The God Delusion was first published in 2006 and has since become an iconic, international bestseller. Richard Dawkins was catapulted to fame in 1976 with publication of *The Selfish Gene*, which was followed with a string of bestselling books: *The Extended Phenotype*, *The Blind Watchmaker*, *River Out of Eden*, *Climbing Mount Improbable*, *Unweaving the Rainbow*, *The Ancestor's Tale*, *The Greatest Show on Earth*, *The Magic of Reality* and a collection of his shorter writings, *A Devil's Chaplain*. His latest books are the two successive parts of his autobiography, *An Appetite for Wonder* and *Brief Candle in the Dark*.

Dawkins is a Fellow of both the Royal Society and the Royal Society of Literature. He is the recipient of numerous honours and awards, including the Royal Society of Literature Award (1987), the Michael Faraday Award of the Royal Society (1990), the International Cosmos Prize for Achievement in Human Science (1997), the Kistler Prize (2001), the Shakespeare Prize (2005), the Lewis Thomas Prize for Writing about Science (2006) and the Galaxy British Book Awards Author of the Year Award (2007). He retired from his position as the Charles Simonyi Professor for the Public Understanding of Science at Oxford University in 2008 and remains a fellow of New College.

www.penguin.co.uk

'Richard Dawkins is energising a huge unbelieving constituency in this country who feel disenfranchised by the present obsession with religion. All power to him, we say.'
Terry Sanderson, National Secular Society

'The most coherent and devastating indictment of religion I have ever read. The case Dawkins makes could not be clearer. There is no God. All religion is wrong.'
Harry Ritchie, *Mail on Sunday*

'Could not be more apt for our time – [Dawkins] pulls out all the stops to demonstrate the force of his thesis in this passionate new book . . . lively and highly readable.'
John Cornwell, *Sunday Times*

'*The God Delusion* is a runaway bestseller, a marked testimony to the hunger many people have for someone in a position of prestige and power to speak for them in such an eloquent voice. Dawkins's latest book deserves multiple readings, not just as an important work of science, but as a great work of literature.'
Steven Weinberg, Nobel Laureate, *Times Literary Supplement*

'I see this as a book for a new millennium, one in which we may be released from lives dominated by the supernatural.'
Brian Eno

'[A] passionately argued book. [Dawkins] boots all the crazy incoherencies put forward by the vicar and his ilk firmly into touch.'
Arena

'Dawkins is Britain's most famous atheist and in *The God Delusion* he gives eloquent vent to his uncompromising views . . . if you want an understanding of evolution or an argument for atheism, there are few better guides than Richard Dawkins.'
Sunday Telegraph

'He is brilliant, articulate, impassioned, and impolite . . . *The God Delusion* is a fine and significant book . . . Dawkins' irreverent and penetrating work will seem a breath of fresh air.'
San Francisco Chronicle

'This thoroughly engaging thesis on atheism cajoles, bullies, persuades and dazzles . . . Some of it is hard to disagree with, some of it will make you hopping mad. Perfect, really.'
***Sunday Times,* Perth**

'Passionate religious irrationality too often poses serious obstacles to human betterment. To oppose it effectively, the world needs equally passionate rationalists unafraid to challenge long accepted beliefs. Richard Dawkins so stands out through the cutting intelligence of *The God Delusion*.'
James D. Watson, Nobel Laureate and co-discoverer of DNA

'He is surely one of the finest living writers in the English language.'
Irish Times

'*The God Delusion* is smart, compassionate, and true . . . If this book doesn't change the world, we're all screwed.'
Penn & Teller

Also by Richard Dawkins

The Selfish Gene
The Extended Phenotype
The Blind Watchmaker
River Out of Eden
Climbing Mount Improbable
Unweaving the Rainbow
A Devil's Chaplain
The Ancestor's Tale (with Yan Wong)
The Greatest Show on Earth
The Magic of Reality (with Dave McKean)
An Appetite for Wonder
Brief Candle in the Dark

THE GOD DELUSION

Richard Dawkins

BLACK SWAN

TRANSWORLD PUBLISHERS
61–63 Uxbridge Road, London W5 5SA
www.penguin.co.uk

Transworld is part of the Penguin Random House group of companies
whose addresses can be found at global.penguinrandomhouse.com

First published in Great Britain in 2006 by Bantam Press
an imprint of Transworld Publishers
Black Swan edition published 2007
Black Swan edition reissued 2016

A CIP catalogue record for this book
is available from the British Library.

ISBN
9781784161927 (B format)
9781784161934 (A format)

Typeset in 11/13pt Minion by Falcon Oast Graphic Art Ltd.
Printed and bound in Great Britain by Clays Ltd, Bungay, Suffolk.

7 9 10 8

In Memoriam
Douglas Adams (1952–2001)

'Isn't it enough to see that a garden is beautiful without having to believe that there are fairies at the bottom of it too?'

Contents

A new introduction for the 10th anniversary edition

The fact that you exist should brim you over with astonishment. You and I, and every other living creature, are machines of ineffable complexity, complexity of a magnitude to challenge credulity. Complexity here means statistical improbability in a non-random direction, the direction of seeming designed for a purpose. The ultimate purpose (gene survival) hides behind a more up-front 'design', details of which vary from species to species. Whatever its specialism – wings for flying, tails for swimming, hands for climbing or digging, galloping legs for prey-catching or predator-escaping – every animal embodies a statistically improbable complexity of detail which approaches (but revealingly falls short of) perfection as an engineer might judge it. 'Statistically improbable' means 'unlikely to have come about by chance'. The God Temptation here is the temptation to evade, by invoking a designer, the responsibility to explain. The point is that the designer himself, in order to be capable of designing, would have to be another complex entity of the kind that, in his turn, needs the same kind of explanation. It's an evasion of responsibility because it invokes the very thing it is supposed to be explaining.

I'm a biologist, so I speak first of the biological version of the God Temptation, the false argument destroyed by Darwin. There is also a cosmological version, which lies outside the Darwinian domain and precedes it by ten billion years. The cosmos may not look so obviously designed as a peacock or its eye. But the laws and constants of physics are fine-tuned in such a way as to set up the conditions under which, in the fullness of time, eyes and peacocks, humans and their brains, will come into existence. The God Temptation here is to invoke an

Intelligent Knob-Twiddler who adjusts the dials of the physical constants so that they have the exquisitely precise values required to bring evolution, and eventually us, into being.

To succumb to the God Temptation in either of those guises, biological or cosmological, is an act of intellectual capitulation. If you are trying to explain something improbable, it can never suffice to invoke an entity that is, in itself, at least as improbable. If you'll stoop to magicking into existence an unexplained peacock-designer, you might as well magic an unexplained peacock and cut out the middleman.

Nevertheless, it's hard not to feel sympathy for such capitulation. The complexity of a living body, indeed of every one of its trillion cells, is so mind-shattering to anyone who truly grasps it (not all do) that the temptation to buckle at the knees and succumb to a non-explanation is almost overwhelming. Even a magic trick can draw the same reaction. There's an old card trick where the conjuror invites a member of the audience to pick a card and show it to the audience. He then burns the card, grinds the ash to powder and rubs it on his forearm. The image of the card appears on his arm, picked out in ash. A conjuror recently told me he performed the trick to a band of Arabs round a camp fire. The tribesmen's reaction made him fear for his life. They sprang up and reached for their guns, thinking he was a djinn. You can see why. You have to smack yourself and shout, 'No! However loudly my senses and my instincts are screaming "Miracle!", it really isn't. There really is a rational explanation. The conjuror prepared the ground in some unknown way before the trick started, and then did some clever prestidigitation while he cunningly distracted my attention.' It's almost as though you have to have 'faith' that it really is only a trick. Faith that nothing supernatural has happened. The laws of physics have not been suspended.

In the case of conjurors we know this to be the case because the best and most honest ones, like Jamy Ian Swiss, or James Randi, or Penn and Teller, or Derren Brown (as opposed to spoonbending charlatans) assure us it is so.* Even if they didn't, the rational thinker falls back on the elegant parsimony of the eighteenth-century philosopher David Hume. Which should surprise you more – that you have been fooled by a trick, or that the laws of physics really have been violated?

When we contemplate the vertebrate eye, or the fine structure of a cell, once again our instincts scream 'Miracle!' and once again we need to smack ourselves. Darwin plays a role akin to the honest conjuror – but he goes further. The honest conjuror tells us it is only a trick but risks expulsion from the Magic Circle if he reveals how it's done. Darwin patiently tells us exactly how the Trick of Life works: cumulative natural selection.

Admittedly that isn't (or probably isn't) how the Cosmological Trick is done. Natural selection explains the miracle of life but it doesn't explain the apparent fine-tuning of the laws and constants of physics – unless you count as a version of natural selection the multiverse theory: there are billions of universes having different laws and constants; with anthropic hindsight we could only find ourselves in one of the minority of universes whose laws and constants happen to be propitious to our evolution. There is a weak sense in which you could regard that as a kind of Darwinism: anthropic *post hoc* selection among universes. The physicist Lee Smolin has provocatively suggested a stronger analogy in which universes give birth to daughter universes with mutated laws and constants.

In any case, Darwin can fairly be said to have done the heavy lifting. Before he came along, any impartial judge would have agreed with Archdeacon William Paley (1743–1805) that the apparent design of physics would be a doddle to explain compared with almost any biological organ, let alone the whole magnificent diversity of purpose-ridden life. Both these

---

* Jamy Ian Swiss ends his emails with a quotation from the celebrated illusionist Karl Germain: 'Conjuring is the only absolutely honest profession – the conjuror promises to deceive, and does.'

versions of the God Temptation are logically fallacious but one of them – the biological one – was so eloquently strong before Darwin, it would tempt one to defy even logic itself. The fact that Darwin solved it so convincingly should now stiffen our confidence to reject the much weaker cosmological version too. Darwin is a role model to inspire all who follow the logical and courageous compulsion to explain complex things in the only legitimate way, which is in terms of simpler things and their interactions.

The publication of *The God Delusion* provoked a swarm of what I came to call 'fleas': religious books with plagiaristic jacket designs and parasitic titles like *The Dawkins Delusion, Deluded by Dawkins, God is no Delusion, The Atheist Delusion, Atheist Delusions, The Devil's Delusion, The God Solution, The God Delusion Revisited, Is God an Illusion?* The 'flea' name came from a line of W. B. Yeats: 'But was there ever dog that praised his fleas?' (Incidentally, there was even a book published called *The Dog Delusion*, with the same cover design as mine, but it isn't a religious book and doesn't qualify as a flea.) Not only the fleas but other religious critics of my book homed in on what they rightly saw as its central and most damaging point, the one briefly reiterated above, dubbed in the book 'the Ultimate Boeing 747'. Grasping at straws, they tried to deny that a god capable of designing something complex must himself be complex.

Dawkins may know something about science, they bent over backwards to concede, but he's no theologian, and theologians are the professionals qualified to tell us what God is like. Some of our best theologians have told us that, far from being complex, God is supremely simple. No less a theologian than Richard Swinburne, then Oxford's Nolloth Professor of the Philosophy of the Christian Religion, said it beautifully clearly. In his 1996 book *Is There a God?*, reissued in 2010, Swinburne endears himself to scientists by praising the virtue of simple explanations but then, astonishingly, goes on to claim that God is the ultimate simple explanation for everything:

> Theism claims that every other object which exists is caused to exist and kept in existence by just one substance, God. And it claims that every property which every substance has is due to God causing or permitting it to exist. It is a hallmark of a simple explanation to postulate few causes. There could in this respect be no simpler explanation than one which postulated only one cause. Theism is simpler than polytheism. And theism postulates for its one cause, a person [with] infinite power (God can do anything logically possible), infinite knowledge (God knows everything logically possible to know), and infinite freedom . . . (p. 43)

As I wrote in my review of the book for the *Sunday Times*,

> Swinburne generously concedes that God cannot accomplish feats that are *logically* impossible, and one feels grateful for this forbearance. That said, there is no limit to the explanatory purposes to which God's infinite power is put. Is science having a little difficulty explaining X? No problem. Don't give X another glance. God's infinite power is effortlessly wheeled in to explain X (along with everything else), and it is always a supremely *simple* explanation because, after all, there is only one God. What could be simpler than that?

Swinburne's is the clearest expression of the remarkably feeble point made later by fleas and others, in reply to my 'Ultimate 747' argument. By theological fiat, God is *declared* to be 'simple'. Theologians are the ones who can speak with authority on God, theologians decree that God is simple, therefore God is simple, therefore the Ultimate 747 argument fails. Such brazen sophistry is beyond astounding. It calls to mind Julian Huxley's satire of Bergsonian vitalism: to postulate a mysterious *élan vital* in explanation for life is like saying that a railway engine is powered by *élan locomotif.* Actually Swinburne's sophistry is worse, because at least Bergson could claim that his *élan vital* was complex, as might be revealed by further investigation. Swinburne, and the fleas that agree with him, have the effrontery to decree by fiat that their *élan théologique* (as we might term it) is not only mysterious (like Bergson's *élan*) but also simple.

God is simple, for Swinburne, because there is only one of him. Polytheism, he states explicitly, is less simple than (mono)theism. Yet that one entity, unitary though he be, has to be clever enough to calculate, with exquisite and prophetic precision, the exact values of the physical constants that would fine-tune a universe to yield, 13.8 billion years later, a species capable of worshipping him. You call that simple? At the same time, in his singular simplicity, he had to foresee that the nuclear force must be set 10^{36} times stronger than gravity; and he had to calculate with similar exactitude the precisely requisite values of half a dozen critical numbers, the fundamental constants of physics.* You and I both possess prodigiously complex brains evolved over hundreds of millions of years, but do you understand quantum mechanics? I certainly don't. Yet God, that paragon of ultimate pure simplicity, not only understands it but invented it. Plus Special and General Relativity. Plus the Higgs boson and dark matter. Finally, the icing on the cake: on top of being the ultimate mathematics and physics genius, God has enough bandwidth to listen to the prayers and praises of billions of people simultaneously (plus how many on other planets, and how many already dead and in Heaven?). He hears their confessed sins and decides which ones should be forgiven, weighs up which cancer patients shall recover, which tsunami or earthquake victims shall be spared, even (according

* A beautifully clear exposition of the fundamental constants is *Just Six Numbers* (1999) by Martin Rees, the Astronomer Royal. Rees does not, of course, draw any theological conclusion. He favours the multiverse.

to his more naïve but still numerous followers) who shall win Wimbledon and who shall be vouchsafed a parking space. Yes, I'm straying into sarcasm territory now, which might seem unfair, but the point about simplicity remains. God may be almighty, all-seeing, all-knowing, all-powerful, all-loving, but the one thing he cannot be, if he is to match up even minimally to his job description, is *simple.* The 'Ultimate 747' argument remains intact and inescapably devastating.

So much for the two scientific versions of the God Temptation. There is also a moral version. Without God, it is said, where is the inducement to be good, what are the sanctions against bad behaviour? How do we even know what is good and what bad? The temptation here is to abdicate the responsibility to think clearly and consistently about morality, and instead take the lazy route of slavishly following an ancient book of rules, rules invented by fallible men (and they *were* men) and tailored to very different times and conditions. Or, worse, to base our moral decisions on the fear that our every move is watched by a great surveillance camera in the sky and so we need to suck up to an obsessively vigilant God, a celestial Nosy Parker, inexhaustibly interested in what goes on in our beds – and even our minds. As for the suggestion that we can't define good and bad without God, it is falsified by the honourable and sophisticated discipline of moral philosophy. But even if it were true, the fact would have no bearing on whether God exists. Maybe there really is no ultimate way to define good and bad. Why should there be, any more than an ultimate way to define beauty?

God also presents a temptation to lazy and sloppy use of language when defining our allegiances.

> 'I'm Christian. Well of course I don't believe any of that supernatural stuff but I was baptized, we go to church at Christmas and I'm certainly not Jewish or Muslim, so I guess that makes me Christian.'

> 'I'm Catholic.'
> 'Ah, I see, so you believe wine turns literally into blood, bread into body, and Mary never died but was assumed bodily into Heaven.'
> 'No, that's just ridiculous, of course I don't believe that.'
> 'Oh, so you're not Catholic after all.'
> 'Well, my family has been Catholic for generations, so doesn't that make me Catholic?'
> 'That's lazy, sloppy abuse of language. My family has been farming for generations but that doesn't define me as a farmer.'
>
> 'By the year 2050 the population of Ruritania will be predominantly Muslim. It's simple demography. Just compare the birth rates of the different communities in Ruritania.'
> 'But you're making the lazy assumption that a child of Muslim parents is defined as a Muslim. Would you define a child of logical positivist parents as a logical positivist? Would you define a child of Keynesian parents as a Keynesian?'

This kind of laziness was documented by a 2011 opinion poll in which I was involved. The decennial UK National Census fell in March 2011. As with previous censuses, everyone was asked to specify their religion ('no religion' was one option). It seemed likely that many of those who ticked 'Christian' had in fact succumbed to the 'lazy temptation': baptized Christian, perhaps, but what did they really believe? This was where I, or rather my charitable foundation, came in. Publication of *The God Delusion* in 2006 had generated two foundations, one in Britain, one in America, sharing the same name, the Richard Dawkins Foundation for Reason and Science (RDFRS). In 2011, the UK Foundation commissioned the respected polling organization Ipsos MORI to survey the actual religious opinions held by those who self-identified as Christian, and we deliberately chose to do it in the very week following the census. A summary of the findings can be seen here: https://www.ipsos-mori.com/Assets/Docs/Polls/ipsos-mori-religious-and-social-attitudes-topline-2012.pdf.

Only people who ticked the 'Christian' box in the census were sampled. 1,136 of them responded. Given that they self-identified as Christian, I was especially surprised – even shocked, to be honest – by their ignorance of the Bible. 'What is the first book of the New Testament?' they were asked. They didn't have to name 'Matthew'; the task was easier then that. They were given a choice of only four: Matthew, Genesis, Acts of the Apostles or Psalms. Only 35 per cent chose Matthew. A massive 39 per cent ticked 'Don't know'. Ignorance of the Bible doesn't rule out sincere religious belief, but it is sobering and it indirectly supports the case for 'laziness' where declarations of religious affiliation are concerned. These were not just any UK citizens, remember. They were specifically those who had self-identified as Christian in the official census a week earlier.

When asked to choose the single main reason for thinking of themselves as belonging to their religion, only 18 per cent said it was because they believed in its teachings. The most popular answer to the question was 'I was christened/baptized into this religion'. That accounted for 46 per cent. Other reasons given included 'One or both of my parents are/were members of this religion' and 'As a child I went to a Sunday school run by this religion'. Compatibly, 17 per cent of these self-described Christians attended church weekly, while 55 per cent never did, or only at Christmas or Easter. Just 35 per cent prayed at least once a week, whether or not they bothered to go to church; 43 per cent prayed never, or less than once per year.

These figures, and the answers to other questions in our survey, seem to support the 'lazy' version of the God Temptation. It is therefore a matter of some interest to know why, given their evident lack of genuine religious belief, these respondents had chosen to tick the 'Christian' box. What did they think it means to call yourself a Christian? The answers to this question are revealing:

Which is the ONE statement that BEST describes what being a Christian means to you personally?

	%
I try to be a good person	40
It's how I was brought up	24
I have accepted Jesus as my Lord and Saviour	15
I believe in the teachings of Jesus	7
It's a British tradition	4
It gives me hope in an afterlife	3
Something else	0
Don't know	2
Prefer not to say	4

Only 22 per cent cited belief in the teachings of Jesus or accepting Jesus as their Lord and Saviour as their primary understanding of what it means to be a Christian. The 28 per cent who chose 'It's how I was brought up' or 'It's a British tradition' exemplify the lazy misuse of language that I have been talking about. So does the answer which I was shocked to see was the most popular: 'I try to be a good person' – and that also illustrates the moral version of the God Temptation which I mentioned earlier.

I do not, of course, deny that many Christians are good people and many good people are Christians. There is no persuasive evidence for any significant association between the two, however. Indeed, it wouldn't surprise me to discover that atheists are more likely to be good than religious people. There is some weak evidence pointing in that direction.* But what distresses me is that at least 40 per cent of our respondents seemed to take the alleged positive association between

* See http://www.cell.com/current-biology/fulltext/S0960-9822%2815%2901167-7.

'Christian' and 'good' for granted, almost as a matter of definition. In the same vein, when thanking somebody for doing a good turn, people often say something like: 'That's very Christian of you.' Or: 'You, Sir, are a gentleman and a Christian.' It is in the same spirit that many people refuse to vote for an atheist in political elections because they think a believer in a higher power – any higher power, it doesn't matter which one – is more likely to take moral decisions.

We chose the very week following the census to run our poll because we suspected – rightly, as it turned out – that the official census would give results that could be misleading. We also feared that such misleading inferences would be used by interested parties, as they undoubtedly were after the 2001 census, to influence policy, perhaps to argue for government-supported faith schools, or bishops in the House of Lords. 'Mr Speaker, Honourable Members, the census shows that Britain is a predominantly Christian country, therefore it's only right that . . .' You can see how it might play.

Actually, the census showed a dramatic drop in the number of professed Christians, from 71.7 per cent in 2001 to 59.3 per cent in 2011. The number professing 'No religious belief' increased over the decade from 14.8 per cent to 25.1 per cent. Parallel counts in the United States show the same trends, but the absolute numbers of religious believers are higher. American religiosity has become legendary. I suspect it is inflated by the same lazy temptation as our UK poll demonstrated, powerfully reinforced by what could be called, with scarcely any exaggeration, intimidation. 'What church do you go to?' The question is presumptuous to the point of rudeness – or so a European would find it. Yet I'm told, again and again, that it is likely to be thrown at newcomers to a neighbourhood in certain states of America as casually and automatically as a solicitous inquiry after health, or a comment on the weather. That the newcomer might not attend a place of worship at all often doesn't cross the Friendly Neighborhood Mind.

It doesn't cross the mind of a typical American politician

either, which is why so many of them drag God into every speech, and why they bend over backwards to appease religious lobbies when talking about important issues such as abortion, stem cell research and the teaching of evolution. And this brings me to my American Foundation, that other spinoff from the publication of *The God Delusion.*

RDFRS (US) has a number of projects going (see www.RichardDawkins.net) and the number has increased since our recent happy union with the Center for Inquiry (CFI). The one I want to focus on here is Openly Secular, a campaign launched with three other important American secular organizations, the Stiefel Freethought Foundation, the Secular Student Alliance and the Secular Coalition for America. Openly Secular is a sophisticated marketing and public awareness campaign designed to raise consciousness of the fact that America is not nearly as religious as politicians, and most other people, think it is. Politicians themselves cannot possibly be as monolithically religious as they pretend. Of the 535 members of the combined houses of Congress, not a single one admits to not believing in a higher power. That is statistically beyond implausible, verging on impossible. More than 20 per cent of the US population at large now tell pollsters they have no religious belief. The figure swells to 40 per cent for American scientists and to 90 per cent for elite scientists (those elected to the National Academy of Sciences), and is probably similar for other educated elites such as top philosophers, historians and other scholars including (for sure) many biblical scholars. In the face of these numbers, how is it even remotely plausible that the figure for the US Congress is zero? The conclusion is glaring: a substantial number of US politicians are simply lying when they profess religious faith.* To put it in starkly cynical terms, a successful US politician is either religious or a well-educated and intelligent liar.

---

* We have off-the-record private information from many of them that this is indeed the case.

It's hard to blame them for lying. They have bought into the widely held belief that it is impossible for a non-believer to win election to high office. None seem to take courage from the fact that the one Congressman who did dare to come out as an atheist, Representative Pete Stark, was repeatedly re-elected (he eventually lost his seat in 2012, aged eighty, for unrelated reasons).* Have US politicians simply not caught up with the fact that the 'nones' (having no religious affiliation), at 23 per cent of Americans, now outnumber Roman Catholics (21 per cent)? Of course most 'nones' would not go so far as to claim the dread word 'atheist'. But avowed atheists combined with agnostics are more than three times as numerous as adherents to the Jewish faith.† And American Jews, to their credit, command attention in the corridors of power.

Our Openly Secular campaign, encouraged by the earlier success of the gay movement, aims to raise consciousness (not least among politicians) by inviting people in all walks of life to 'come out' as non-believers. Those who have made short videos for us include not only celebrities such as Bill Maher, Julia Sweeney, Penn & Teller, John Davidson, Killah Priest, Chris Kluwe, Arian Foster, John de Lancie and principals of the band Nightwish but, of equal importance, 'ordinary nice people from next door'. While resolutely opposed to 'outing' people against their wishes, we hope to give them the courage to come forward and take a stand for truth, in the same way as the Gay Pride advocates of an earlier decade did. It is our (plausible) hope that a tipping point will be reached, whereupon floodgates will be opened and even politicians will realize that they no longer need to vote against their consciences in order to suck up to religious lobbies.

The legendarily high religiosity of the United States is at least

* Representative Barney Frank, publicly 'out' as gay (an extreme rarity in Congress), was repeatedly re-elected by his liberal and sophisticated constituency. Nevertheless he withheld his atheism until after he retired.

† http://www.religionnews.com/2015/05/12/christians-lose-ground-nones-soar-new-portrait-u-s-religion/.

partly a manifestation of succumbing to the 'lazy' temptation uncovered in Britain by the RDFRS UK poll. But there does seem to be a real difference as well. America really is more religious than Britain and western Europe. Why this should be is one of the commonest questions I am asked, especially by American audiences.

I used to answer with a paradox. Britain, Scandinavia and other countries with an established church are the least religious. Religion has become boring. Churches are places you visit only to be baptized, married or buried, perhaps also to carol some Christmas nostalgia. But in America, precisely because the constitution bans the establishment of religion, religion has become free-enterprise, entrepreneurial, competitive, charismatic, exciting, a vibrant and lucrative branch of show business.

I still find that argument somewhat plausible, especially when I look at megachurch televangelists with their mansions and Lear jets, preaching a 'Jesus wants you to be rich' gospel of self-interest. I suspect that it represents a part of the truth. However, Gregory Paul, Jerry Coyne and others have persuaded me of a different hypothesis, for which there is positive statistical evidence. This has been called the Existential Security Hypothesis. Coyne attributes it originally to Karl Marx, who famously stated in the 1840s:

> Religion is the sigh of the oppressed creature, the heart of a heartless world, and the soul of soulless conditions. It is the opium of the people.

Marx's recognition of this aspect of the God Temptation is tinged with sympathy for its victims and in this, if in nothing else, I join him. No wonder religion was popular among the slaves of America (and remains so among their descendants to this day). No wonder the pacifying opium of religion was actively pushed by oppressors, then and down the ages.

Greg Paul's exhaustive research looks across countries and finds that religiosity increases with various indices of social malaise – measures, we might say, of 'existential insecurity'. The correlation of religiosity with income inequality, for example, is 0.707. With infant mortality it is 0.746; with abortion rates among teenagers it is 0.825; with rates of gonorrhoea infection it is 0.643. The idea is that people tend to resort to religion in countries where they feel insecure in their lives, unsupported by health care and other social welfare provisions; more at risk in this world, so more tempted to place their forlorn hopes in a mythical next world.

Correlation can suggest causation but cannot prove it, and it doesn't tell us which way, if any, the causal arrow points. Does social malaise cause religiosity or the other way around? Coyne favours the former, on persuasive grounds.* The evidence is interesting. When income inequality changes, religiosity changes too, in the predicted direction but with a one-year time lag. Societal despair provides a climate in which religion flourishes. So, while I am still committed to persuading people on intellectual grounds that God is indeed a Delusion, it can be argued that a better route to killing religion is to abolish poverty and especially inequality. Working to improve education fosters both routes.

Jerry Coyne's book *Faith versus Fact* should be added to the list of so-called 'horseman' books, as should Lawrence Krauss's *A Universe from Nothing*. One of the achievements of RDFRS (US) was to get the original 'Four Horsemen of Atheism' (Sam Harris, Dan Dennett, Christopher Hitchens and me) together under one roof (Christopher's) for an unchaired and unscripted filmed conversation.† I think the lack of a chairman improved the conversation, as it usually does. I don't object to the horseman label, by the way. I'm less keen on 'new atheist': it isn't clear to me how we differ from old atheists.

---

* https://whyevolutionistrue.wordpress.com/2015/09/15/will-nonbelief-replace-religion-within-25-years/. Coyne's main source is the 2004 book by Norris and Inglehart, *Sacred and Secular*.

† https://richarddawkins.net/2009/02/the-four-horsemen-hd-hour-1-of-2-discussions-with-richard-dawkins-ep-1-2/.

Another successful enterprise of RDFRS (US) has been the Clergy Project (TCP). Dan Barker (*Losing Faith in Faith*) made us aware that there are clergy who have become atheists but feel, for evident reasons, unable to come clean. Dan himself concealed his apostasy for a year, even continuing to write hymns (he's a talented musician) before finally breaking free and joining his now wife Annie Laurie Gaylor in the Freedom from Religion Foundation (FFRF). From the inception of RDFRS it was my dream to find a way of rescuing these honourable renegades. My original hope was to finance scholarships for atheistic clergymen and women, retraining them to make their living in a more reputable career – carpentry, perhaps. It soon became clear that we couldn't raise enough money to do this on anything other than a token scale. The Clergy Project was something we could do: a small but significant step in the right direction. We provided a website where atheistic clergy could meet each other, under false names, in conditions of complete confidentiality. They could discuss their shared problems, getting advice from each other and from Dan Barker and others who had already 'come out'. Membership of TCP is strictly controlled. New members are carefully vetted for fear of fifth columnists who might 'out' people before they are ready to face the world and risk losing friends, family, their livelihood and the respect of their community. TCP has now constituted itself as a charity, independent of its parent organizations (RDFRS and FFRF) and entirely governed by its members.

From a handful of founder members known to each other only by *noms de guerre* like 'Adam' and 'Chris', TCP has ballooned to the point where in 2016 membership stands at nearly 700, mostly ex-Christians but including people from all the other major religions too. That's an impressive number. If there are 700 clergy who have actually heard about the secret club for atheist clergy and are prepared to take the risk of joining it, think how much larger must be the total number out there. There are probably many who scarcely dare admit their nonbelief even to themselves, and need only the reassurance of knowing they are not alone to allow themselves to do so. I take great encouragement from that thought.

For an insightful account and analysis of what makes these apostates tick (their stories are quite variable), see *Caught in the Pulpit* by Daniel Dennett and Linda LaScola.* Also available and listed on the TCP website† are personal memoirs by TCP alumni who have made the courageous leap and 'come out'. Catherine Dunphy, for example, in *From Apostle to Apostate*, well conveys the harrowing personal difficulties faced by her and her colleagues, and she gives an especially well-informed history of TCP.

One other spin-off from *The God Delusion* is unusual enough to be worth mentioning. The book has been widely denounced (mostly by critics who haven't read it but only read other critics) as 'strident' and 'shrill'. One critic, who cannot possibly have read it, went so far as to suspect me of 'Tourette's Syndrome'. In fact the book's tone is mostly rather mild – certainly milder than many of the 'fleas' that responded to it. The illusion of stridency arises because of the long-standing convention, observed by believers and non-believers alike, that you are not allowed to criticize religion in polite society. 'Why not [to quote Douglas Adams]? Because you're not.' The result is that people literally hear mildly expressed criticism as strident, even though it isn't by the normal standards of other fields such as politics, journalism or (as I pointed out in the preface to the paperback edition) restaurant criticism.

But there is one passage whose tone could fairly be heard as strong by any standards: the opening sentence of chapter 2.

> The God of the Old Testament is arguably the most unpleasant character in all fiction: jealous and proud of it; a petty, unjust, unforgiving control-freak; a

---

* This book has been turned into a play by Martin Gazzaniga, sponsored by RDFRS (US).

† http://clergyproject.org/booksandblogs/.

> vindictive, bloodthirsty ethnic cleanser; a misogynistic, homophobic, racist, infanticidal, genocidal, filicidal, pestilential, megalomaniacal, sadomasochistic, capriciously malevolent bully.

The sentence even provoked an accusation of anti-semitism by Britain's most senior rabbi. The accusation was made in the heat of a moment and he soon withdrew it with a characteristically gracious and charming apology. He had misunderstood me as singling out the (Jewish) Old Testament God by contrast with the God of the Christian New Testament, whereas I actually think the central 'atonement' dogma of Christianity (due to Paul rather than Jesus) is obnoxious even by the elevated standard set in the Old Testament (see chapter 7).

Strong though it sounds, that sentence from the beginning of chapter 2 can be amply justified, word for word – every single one of them – from the Bible itself. Repeated accusations of stridency against my infamous sentence provoked me to plan a lecture in which I would cite chapter and verse for every one of my seventeen adjectives and three nouns. It soon became clear that the sheer number of verses cited would prolong the lecture far beyond a mere hour. The material was rich enough for a whole book, and I knew just the man to write it: Dan Barker, the ex-preacher who, by his own account, had been the sort of zealot you wouldn't want to sit next to on a bus: the sort of preacher who would thrust a bible in the face of a perfect stranger and ask if he were saved. Dan knows his Bible like a London cabbie's hippocampus knows 'The Knowledge'. I put it to him. He jumped at the idea and the result is his splendid book *God: The Most Unpleasant Character in All Fiction.* There's a chapter for every one of the words in my list, each chapter filled with verses from scripture, interspersed with Dan's entertaining and well-informed commentary.

The God Delusion has sold more than three million copies so far, a paltry number compared to the Bible's five billion, or the slightly lower figure for the Qur'an. I like to think most of my three million copies have actually been read. If only we could say the same of the billions of Bibles and Qur'ans, the need for my book might be sensibly diminished. For, on the face of it, you could plausibly argue that the best antidote against all three of the Abrahamic religions is a thorough reading of their holy books. The nasty bits are seldom mentioned in churches or Sunday schools, and many devout believers are blissfully unaware of their existence.* Even when their existence is admitted, they are bowdlerized by a piece of intellectual sleight of hand, and this is yet another version of The God Temptation that I wish to mention. The wolfish horror of the worst scriptural verses is cloaked under various forms of sheep's clothing: the words are not meant to be taken literally, they are 'metaphorical'.

'Not meant' by whom? Nobody knows who originally invented the myth of Abraham's cruel abuse of Isaac (Ishmael in the Islamic version). Modern theologians don't take this shocking story literally but excuse it as a parable admonishing the Israelites to stop sacrificing children.† Did the anonymous scribe who first turned oral legend into writing believe it was literally true? We don't know. But no reasonable person could deny that the vast majority of ordinary followers of the religions concerned have down the centuries believed it literally happened. And still persisted in worshipping the psychotic monster who, they sincerely believed, gave Abraham his orders. And the same for the other horror stories of scripture.

The claim of some modern apologists that literal interpretation of the Bible is a recent phenomenon is unpersuasive, to say

---

* For an entertaining illustration of this, see https://www.youtube.com/watch?v=zEnWw_lH4tQ. A pair of Dutch investigators wrapped a bible in a Qur'an cover, and read selected horrible verses to random people in the street. Their victims were flabbergasted when they eventually discovered the truth.

† By the way, why express such an important lesson in a parable? Why not just say, more clearly and directly: 'Stop sacrificing children, chaps. Do sheep instead.'

the least. Take just one highly influential example. Archbishop Ussher's 1650 calculation of the date of Creation (4004 BC) is based on a literalistic adding up of the ages cited in the list of 'begats'. Metaphorical? Er, no, you don't meticulously add up a lot of *metaphorical* numbers to come up with the *actual* date of an alleged historical event. Ussher's ludicrous calculation also incidentally shows up the presumptuous arrogance typical of the theological mind. Not only did Creation fall in 4004 BC (as opposed to 4003 or 4005). It was Saturday, 22nd October if you please, not the 21st or 23rd. The archbishop's effrontery is matched only by that of modern theologians who claim that biblical literalism is a recent aberration.

In my personal view, illustrated by the final chapter of this book, the saddest version of the God Temptation is the temptation to forgo the spiritual – yes, spiritual – joy of a rational, scientific understanding of life, and of the universe in which life finds itself, in favour of a primitive, superstitious supernaturalism. And to tempt children down the same barren path is a sin so grievous that, if wilfully and knowingly perpetrated, it brings millstones fleetingly to mind.

Preface to the paperback edition

The God Delusion in the hardback edition was widely described as the surprise bestseller of 2006. It was warmly received by the great majority of those who sent in their personal reviews to Amazon (more than 1,000 at the time of writing). Approval was less overwhelming in the printed reviews, however. A cynic might put this down to an unimaginative reflex of reviews editors: It has 'God' in the title, so send it to a known faith-head. That would be too cynical, however. Several unfavourable reviews began with the phrase, which I long ago learned to treat as ominous, 'I'm an atheist BUT . . .' As Daniel Dennett noted in *Breaking the Spell*, a bafflingly large number of intellectuals 'believe in belief' even though they lack religious belief themselves. These vicarious second-order believers are often more zealous than the real thing, their zeal pumped up by ingratiating broad-mindedness: 'Alas, I can't share your faith but I *respect* and sympathize with it.'

'I'm an atheist, BUT . . .' The sequel is nearly always unhelpful, nihilistic or – worse – suffused with a sort of exultant negativity. Notice, by the way, the distinction from another favourite genre: 'I *used* to be an atheist, but . . .' That is one of the oldest tricks in the book, much favoured by religious apologists from C. S. Lewis to the present day. It serves to establish some sort of street cred up front, and it is amazing how often it works. Look out for it.

I wrote an article for the website RichardDawkins.net called 'I'm an atheist BUT . . .' and I have borrowed from it in the following list of critical or otherwise negative points from reviews of the hardback. That website, conducted by the inspired Josh Timonen, has attracted an enormous number of contributors who have effectively eviscerated all these criticisms, but in less guarded, more outspoken tones than my own, or than those of my academic colleagues A. C. Grayling, Daniel Dennett, Paul Kurtz, Steven Weinberg and others who have done so in print (and whose comments are reproduced on the same website).

You can't criticize religion without a detailed analysis of learned books of theology.

Surprise bestseller? If I'd gone to town, as one self-consciously intellectual critic wished, on the epistemological differences between Aquinas and Duns Scotus; if I'd done justice to Eriugena on subjectivity, Rahner on grace or Moltmann on hope (as he vainly hoped I would), my book would have been more than a *surprise* bestseller: it would have been a miraculous one. But that is not the point. Unlike Stephen Hawking (who accepted advice that every formula he published would halve his sales), I would happily have forgone bestsellerdom if there had been the slightest hope of Duns Scotus illuminating my central question of whether God exists. The vast majority of theological writings simply assume that he does, and go on from there. For my purposes, I need consider only those theologians who take seriously the possibility that God does not exist and argue that he does. This I think Chapter 3 achieves, with what I hope is good humour and sufficient comprehensiveness.

When it comes to good humour, I cannot improve on the splendid 'Courtier's Reply', published by P. Z. Myers on his 'Pharyngula' website.

> I have considered the impudent accusations of Mr Dawkins with exasperation at his lack of serious scholarship. He has apparently not read the detailed discourses of Count Roderigo of Seville on the exquisite and exotic leathers of the Emperor's boots, nor does he give a moment's consideration to Bellini's masterwork, *On the Luminescence of the Emperor's Feathered Hat.* We have entire schools

> dedicated to writing learned treatises on the beauty of the Emperor's raiment, and every major newspaper runs a section dedicated to imperial fashion . . . Dawkins arrogantly ignores all these deep philosophical ponderings to crudely accuse the Emperor of nudity . . . Until Dawkins has trained in the shops of Paris and Milan, until he has learned to tell the difference between a ruffled flounce and a puffy pantaloon, we should all pretend he has not spoken out against the Emperor's taste. His training in biology may give him the ability to recognize dangling genitalia when he sees it, but it has not taught him the proper appreciation of Imaginary Fabrics.

To expand the point, most of us happily disavow fairies, astrology and the Flying Spaghetti Monster, without first immersing ourselves in books of Pastafarian theology etc.

The next criticism is a related one: the great 'straw man' offensive.

You always attack the worst of religion and ignore the best.

'You go after crude, rabble-rousing chancers like Ted Haggard, Jerry Falwell and Pat Robertson, rather than sophisticated theologians like Tillich or Bonhoeffer who teach the sort of religion I believe in.'

If only such subtle, nuanced religion predominated, the world would surely be a better place, and I would have written a different book. The melancholy truth is that this kind of understated, decent, revisionist religion is numerically negligible. To the vast majority of believers around the world, religion all too closely resembles what you hear from the likes of Robertson, Falwell or Haggard, Osama bin Laden or the Ayatollah Khomeini. These are not straw men, they are all too influential, and everybody in the modern world has to deal with them.

I'm an atheist, but I wish to dissociate myself from your shrill, strident, intemperate, intolerant, ranting language. Actually, if you look at the language of *The God Delusion*, it is rather less shrill or intemperate than we regularly take in our stride – when listening to political commentators for example, or theatre, art or book critics. Here are some samples of recent restaurant criticism from leading London newspapers:

> 'It is difficult, if not impossible, to imagine anyone conjuring up a restaurant, even in their sleep, where the food in its mediocrity comes so close to inedible.'
>
> 'All things considered, quite the worst restaurant in London, maybe the world . . . serves horrendous food, grudgingly, in a room that is a museum to Italian waiters' taste circa 1976.'
>
> 'The worst meal I've ever eaten. Not by a small margin. I mean the worst! The most unrelievedly awful!'
>
> '[What] looked like a sea mine in miniature was the most disgusting thing I've put in my mouth since I ate earthworms at school.'

The strongest language to be found in *The God Delusion* is tame and measured by comparison. If it sounds intemperate, it is only because of the weird convention, almost universally accepted (see the quotation from Douglas Adams on pages 42–3), that religious faith is uniquely privileged: above and beyond criticism. Insulting a restaurant might seem trivial compared to insulting God. But restaurateurs and chefs really exist and they have feelings to be hurt, whereas blasphemy, as the witty bumper sticker puts it, is a victimless crime.

In 1915, the British Member of Parliament Horatio Bottomley recommended that, after the war, 'If by chance you should discover one day in a restaurant you are being served by a German waiter, you will throw the soup in his foul face; if you find yourself sitting at the side of a German clerk, you will spill the inkpot over his foul head.' Now that's strident and intolerant (and, I should have thought, ridiculous and ineffective as rhetoric even in its own time). Contrast it with the opening sentence of Chapter 2, which is the passage most often quoted as 'strident' or 'shrill'. It is not for me to say whether I succeeded, but my intention was closer to robust but humorous broadside than shrill polemic. In public readings of *The God Delusion* this is the one passage that is guaranteed to get a good-natured laugh, which is why my wife and I invariably use it as the warm-up act to break the ice with a new audience. If I could venture to suggest why the humour works, I think it is the incongruous mismatch between a subject that *could* have been stridently or vulgarly expressed, and the actual expression in a drawn-out list of Latinate or pseudo-scholarly words ('filicidal', 'megalomaniacal', 'pestilential'). My model here was one of the funniest writers of the twentieth century, and nobody could call Evelyn Waugh shrill or strident (I even gave the game away by mentioning his name in the anecdote that immediately follows, on page 51).

Book critics or theatre critics can be derisively negative and gain delighted praise for the trenchant wit of their review. But in criticisms of religion even *clarity* ceases to be a virtue and sounds like aggressive hostility. A politician may attack an opponent scathingly across the floor of the House and earn plaudits for his robust pugnacity. But let a soberly reasoning critic of religion employ what would in other contexts sound merely direct or forthright, and it will be described as a 'rant'. Polite society will purse its lips and shake its head: even secular polite society, and especially that part of secular society that loves to announce, 'I'm an atheist, BUT . . .'

You are only preaching to the choir. What's the point?

'Converts' Corner' on RichardDawkins.net gives the lie to this premise, but even taking it at face value there are good answers.

One is that the non-believing choir is a lot bigger than many people think, especially in America. But, again especially in America, it is largely a closet choir, and it desperately needs encouragement to come out. Judging by the thanks I received all over North America on my book tour, the encouragement that people like Sam Harris, Dan Dennett, Christopher Hitchens and me are able to give is greatly appreciated.

A more subtle reason for preaching to the choir is the need to raise consciousness. When the feminists raised our consciousness about sexist pronouns, they would have been preaching to the choir where the more substantive issues of the rights of women and the evils of discrimination against them were concerned. But that decent, liberal choir still needed its consciousness raised with respect to everyday language. However right-on we may have been on the political issues of rights and discrimination, we nevertheless still unconsciously bought into linguistic conventions that made half the human race feel excluded.

There are other linguistic conventions that need to go the same way as sexist pronouns, and the atheist choir is not exempt. We all need our consciousness raised. Atheists as well as theists unconsciously observe society's convention that we must be especially polite and respectful to faith. And I never tire of drawing attention to society's tacit acceptance of the labelling of small children with the religious opinions of their parents. Atheists need to raise their own consciousness of the anomaly: religious opinion is the one kind of parental opinion that – by almost universal consent – can be fastened upon children who are, in truth, too young to know what their opinion really is. There is no such thing as a Christian child: only a child of Christian parents. Seize every opportunity to ram it home.

You are just as much of a fundamentalist as those you criticize.

No, please, it is all too easy to mistake passion that can change its mind for fundamentalism, which never will. Fundamentalist Christians are passionately opposed to evolution and I am passionately in favour of it. Passion for passion, we are evenly matched. And that, according to some, means we are equally fundamentalist. But, to borrow an aphorism whose source I am unable to pin down, when two opposite points of view are expressed with equal force, the truth does not necessarily lie midway between them. It is possible for one side to be simply wrong. And that justifies passion on the other side.

Fundamentalists know what they believe and they know that nothing will change their minds. The quotation from Kurt Wise on page 323 says it all: '. . . if all the evidence in the universe turns against creationism, I would be the first to admit it, but I would still be a creationist because that is what the Word of God seems to indicate. Here I must stand.' It is impossible to overstress the difference between such a passionate commitment to biblical fundamentals and the true scientist's equally passionate commitment to evidence. The fundamentalist Kurt Wise proclaims that all the evidence in the universe would not change his mind. The true scientist, however passionately he may 'believe' in evolution, knows exactly what it would take to change his mind: Evidence. As J. B. S. Haldane said when asked what evidence might contradict evolution, 'Fossil rabbits in the Precambrian.' Let me coin my own opposite version of Kurt Wise's manifesto: 'If all the evidence in the universe turned in favour of creationism, I would be the first to admit it, and I would immediately change my mind. As things stand, however, all available evidence (and there is a vast amount of it) favours evolution. It is for this reason and this reason alone that I argue for evolution with a passion that matches the passion of those who argue against it. My passion is based on evidence. Theirs, flying in the face of evidence as it does, is truly fundamentalist.'

I'm an atheist myself, but religion is here to stay. Live with it.

'You want to get rid of religion? Good luck to you! You think you can get rid of religion? What planet are you living on? Religion is a fixture. Get over it!'

I could bear any of these downers, if they were uttered in something approaching a tone of regret or concern. On the contrary. The tone of voice is sometimes downright gleeful. I don't think it's masochism. More probably, we can put it down to 'belief in belief' again. These people may not be religious themselves, but they love the idea that other people are religious. This brings me to my final category of naysayers.

I'm an atheist myself, but people need religion.

'What are you going to put in its place? How are you going to comfort the bereaved? How are you going to fill the need?'

What patronizing condescension! 'You and I, of course, are much too intelligent and well educated to need religion. But ordinary people, *hoi polloi*, the Orwellian proles, the Huxleian Deltas and Epsilon semi-morons, need religion.' I am reminded of an occasion when I was lecturing at a conference on the public understanding of science, and I briefly inveighed against 'dumbing down'. In the question and answer session at the end, one member of the audience stood up and suggested that dumbing down might be necessary 'to bring minorities and women to science'. His tone of voice told that he genuinely thought he was being liberal and progressive. I can just imagine what the women and 'minorities' in the audience thought about it.

Returning to humanity's need for comfort, it is, of course, real, but isn't there something childish in the belief that the universe owes us comfort, as of right? Isaac Asimov's remark about the infantilism of pseudoscience is just as applicable to religion: 'Inspect every piece of pseudoscience and you will find a security blanket, a thumb to suck, a skirt to hold.' It is astonishing, moreover, how many people are unable to understand that 'X is comforting' does not imply 'X is true'.

A related plaint concerns the need for a 'purpose' in life. To quote one Canadian critic:

> The atheists may be right about God. Who knows?
> But God or no God, it's clear that something in the

> human soul requires a belief that life has a purpose that transcends the material plane. One would think that a more-rational-than-thou empiricist such as Dawkins would recognize this unchanging aspect of human nature . . . does Dawkins really think this world would be a more humane place if we all looked to *The God Delusion* instead of The Bible for truth and comfort?

Actually yes, since you mention 'humane', yes I do, but I must repeat, yet again, that the consolation-content of a belief does not raise its truth-value. Of course I cannot deny the need for emotional comfort, and I cannot claim that the world-view adopted in this book offers any more than moderate comfort to, for example, the bereaved. But if the comfort that religion seems to offer is founded on the neurologically highly implausible premise that we survive the death of our brains, do you really want to defend it? In any case, I don't think I have ever met anyone at a funeral who dissents from the view that the non-religious parts (eulogies, the deceased's favourite poems or music) are more moving than the prayers.

Having read *The God Delusion*, Dr David Ashton, a British consultant physician, wrote to me on the unexpected death, on Christmas Day 2006, of his beloved seventeen-year-old son, Luke. Shortly before Luke's death, the two of them had talked appreciatively of the charitable foundation that I am setting up to encourage reason and science. At Luke's funeral on the Isle of Man, his father suggested to the congregation that, if they wished to make any kind of contribution in Luke's memory, they should send it to my foundation, as Luke would have wished. The thirty cheques received amounted to more than £2,000, including more than £600 from a whip-round in the local village pub. This boy was obviously much loved. When I read the Order of Service for the funeral ceremony, I literally wept (although I had never met Luke), and I asked for permission to reproduce it at RichardDawkins.net. A lone piper played the Manx lament 'Ellan Vannin'. Two friends spoke eulogies. Dr Ashton himself recited Dylan Thomas's beautiful poem 'Fern Hill' ('Now as I was young and easy, under the apple boughs' – so achingly evocative of lost youth). And then, I catch my breath to report, he read the opening lines of my own *Unweaving the Rainbow*, lines that I have long earmarked for my own funeral.

Obviously there are exceptions, but I suspect that for many people the main reason they cling to religion is not that it is consoling, but that they have been let down by our educational system and don't realize that non-belief is even an option. This is certainly true of most people who think they are creationists. They have simply not been properly taught Darwin's astounding alternative. Probably the same is true of the belittling myth that people 'need' religion. At a recent conference in 2006, an anthropologist (and prize specimen of I'm-an-atheist-buttery) quoted Golda Meir when asked whether she believed in God: 'I believe in the Jewish people, and the Jewish people believe in God.' Our anthropologist substituted his own version: 'I believe in people, and people believe in God.' I prefer to say that I believe in people, and people, when given the right encouragement to think for themselves about all the information now available, very often turn out *not* to believe in God and to lead fulfilled and satisfied – indeed, *liberated* – lives.

In this new paperback edition I have taken the opportunity to make a few minor improvements, and correct some small errors that readers of the hardback have kindly drawn to my attention.

Preface to the first edition

As a child, my wife hated her school and wished she could leave. Years later, when she was in her twenties, she disclosed this unhappy fact to her parents, and her mother was aghast: 'But darling, why didn't you come to us and tell us?' Lalla's reply is my text for today: 'But I didn't know I could.'

I didn't know I could.

I suspect – well, I am sure – that there are lots of people out there who have been brought up in some religion or other, are unhappy in it, don't believe it, or are worried about the evils that are done in its name; people who feel vague yearnings to leave their parents' religion and wish they could, but just don't realize that leaving is an option. If you are one of them, this book is for you. It is intended to raise consciousness – raise consciousness to the fact that to be an atheist is a realistic aspiration, and a brave and splendid one. You can be an atheist who is happy, balanced, moral, and intellectually fulfilled. That is the first of my consciousness-raising messages. I also want to raise consciousness in three other ways, which I'll come on to.

In January 2006 I presented a two-part television documentary on British television (Channel Four) called *Root of All Evil?* From the start, I didn't like the title and fought it hard. Religion is not the root of *all* evil, for no one thing is the root of all anything. But I was delighted with the advertisement that Channel Four put in the national newspapers. It was a picture of the Manhattan skyline with the caption 'Imagine a world without religion.' What was the connection? The twin towers of the World Trade Center were conspicuously present.

Imagine, with John Lennon, a world with no religion. Imagine no suicide bombers, no 9/11, no 7/7, no Crusades, no witch-hunts, no Gunpowder Plot, no Indian partition, no

Israeli/Palestinian wars, no Serb/Croat/Muslim massacres, no persecution of Jews as 'Christ-killers', no Northern Ireland 'troubles', no 'honour killings', no shiny-suited bouffant-haired televangelists fleecing gullible people of their money ('God wants you to give till it hurts'). Imagine no Taliban to blow up ancient statues, no public beheadings of blasphemers, no flogging of female skin for the crime of showing an inch of it. Incidentally, my colleague Desmond Morris informs me that John Lennon's magnificent song is sometimes performed in America with the phrase 'and no religion too' expurgated. One version even has the effrontery to change it to 'and *one* religion too'.

Perhaps you feel that agnosticism is a reasonable position, but that atheism is just as dogmatic as religious belief? If so, I hope Chapter 2 will change your mind, by persuading you that 'the God Hypothesis' is a scientific hypothesis about the universe, which should be analysed as sceptically as any other. Perhaps you have been taught that philosophers and theologians have put forward good reasons to believe in God. If you think that, you might enjoy Chapter 3 on 'Arguments for God's existence' – the arguments turn out to be spectacularly weak. Maybe you think it is obvious that God must exist, for how else could the world have come into being? How else could there be life, in all its rich diversity, with every species looking uncannily as though it had been 'designed'? If your thoughts run along those lines, I hope you will gain enlightenment from Chapter 4 on 'Why there almost certainly is no God'. Far from pointing to a designer, the illusion of design in the living world is explained with far greater economy and with devastating elegance by Darwinian natural selection. And, while natural selection itself is limited to explaining the living world, it raises our consciousness to the likelihood of comparable explanatory 'cranes' that may aid our understanding of the cosmos itself. The power of cranes such as natural selection is the second of my four consciousness-raisers.

Perhaps you think there must be a god or gods because anthropologists and historians report that believers dominate every human culture. If you find that convincing, please refer to Chapter 5, on 'The roots of religion', which explains why belief is so ubiquitous. Or do you think that religious belief is necessary in order for us to have justifiable morals? Don't we need God, in order to be good? Please read Chapters 6 and 7 to see why this is not so. Do you still have a soft spot for religion as a good thing for the world, even if you yourself have lost your faith? Chapter 8 will invite you to think about ways in which religion is not such a good thing for the world.

If you feel trapped in the religion of your upbringing, it would be worth asking yourself how this came about. The answer is usually some form of childhood indoctrination. If you are religious at all it is overwhelmingly probable that your religion is that of your parents. If you were born in Arkansas and you think Christianity is true and Islam false, knowing full well that you would think the opposite if you had been born in Afghanistan, you are the victim of childhood indoctrination. *Mutatis mutandis* if you were born in Afghanistan.

The whole matter of religion and childhood is the subject of Chapter 9, which also includes my third consciousness-raiser. Just as feminists wince when they hear 'he' rather than 'he or she', or 'man' rather than 'human', I want everybody to flinch whenever we hear a phrase such as 'Catholic child' or 'Muslim child'. Speak of a 'child of Catholic parents' if you like; but if you hear anybody speak of a 'Catholic child', stop them and politely point out that children are too young to know where they stand on such issues, just as they are too young to know where they stand on economics or politics. Precisely because my purpose is consciousness-raising, I shall not apologize for mentioning it here in the Preface as well as in Chapter 9. You can't say it too often. I'll say it again. That is not a Muslim child, but a child of Muslim parents. That child is too young to know whether it is a Muslim or not. There is no such thing as a Muslim child. There is no such thing as a Christian child.

Chapters 1 and 10 top and tail the book by explaining, in their different ways, how a proper understanding of the magnificence of the real world, while never becoming a religion, can fill the inspirational role that religion has historically – and inadequately – usurped.

My fourth consciousness-raiser is atheist pride. Being an atheist is nothing to be apologetic about. On the contrary, it is something to be proud of, standing tall to face the far horizon, for atheism nearly always indicates a healthy independence of mind and, indeed, a healthy mind. There are many people who know, in their heart of hearts, that they are atheists, but dare not admit it to their families or even, in some cases, to themselves. Partly, this is because the very word 'atheist' has been assiduously built up as a terrible and frightening label. Chapter 9 quotes the comedian Julia Sweeney's tragi-comic story of her parents' discovery, through reading a newspaper, that she had become an atheist. Not believing in God they could just about take, but an atheist! An *ATHEIST*? (The mother's voice rose to a scream.)

I need to say something to American readers in particular at this point, for the religiosity of today's America is something truly remarkable. The lawyer Wendy Kaminer was exaggerating only slightly when she remarked that making fun of religion is as risky as burning a flag in an American Legion Hall.[1] The status of atheists in America today is on a par with that of homosexuals fifty years ago. Now, after the Gay Pride movement, it is possible, though still not very easy, for a homosexual to be elected to public office. A Gallup poll taken in 1999 asked Americans whether they would vote for an otherwise well-qualified person who was a woman (95 per cent would), Roman Catholic (94 per cent would), Jew (92 per cent), black (92 per cent), Mormon (79 per cent), homosexual (79 per cent) or atheist (49 per cent). Clearly we have a long way to go. But atheists are a lot more numerous, especially among the educated elite, than many realize. This was so even in the nineteenth century, when John Stuart Mill was already able to say: 'The world would be astonished if it knew how great a proportion of its brightest ornaments, of those most distinguished even in popular estimation for wisdom and virtue, are complete sceptics in religion.'

This must be even truer today and, indeed, I present evidence for it in Chapter 3. The reason so many people don't notice atheists is that many of us are reluctant to 'come out'. My dream is that this book may help people to come out. Exactly as in the case of the gay movement, the more people come out, the easier it will be for others to join them. There may be a critical mass for the initiation of a chain reaction.

American polls suggest that atheists and agnostics far outnumber religious Jews, and even outnumber most other particular religious groups. Unlike Jews, however, who are notoriously one of the most effective political lobbies in the United States, and unlike evangelical Christians, who wield even greater political power, atheists and agnostics are not organized and therefore exert almost zero influence. Indeed, organizing atheists has been compared to herding cats, because they tend to think independently and will not conform to authority. But a good first step would be to build up a critical mass of those willing to 'come out', thereby encouraging others to do so. Even if they can't be herded, cats in sufficient numbers can make a lot of noise and they cannot be ignored.

The word 'delusion' in my title has disquieted some psychiatrists who regard it as a technical term, not to be bandied about. Three of them wrote to me to propose a special technical term for religious delusion: 'relusion'.[2] Maybe it'll catch on. But for now I am going to stick with 'delusion', and I need to justify my use of it. The *Penguin English Dictionary* defines a delusion as 'a false belief or impression'. Surprisingly, the illustrative quotation the dictionary gives is from Phillip E. Johnson: 'Darwinism is the story of humanity's liberation from the delusion that its destiny is controlled by a power higher than itself.' Can that be the same Phillip E. Johnson who leads the creationist charge against Darwinism in America today? Indeed it is, and the quotation is, as we might guess, taken out of context. I hope the fact that I have stated as much will be noted, since the same courtesy has not been extended to me in numerous creationist quotations of my works, deliberately and misleadingly taken out of context. Whatever Johnson's own meaning, his sentence as it stands is one that I would be happy to endorse. The dictionary supplied with Microsoft Word defines a delusion as 'a persistent false belief held in the face of strong contradictory evidence, especially as a symptom of psychiatric disorder'. The first part captures religious faith perfectly. As to whether it is a symptom of a psychiatric disorder, I am inclined to follow Robert M. Pirsig, author of *Zen and the Art of Motorcycle Maintenance*: 'When one person suffers from a delusion, it is called insanity. When many people suffer from a delusion it is called Religion.'

If this book works as I intend, religious readers who open it will be atheists when they put it down. What presumptuous optimism! Of course, dyed-in-the-wool faith-heads are immune to argument, their resistance built up over years of childhood indoctrination using methods that took centuries to mature (whether by evolution or design). Among the more effective immunological devices is a dire warning to avoid even opening a book like this, which is surely a work of Satan. But I believe there are plenty of open-minded people out there: people whose childhood indoctrination was not too insidious, or for other reasons didn't 'take', or whose native intelligence is strong enough to overcome it. Such free spirits should need only a little encouragement to break free of the vice of religion altogether. At very least, I hope that nobody who reads this book will be able to say, 'I didn't know I could.'

For help in the preparation of this book, I am grateful to many friends and colleagues. I cannot mention them all, but they include my literary agent John Brockman, and my editors, Sally Gaminara (for Transworld) and Eamon Dolan (for Houghton Mifflin), both of whom read the book with sensitivity and intelligent understanding, and gave me a helpful mixture of criticism and advice. Their whole-hearted and enthusiastic belief in the book was very encouraging to me. Gillian

Somerscales has been an exemplary copy editor, as constructive with her suggestions as she was meticulous with her corrections. Others who criticized various drafts, and to whom I am very grateful, are Jerry Coyne, J. Anderson Thomson, R. Elisabeth Cornwell, Ursula Goodenough, Latha Menon and especially Karen Owens, critic *extraordinaire*, whose acquaintance with the stitching and unstitching of every draft of the book has been almost as detailed as my own.

The book owes something (and vice versa) to the two-part television documentary *Root of All Evil?*, which I presented on British television (Channel Four) in January 2006. I am grateful to all who were involved in the production, including Deborah Kidd, Russell Barnes, Tim Cragg, Adam Prescod, Alan Clements and Hamish Mykura. For permission to use quotations from the documentary I thank IWC Media and Channel Four. *Root of All Evil?* achieved excellent ratings in Britain, and it has also been taken by the Australian Broadcasting Corporation. It remains to be seen whether any US television channel will dare to show it.*

This book has been developing in my mind for some years. During that time, some of the ideas inevitably found their way into lectures, for example my Tanner Lectures at Harvard, and articles in newspapers and magazines. Readers of my regular column in *Free Inquiry*, especially, may find certain passages familiar. I am grateful to Tom Flynn, the Editor of that admirable magazine, for the stimulus he gave me when he commissioned me to become a regular columnist. After a temporary hiatus during the finishing of the book, I hope now to resume my column, and will no doubt use it to respond to the aftermath of the book.

For a variety of reasons I am grateful to Dan Dennett, Marc Hauser, Michael Stirrat, Sam Harris, Helen Fisher, Margaret Downey, Ibn Warraq, Hermione Lee, Julia Sweeney, Dan

---

*As the paperback goes to press, the answer is still no. DVDs, however, are now available from http://richarddawkins.net/store.

Barker, Josephine Welsh, Ian Baird and especially George Scales. Nowadays, a book such as this is not complete until it becomes the nucleus of a living website, a forum for supplementary materials, reactions, discussions, questions and answers – who knows what the future may bring? I hope that www.richarddawkins.net, the website of the Richard Dawkins Foundation for Reason and Science, will come to fill that role, and I am extremely grateful to Josh Timonen for the artistry, professionalism and sheer hard work that he is putting into it.

Above all, I thank my wife Lalla Ward, who has coaxed me through all my hesitations and self-doubts, not just with moral support and witty suggestions for improvement, but by reading the entire book aloud to me, at two different stages in its development, so I could apprehend very directly how it might seem to a reader other than myself. I recommend the technique to other authors, but I must warn that for best results the reader must be a professional actor, with voice and ear sensitively tuned to the music of language.

CHAPTER 1

A deeply religious non-believer

I don't try to imagine a personal God; it suffices to stand in awe at the structure of the world, insofar as it allows our inadequate senses to appreciate it.

ALBERT EINSTEIN

DESERVED RESPECT

The boy lay prone in the grass, his chin resting on his hands. He suddenly found himself overwhelmed by a heightened awareness of the tangled stems and roots, a forest in microcosm, a transfigured world of ants and beetles and even – though he wouldn't have known the details at the time – of soil bacteria by the billions, silently and invisibly shoring up the economy of the micro-world. Suddenly the micro-forest of the turf seemed to swell and become one with the universe, and with the rapt mind of the boy contemplating it. He interpreted the experience in religious terms and it led him eventually to the priesthood. He was ordained an Anglican priest and became a chaplain at my school, a teacher of whom I was fond. It is thanks to decent liberal clergymen like him that nobody could ever claim that I had religion forced down my throat.*

In another time and place, that boy could have been me under the stars, dazzled by Orion, Cassiopeia and Ursa Major,

---

*Our sport during lessons was to sidetrack him away from scripture and towards stirring tales of Fighter Command and the Few. He had done war

tearful with the unheard music of the Milky Way, heady with the night scents of frangipani and trumpet flowers in an African garden. Why the same emotion should have led my chaplain in one direction and me in the other is not an easy question to answer. A quasi-mystical response to nature and the universe is common among scientists and rationalists. It has no connection with supernatural belief. In his boyhood at least, my chaplain was presumably not aware (nor was I) of the closing lines of *The Origin of Species* – the famous 'entangled bank' passage, 'with birds singing on the bushes, with various insects flitting about, and with worms crawling through the damp earth'. Had he been, he would certainly have identified with it and, instead of the priesthood, might have been led to Darwin's view that all was 'produced by laws acting around us':

> Thus, from the war of nature, from famine and death, the most exalted object which we are capable of conceiving, namely, the production of the higher animals, directly follows. There is grandeur in this view of life, with its several powers, having been originally breathed into a few forms or into one; and that, whilst this planet has gone cycling on according to the fixed law of gravity, from so simple a beginning endless forms most beautiful and most wonderful have been, and are being, evolved.

Carl Sagan, in *Pale Blue Dot*, wrote:

> How is it that hardly any major religion has looked

---

service in the RAF and it was with familiarity, and something of the affection that I still retain for the Church of England (at least by comparison with the competition), that I later read John Betjeman's poem:

> Our padre is an old sky pilot,
> Severely now they've clipped his wings,
> But still the flagstaff in the Rect'ry garden
> Points to Higher Things . . .

> at science and concluded, 'This is better than we thought! The Universe is much bigger than our prophets said, grander, more subtle, more elegant'? Instead they say, 'No, no, no! My god is a little god, and I want him to stay that way.' A religion, old or new, that stressed the magnificence of the Universe as revealed by modern science might be able to draw forth reserves of reverence and awe hardly tapped by the conventional faiths.

All Sagan's books touch the nerve-endings of transcendent wonder that religion monopolized in past centuries. My own books have the same aspiration. Consequently I hear myself often described as a deeply religious man. An American student wrote to me that she had asked her professor whether he had a view about me. 'Sure,' he replied. 'He's positive science is incompatible with religion, but he waxes ecstatic about nature and the universe. To me, that *is* religion!' But is 'religion' the right word? I don't think so. The Nobel Prize-winning physicist (and atheist) Steven Weinberg made the point as well as anybody, in *Dreams of a Final Theory*:

> Some people have views of God that are so broad and flexible that it is inevitable that they will find God wherever they look for him. One hears it said that 'God is the ultimate' or 'God is our better nature' or 'God is the universe.' Of course, like any other word, the word 'God' can be given any meaning we like. If you want to say that 'God is energy,' then you can find God in a lump of coal.

Weinberg is surely right that, if the word God is not to become completely useless, it should be used in the way people have generally understood it: to denote a supernatural creator that is 'appropriate for us to worship'.

Much unfortunate confusion is caused by failure to distinguish what can be called Einsteinian religion from supernatural religion. Einstein sometimes invoked the name of God (and he is not the only atheistic scientist to do so), inviting misunderstanding by supernaturalists eager to misunderstand and claim so illustrious a thinker as their own. The dramatic (or was it mischievous?) ending of Stephen Hawking's *A Brief History of Time*, 'For then we should know the mind of God', is notoriously misconstrued. It has led people to believe, mistakenly of course, that Hawking is a religious man. The cell biologist Ursula Goodenough, in *The Sacred Depths of Nature*, sounds more religious than Hawking or Einstein. She loves churches, mosques and temples, and numerous passages in her book fairly beg to be taken out of context and used as ammunition for supernatural religion. She goes so far as to call herself a 'Religious Naturalist'. Yet a careful reading of her book shows that she is really as staunch an atheist as I am.

'Naturalist' is an ambiguous word. For me it conjures my childhood hero, Hugh Lofting's Doctor Dolittle (who, by the way, had more than a touch of the 'philosopher' naturalist of HMS *Beagle* about him). In the eighteenth and nineteenth centuries, naturalist meant what it still means for most of us today: a student of the natural world. Naturalists in this sense, from Gilbert White on, have often been clergymen. Darwin himself was destined for the Church as a young man, hoping that the leisurely life of a country parson would enable him to pursue his passion for beetles. But philosophers use 'naturalist' in a very different sense, as the opposite of *supernaturalist*. Julian Baggini explains in *Atheism: A Very Short Introduction* the meaning of an atheist's commitment to naturalism: 'What most atheists do believe is that although there is only one kind of stuff in the universe and it is physical, out of this stuff come minds, beauty, emotions, moral values – in short the full gamut of phenomena that gives richness to human life.'

Human thoughts and emotions *emerge* from exceedingly complex interconnections of physical entities within the brain.

An atheist in this sense of philosophical naturalist is somebody who believes there is nothing beyond the natural, physical world, no *super*natural creative intelligence lurking behind the observable universe, no soul that outlasts the body and no miracles – except in the sense of natural phenomena that we don't yet understand. If there is something that appears to lie beyond the natural world as it is now imperfectly understood, we hope eventually to understand it and embrace it within the natural. As ever when we unweave a rainbow, it will not become less wonderful.

Great scientists of our time who sound religious usually turn out not to be so when you examine their beliefs more deeply. This is certainly true of Einstein and Hawking. The present Astronomer Royal and President of the Royal Society, Martin Rees, told me that he goes to church as an 'unbelieving Anglican . . . out of loyalty to the tribe'. He has no theistic beliefs, but shares the poetic naturalism that the cosmos provokes in the other scientists I have mentioned. In the course of a recently televised conversation, I challenged my friend the obstetrician Robert Winston, a respected pillar of British Jewry, to admit that his Judaism was of exactly this character and that he didn't really believe in anything supernatural. He came close to admitting it but shied at the last fence (to be fair, he was supposed to be interviewing me, not the other way around).[3] When I pressed him, he said he found that Judaism provided a good discipline to help him structure his life and lead a good one. Perhaps it does; but that, of course, has not the smallest bearing on the truth value of any of its supernatural claims. There are many intellectual atheists who proudly call themselves Jews and observe Jewish rites, perhaps out of loyalty to an ancient tradition or to murdered relatives, but also because of a confused and confusing willingness to label as 'religion' the pantheistic reverence which many of us share with its most distinguished exponent, Albert Einstein. They may not believe but, to borrow a phrase from the philosopher Daniel Dennett, they 'believe in belief'.[4]

One of Einstein's most eagerly quoted remarks is 'Science without religion is lame, religion without science is blind.' But Einstein also said,

> It was, of course, a lie what you read about my religious convictions, a lie which is being systematically repeated. I do not believe in a personal God and I have never denied this but have expressed it clearly. If something is in me which can be called religious then it is the unbounded admiration for the structure of the world so far as our science can reveal it.

Does it seem that Einstein contradicted himself? That his words can be cherry-picked for quotes to support both sides of an argument? No. By 'religion' Einstein meant something entirely different from what is conventionally meant. As I continue to clarify the distinction between supernatural religion on the one hand and Einsteinian religion on the other, bear in mind that I am calling only *supernatural* gods delusional.

Here are some more quotations from Einstein, to give a flavour of Einsteinian religion.

> I am a deeply religious nonbeliever. This is a somewhat new kind of religion.

> I have never imputed to Nature a purpose or a goal, or anything that could be understood as anthropomorphic. What I see in Nature is a magnificent structure that we can comprehend only very imperfectly, and that must fill a thinking person with a feeling of humility. This is a genuinely religious feeling that has nothing to do with mysticism.

> The idea of a personal God is quite alien to me and seems even naive.

In greater numbers since his death, religious apologists understandably try to claim Einstein as one of their own. Some of his religious contemporaries saw him very differently. In 1940 Einstein wrote a famous paper justifying his statement 'I do not believe in a personal God.' This and similar statements provoked a storm of letters from the religiously orthodox, many of them alluding to Einstein's Jewish origins. The extracts that follow are taken from Max Jammer's book *Einstein and Religion* (which is also my main source of quotations from Einstein himself on religious matters). The Roman Catholic Bishop of Kansas City said: 'It is sad to see a man, who comes from the race of the Old Testament and its teaching, deny the great tradition of that race.' Other Catholic clergymen chimed in: 'There is no other God but a personal God . . . Einstein does not know what he is talking about. He is all wrong. Some men think that because they have achieved a high degree of learning in some field, they are qualified to express opinions in all.' The notion that religion is a proper *field*, in which one might claim *expertise*, is one that should not go unquestioned. That clergyman presumably would not have deferred to the expertise of a claimed 'fairyologist' on the exact shape and colour of fairy wings. Both he and the bishop thought that Einstein, being theologically untrained, had misunderstood the nature of God. On the contrary, Einstein understood very well exactly what he was denying.

An American Roman Catholic lawyer, working on behalf of an ecumenical coalition, wrote to Einstein:

> We deeply regret that you made your statement . . . in which you ridicule the idea of a personal God. In the past ten years nothing has been so calculated to make people think that Hitler had some reason to expel the Jews from Germany as your statement. Conceding your right to free speech, I still say that your statement constitutes you as one of the greatest sources of discord in America.

A New York rabbi said: 'Einstein is unquestionably a great scientist, but his religious views are diametrically opposed to Judaism.'

'But'? '*But*'? Why not 'and'?

The president of a historical society in New Jersey wrote a letter that so damningly exposes the weakness of the religious mind, it is worth reading twice:

> We respect your learning, Dr Einstein; but there is one thing you do not seem to have learned: that God is a spirit and cannot be found through the telescope or microscope, no more than human thought or emotion can be found by analyzing the brain. As everyone knows, religion is based on Faith, not knowledge. Every thinking person, perhaps, is assailed at times with religious doubt. My own faith has wavered many a time. But I never told anyone of my spiritual aberrations for two reasons: (1) I feared that I might, by mere suggestion, disturb and damage the life and hopes of some fellow being; (2) because I agree with the writer who said, 'There is a mean streak in anyone who will destroy another's faith.' . . . I hope, Dr Einstein, that you were misquoted and that you will yet say something more pleasing to the vast number of the American people who delight to do you honor.

What a devastatingly revealing letter! Every sentence drips with intellectual and moral cowardice.

Less abject but more shocking was the letter from the Founder of the Calvary Tabernacle Association in Oklahoma:

> Professor Einstein, I believe that every Christian in America will answer you, 'We will not give up our belief in our God and his son Jesus Christ, but we

> invite you, if you do not believe in the God of the people of this nation, to go back where you came from.' I have done everything in my power to be a blessing to Israel, and then you come along and with one statement from your blasphemous tongue, do more to hurt the cause of your people than all the efforts of the Christians who love Israel can do to stamp out anti-Semitism in our land. Professor Einstein, every Christian in America will immediately reply to you, 'Take your crazy, fallacious theory of evolution and go back to Germany where you came from, or stop trying to break down the faith of a people who gave you a welcome when you were forced to flee your native land.'

The one thing all his theistic critics got right was that Einstein was not one of them. He was repeatedly indignant at the suggestion that he was a theist. So, was he a deist, like Voltaire and Diderot? Or a pantheist, like Spinoza, whose philosophy he admired: 'I believe in Spinoza's God who reveals himself in the orderly harmony of what exists, not in a God who concerns himself with fates and actions of human beings'?

Let's remind ourselves of the terminology. A theist believes in a supernatural intelligence who, in addition to his main work of creating the universe in the first place, is still around to oversee and influence the subsequent fate of his initial creation. In many theistic belief systems, the deity is intimately involved in human affairs. He answers prayers; forgives or punishes sins; intervenes in the world by performing miracles; frets about good and bad deeds, and knows when we do them (or even *think* of doing them). A deist, too, believes in a supernatural intelligence, but one whose activities were confined to setting up the laws that govern the universe in the first place. The deist God never intervenes thereafter, and certainly has no specific interest in human affairs. Pantheists don't believe in a supernatural God at all, but use the word God as a non-supernatural synonym for Nature, or for the Universe, or for the lawfulness that governs its workings. Deists differ from theists in that their God does not answer prayers, is not interested in sins or confessions, does not read our thoughts and does not intervene with capricious miracles. Deists differ from pantheists in that the deist God is some kind of cosmic intelligence, rather than the pantheist's metaphoric or poetic *synonym* for the laws of the universe. Pantheism is sexed-up atheism. Deism is watered-down theism.

There is every reason to think that famous Einsteinisms like 'God is subtle but he is not malicious' or 'He does not play dice' or 'Did God have a choice in creating the Universe?' are pantheistic, not deistic, and certainly not theistic. 'God does not play dice' should be translated as 'Randomness does not lie at the heart of all things.' 'Did God have a choice in creating the Universe?' means 'Could the universe have begun in any other way?' Einstein was using 'God' in a purely metaphorical, poetic sense. So is Stephen Hawking, and so are most of those physicists who occasionally slip into the language of religious metaphor. Paul Davies's *The Mind of God* seems to hover somewhere between Einsteinian pantheism and an obscure form of deism – for which he was rewarded with the Templeton Prize (a very large sum of money given annually by the Templeton Foundation, usually to a scientist who is prepared to say something nice about religion).

Let me sum up Einsteinian religion in one more quotation from Einstein himself: 'To sense that behind anything that can be experienced there is a something that our mind cannot grasp and whose beauty and sublimity reaches us only indirectly and as a feeble reflection, this is religiousness. In this sense I am religious.' In this sense I too am religious, with the reservation that 'cannot grasp' does not have to mean 'forever ungraspable'. But I prefer not to call myself religious because it is misleading. It is destructively misleading because, for the vast majority of people, 'religion' implies 'supernatural'. Carl Sagan put it well: '. . . if by "God" one means the set of physical laws that govern the universe, then clearly there is such a God. This God is emotionally unsatisfying . . . it does not make much sense to pray to the law of gravity.'

Amusingly, Sagan's last point was foreshadowed by the Reverend Dr Fulton J. Sheen, a professor at the Catholic University of America, as part of a fierce attack upon Einstein's 1940 disavowal of a personal God. Sheen sarcastically asked whether anyone would be prepared to lay down his life for the Milky Way. He seemed to think he was making a point against Einstein, rather than for him, for he added: 'There is only one fault with his cosmical religion: he put an extra letter in the word – the letter "s".' There is nothing comical about Einstein's beliefs. Nevertheless, I wish that physicists would refrain from using the word God in their special metaphorical sense. The metaphorical or pantheistic God of the physicists is light years away from the interventionist, miracle-wreaking, thought-reading, sin-punishing, prayer-answering God of the Bible, of priests, mullahs and rabbis, and of ordinary language. Deliberately to confuse the two is, in my opinion, an act of intellectual high treason.

UNDESERVED RESPECT

My title, *The God Delusion*, does not refer to the God of Einstein and the other enlightened scientists of the previous section. That is why I needed to get Einsteinian religion out of the way to begin with: it has a proven capacity to confuse. In the rest of this book I am talking only about *supernatural* gods, of which the most familiar to the majority of my readers will be Yahweh, the God of the Old Testament. I shall come to him in a moment. But before leaving this preliminary chapter I need to deal with one more matter that would otherwise bedevil the whole book. This time it is a matter of etiquette. It is possible that religious readers will be offended by what I have to say, and will find in these pages insufficient *respect* for their own particular beliefs (if not the beliefs that others treasure). It would be a shame if such offence prevented them from reading on, so I want to sort it out here, at the outset.

A widespread assumption, which nearly everybody in our society accepts – the non-religious included – is that religious faith is especially vulnerable to offence and should be protected by an abnormally thick wall of respect, in a different class from the respect that any human being should pay to any other. Douglas Adams put it so well, in an impromptu speech made in Cambridge shortly before his death,[5] that I never tire of sharing his words:

> Religion . . . has certain ideas at the heart of it which we call sacred or holy or whatever. What it means is, 'Here is an idea or a notion that you're not allowed to say anything bad about; you're just not. Why not? – because you're not!' If somebody votes for a party that you don't agree with, you're free to argue about it as much as you like; everybody will have an argument but nobody feels aggrieved by it. If somebody thinks taxes should go up or down you are free to have an argument about it. But on the other hand if somebody says 'I mustn't move a light switch on a Saturday', you say, 'I *respect* that'.
>
> Why should it be that it's perfectly legitimate to support the Labour party or the Conservative party, Republicans or Democrats, this model of economics versus that, Macintosh instead of Windows – but to have an opinion about how the Universe began, about who created the Universe . . . no, that's holy? . . . We are used to not challenging religious ideas but it's very interesting how much of a furore Richard creates when he does it! Everybody gets absolutely frantic about it because you're not allowed to say these things. Yet when you look at it rationally there is no reason why those ideas

> shouldn't be as open to debate as any other, except that we have agreed somehow between us that they shouldn't be.

Here's a particular example of our society's overweening respect for religion, one that really matters. By far the easiest grounds for gaining conscientious objector status in wartime are religious. You can be a brilliant moral philosopher with a prize-winning doctoral thesis expounding the evils of war, and still be given a hard time by a draft board evaluating your claim to be a conscientious objector. Yet if you can say that one or both of your parents is a Quaker you sail through like a breeze, no matter how inarticulate and illiterate you may be on the theory of pacifism or, indeed, Quakerism itself.

At the opposite end of the spectrum from pacifism, we have a pusillanimous reluctance to use religious names for warring factions. In Northern Ireland, Catholics and Protestants are euphemized to 'Nationalists' and 'Loyalists' respectively. The very word 'religions' is bowdlerized to 'communities', as in 'inter-community warfare'. Iraq, as a consequence of the Anglo-American invasion of 2003, degenerated into sectarian civil war between Sunni and Shia Muslims. Clearly a religious conflict – yet in the *Independent* of 20 May 2006 the front-page headline and first leading article both described it as 'ethnic cleansing'. 'Ethnic' in this context is yet another euphemism. What we are seeing in Iraq is religious cleansing. The original usage of 'ethnic cleansing' in the former Yugoslavia is also arguably a euphemism for religious cleansing, involving Orthodox Serbs, Catholic Croats and Muslim Bosnians.[6]

I have previously drawn attention to the privileging of religion in public discussions of ethics in the media and in government.[7] Whenever a controversy arises over sexual or reproductive morals, you can bet that religious leaders from several different faith groups will be prominently represented on influential committees, or on panel discussions on radio or television. I'm not suggesting that we should go out of our way to censor the views of these people. But why does our society beat a path to their door, as though they had some expertise comparable to that of, say, a moral philosopher, a family lawyer or a doctor?

Here's another weird example of the privileging of religion. On 21 February 2006 the United States Supreme Court ruled, in accordance with the Constitution, that a church in New Mexico should be exempt from the law, which everybody else has to obey, against the taking of hallucinogenic drugs.[8] Faithful members of the Centro Espirita Beneficiente Uniao do Vegetal believe that they can understand God only by drinking hoasca tea, which contains the illegal hallucinogenic drug dimethyltryptamine. Note that it is sufficient that they *believe* that the drug enhances their understanding. They do not have to produce evidence. Conversely, there is plenty of evidence that cannabis eases the nausea and discomfort of cancer sufferers undergoing chemotherapy. Yet, again in accordance with the Constitution, the Supreme Court ruled in 2005 that all patients who use cannabis for medicinal purposes are vulnerable to federal prosecution (even in the minority of states where such specialist use is legalized). Religion, as ever, is the trump card. Imagine members of an art appreciation society pleading in court that they 'believe' they need a hallucinogenic drug in order to enhance their understanding of Impressionist or Surrealist paintings. Yet, when a church claims an equivalent need, it is backed by the highest court in the land. Such is the power of religion as a talisman.

Eighteen years ago, I was one of thirty-six writers and artists commissioned by the magazine *New Statesman* to write in support of the distinguished author Salman Rushdie,[9] then under sentence of death for writing a novel. Incensed by the 'sympathy' for Muslim 'hurt' and 'offence' expressed by Christian leaders and even some secular opinion-formers, I drew the following parallel:

> If the advocates of apartheid had their wits about them they would claim – for all I know truthfully – that allowing mixed races is against their religion. A good part of the opposition would respectfully tip-toe away. And it is no use claiming that this is an unfair parallel because apartheid has no rational justification. The whole point of religious faith, its strength and chief glory, is that it does not depend on rational justification. The rest of us are expected to defend our prejudices. But ask a religious person to justify their faith and you infringe 'religious liberty'.

Little did I know that something pretty similar would come to pass in the twenty-first century. The *Los Angeles Times* (10 April 2006) reported that numerous Christian groups on campuses around the United States were suing their universities for enforcing anti-discrimination rules, including prohibitions against harassing or abusing homosexuals. As a typical example, in 2004 James Nixon, a twelve-year-old boy in Ohio, won the right in court to wear a T-shirt to school bearing the words 'Homosexuality is a sin, Islam is a lie, abortion is murder. Some issues are just black and white!'[10] The school told him not to wear the T-shirt – and the boy's parents sued the school. The parents might have had a conscionable case if they had based it on the First Amendment's guarantee of freedom of speech. But they didn't. Instead, the Nixons' lawyers appealed to the constitutional right to freedom of *religion*. Their victorious lawsuit was supported by the Alliance Defense Fund of Arizona, whose business it is to 'press the legal battle for religious freedom'.

The Reverend Rick Scarborough, supporting the wave of similar Christian lawsuits brought to establish religion as a legal justification for discrimination against homosexuals and other groups, has named it the civil rights struggle of the twenty-first century: 'Christians are going to have to take a stand for the right to be Christian.'[11] Once again, if such people took their stand on the right to free speech, one might reluctantly sympathize. But that isn't what it is about. 'The right to be Christian' seems in this case to mean 'the right to poke your nose into other people's private lives'. The legal case in favour of discrimination against homosexuals is being mounted as a counter-suit against alleged religious discrimination! And the law seems to respect this. You can't get away with saying, 'If you try to stop me from insulting homosexuals it violates my freedom of prejudice.' But you can get away with saying, 'It violates my freedom of religion.' What, when you think about it, is the difference? Yet again, religion trumps all.

I'll end the chapter with a particular case study, which tellingly illuminates society's exaggerated respect for religion, over and above ordinary human respect. The case flared up in February 2006 – a ludicrous episode, which veered wildly between the extremes of comedy and tragedy. The previous September, the Danish newspaper *Jyllands-Posten* published twelve cartoons depicting the prophet Muhammad. Over the next three months, indignation was carefully and systematically nurtured throughout the Islamic world by a small group of Muslims living in Denmark, led by two imams who had been granted sanctuary there.[12] In late 2005 these malevolent exiles travelled from Denmark to Egypt bearing a dossier, which was copied and circulated from there to the whole Islamic world, including, importantly, Indonesia. The dossier contained falsehoods about alleged maltreatment of Muslims in Denmark, and the tendentious lie that *Jyllands-Posten* was a government-run newspaper. It also contained the twelve cartoons which, crucially, the imams had supplemented with three additional images whose origin was mysterious but which certainly had no connection with Denmark. Unlike the original twelve, these three add-ons were genuinely offensive – or would have been if they had, as the zealous propagandists alleged, depicted Muhammad. A particularly damaging one of these three was not a cartoon at all but a faxed photograph of a bearded man wearing a fake pig's snout held on with elastic. It has subsequently turned out that this was an Associated Press photograph of a Frenchman entered for a pig-squealing contest at a country fair in France.[13] The photograph had no connection whatsoever with the prophet Muhammad, no connection with Islam, and no connection with Denmark. But the Muslim activists, on their mischief-stirring hike to Cairo, implied all three connections . . . with predictable results.

The carefully cultivated 'hurt' and 'offence' was brought to an explosive head five months after the twelve cartoons were originally published. Demonstrators in Pakistan and Indonesia burned Danish flags (where did they get them from?) and hysterical demands were made for the Danish government to apologize. (Apologize for what? They didn't draw the cartoons, or publish them. Danes just live in a country with a free press, something that people in many Islamic countries might have a hard time understanding.) Newspapers in Norway, Germany, France and even the United States (but, conspicuously, not Britain) reprinted the cartoons in gestures of solidarity with *Jyllands-Posten*, which added fuel to the flames. Embassies and consulates were trashed, Danish goods were boycotted, Danish citizens and, indeed, Westerners generally, were physically threatened; Christian churches in Pakistan, with no Danish or European connections at all, were burned. Nine people were killed when Libyan rioters attacked and burned the Italian consulate in Benghazi. As Germaine Greer wrote, what these people really love and do best is pandemonium.[14]

A bounty of $1 million was placed on the head of 'the Danish cartoonist' by a Pakistani imam – who was apparently unaware that there were twelve different Danish cartoonists, and almost certainly unaware that the three most offensive pictures had never appeared in Denmark at all (and, by the way, where was that million going to come from?). In Nigeria, Muslim protesters against the Danish cartoons burned down several Christian churches, and used machetes to attack and kill (black Nigerian) Christians in the streets. One Christian was put inside a rubber tyre, doused with petrol and set alight. Demonstrators were photographed in Britain bearing banners saying 'Slay those who insult Islam', 'Butcher those who mock Islam', 'Europe you will pay: Demolition is on its way' and 'Behead those who insult Islam'. Fortunately, our political leaders were on hand to remind us that Islam is a religion of peace and mercy.

In the aftermath of all this, the journalist Andrew Mueller interviewed Britain's leading 'moderate' Muslim, Sir Iqbal Sacranie.[15] Moderate he may be by today's Islamic standards, but in Andrew Mueller's account he still stands by the remark he made when Salman Rushdie was condemned to death for writing a novel: 'Death is perhaps too easy for him' – a remark that sets him in ignominious contrast to his courageous predecessor as Britain's most influential Muslim, the late Dr Zaki Badawi, who offered Salman Rushdie sanctuary in his own home. Sacranie told Mueller how concerned he was about the Danish cartoons. Mueller was concerned too, but for a different reason: 'I am concerned that the ridiculous, disproportionate reaction to some unfunny sketches in an obscure Scandinavian newspaper may confirm that . . . Islam and the west are fundamentally irreconcilable.' Sacranie, on the other hand, praised British newspapers for not reprinting the cartoons, to which Mueller voiced the suspicion of most of the nation that 'the restraint of British newspapers derived less from sensitivity to Muslim discontent than it did from a desire not to have their windows broken'.

Sacranie explained that 'The person of the Prophet, peace be upon him, is revered so profoundly in the Muslim world, with a love and affection that cannot be explained in words. It goes beyond your parents, your loved ones, your children. That is part of the faith. There is also an Islamic teaching that one does not depict the Prophet.' This rather assumes, as Mueller observed,

> that the values of Islam trump anyone else's – which is what any follower of Islam does assume, just as any follower of any religion believes that theirs is the sole way, truth and light. If people wish to love a 7th century preacher more than their own families, that's up to them, but nobody else is obliged to take it seriously . . .

Except that if you don't take it seriously and accord it proper respect you are physically threatened, on a scale that no other religion has aspired to since the Middle Ages. One can't help wondering why such violence is necessary, given that, as Mueller notes: 'If any of you clowns are right about anything, the cartoonists are going to hell anyway – won't that do? In the meantime, if you want to get excited about affronts to Muslims, read the Amnesty International reports on Syria and Saudi Arabia.'

Many people have noted the contrast between the hysterical 'hurt' professed by Muslims and the readiness with which Arab media publish stereotypical anti-Jewish cartoons. At a demonstration in Pakistan against the Danish cartoons, a woman in a black burka was photographed carrying a banner reading 'God Bless Hitler'.

In response to all this frenzied pandemonium, decent liberal newspapers deplored the violence and made token noises about free speech. But at the same time they expressed 'respect' and 'sympathy' for the deep 'offence' and 'hurt' that Muslims had 'suffered'. The 'hurt' and 'suffering' consisted, remember, not in any person enduring violence or real pain of any kind: nothing more than a few daubs of printing ink in a newspaper that nobody outside Denmark would ever have heard of but for a deliberate campaign of incitement to mayhem.

I am not in favour of offending or hurting anyone just for the sake of it. But I am intrigued and mystified by the disproportionate privileging of religion in our otherwise secular societies. All politicians must get used to disrespectful cartoons of their faces, and nobody riots in their defence. What is so special about religion that we grant it such uniquely privileged respect? As H. L. Mencken said: 'We must respect the other fellow's religion, but only in the sense and to the extent that we respect his theory that his wife is beautiful and his children smart.'

It is in the light of the unparalleled presumption of respect for religion* that I make my own disclaimer for this book. I shall not go out of my way to offend, but nor shall I don kid gloves to handle religion any more gently than I would handle anything else.

---

* A stunning example of such 'respect' was reported in the *New York Times* while this paperback was in proof. In January 2007, a German Muslim woman had applied for a fast-track divorce on the grounds that her husband, from the very start of the marriage, repeatedly and seriously beat her. While not denying the facts, judge Christa Datz-Winter turned down the application, citing the Qur'an. 'In a remarkable ruling that underlines the tension between Muslim customs and European laws, the judge, Christa Datz-Winter, said that the couple came from a Moroccan cultural milieu, in which she said it was common for husbands to beat their wives. The Koran, she wrote, sanctions such physical abuse' (*New York Times*, 23 March 2007). This incredible story came to light in March 2007 when the unfortunate woman's lawyer disclosed it. To its credit, the Frankfurt court promptly removed Judge Datz-Winter from the case. Nevertheless, the *New York Times* article concludes by quoting a suggestion that the episode will do great damage to other Muslim women suffering domestic abuse: 'Many are already afraid of going to court against their spouses. There have been a string of so-called honor-killings here, in which Turkish Muslim men have murdered women.' Judge Datz-Winter's motivation was put down to 'cultural sensitivity', but there is another name by which you could call it: patronizing insult. 'Of course we Europeans wouldn't dream of behaving like this, but wife-beating is part of "their culture", sanctioned by "their religion", and we should "respect" it.'

CHAPTER 2

The God Hypothesis

The religion of one age is the literary entertainment of the next.

RALPH WALDO EMERSON

The God of the Old Testament is arguably the most unpleasant character in all fiction: jealous and proud of it; a petty, unjust, unforgiving control-freak; a vindictive, bloodthirsty ethnic cleanser; a misogynistic, homophobic, racist, infanticidal, genocidal, filicidal, pestilential, megalomaniacal, sado-masochistic, capriciously malevolent bully. Those of us schooled from infancy in his ways can become desensitized to their horror. A *naïf* blessed with the perspective of innocence has a clearer perception. Winston Churchill's son Randolph somehow contrived to remain ignorant of scripture until Evelyn Waugh and a brother officer, in a vain attempt to keep Churchill quiet when they were posted together during the war, bet him he couldn't read the entire Bible in a fortnight: 'Unhappily it has not had the result we hoped. He has never read any of it before and is hideously excited; keeps reading quotations aloud "I say I bet you didn't know this came in the Bible . . ." or merely slapping his side & chortling "God, isn't God a shit!" '[16] Thomas Jefferson – better read – was of a similar opinion, describing the God of Moses as 'a being of terrific character – cruel, vindictive, capricious and unjust'.

It is unfair to attack such an easy target. The God Hypothesis should not stand or fall with its most unlovely instantiation,

Yahweh, nor his insipidly opposite Christian face, 'Gentle Jesus meek and mild'. (To be fair, this milksop *persona* owes more to his Victorian followers than to Jesus himself. Could anything be more mawkishly nauseating than Mrs C. F. Alexander's 'Christian children all must be / Mild, obedient, good as he'?) I am not attacking the particular qualities of Yahweh, or Jesus, or Allah, or any other specific god such as Baal, Zeus or Wotan. Instead I shall define the God Hypothesis more defensibly: *there exists a superhuman, supernatural intelligence who deliberately designed and created the universe and everything in it, including us.* This book will advocate an alternative view: *any creative intelligence, of sufficient complexity to design anything, comes into existence only as the end product of an extended process of gradual evolution.* Creative intelligences, being evolved, necessarily arrive late in the universe, and therefore cannot be responsible for designing it. God, in the sense defined, is a delusion; and, as later chapters will show, a pernicious delusion.

Not surprisingly, since it is founded on local traditions of private revelation rather than evidence, the God Hypothesis comes in many versions. Historians of religion recognize a progression from primitive tribal animisms, through polytheisms such as those of the Greeks, Romans and Norsemen, to monotheisms such as Judaism and its derivatives, Christianity and Islam.

POLYTHEISM

It is not clear why the change from polytheism to monotheism should be assumed to be a self-evidently progressive improvement. But it widely is – an assumption that provoked Ibn Warraq (author of *Why I Am Not a Muslim*) wittily to conjecture that monotheism is in its turn doomed to subtract one more god and become atheism. The *Catholic Encyclopedia* dismisses polytheism and atheism in the same insouciant breath: 'Formal dogmatic atheism is self-refuting, and has never *de facto* won the reasoned assent of any considerable number of men. Nor can polytheism, however easily it may take hold of the popular imagination, ever satisfy the mind of a philosopher.'[17]

Monotheistic chauvinism was until recently written into the charity law of both England and Scotland, discriminating against polytheistic religions in granting tax-exempt status, while allowing an easy ride to charities whose object was to promote monotheistic religion, sparing them the rigorous vetting quite properly required of secular charities. It was my ambition to persuade a member of Britain's respected Hindu community to come forward and bring a civil action to test this snobbish discrimination against polytheism.

Far better, of course, would be to abandon the promotion of religion altogether as grounds for charitable status. The benefits of this to society would be great, especially in the United States, where the sums of tax-free money sucked in by churches, and polishing the heels of already well-heeled televangelists, reach levels that could fairly be described as obscene. The aptly named Oral Roberts once told his television audience that God would kill him unless they gave him $8 million. Almost unbelievably, it worked. Tax-free! Roberts himself is still going strong, as is 'Oral Roberts University' of Tulsa, Oklahoma. Its buildings, valued at $250 million, were directly commissioned by God himself in these words: 'Raise up your students to hear My voice, to go where My light is dim, where My voice is heard small, and My healing power is not known, even to the uttermost bounds of the Earth. Their work will exceed yours, and in this I am well pleased.'

On reflection, my imagined Hindu litigator would have been as likely to play the 'If you can't beat them join them' card. His polytheism isn't really polytheism but monotheism in disguise. There is only one God – Lord Brahma the creator, Lord Vishnu the preserver, Lord Shiva the destroyer, the goddesses Saraswati, Laxmi and Parvati (wives of Brahma, Vishnu and Shiva), Lord Ganesh the elephant god, and hundreds of others, all are just different manifestations or incarnations of the one God.

Christians should warm to such sophistry. Rivers of medieval ink, not to mention blood, have been squandered over the 'mystery' of the Trinity, and in suppressing deviations such as the Arian heresy. Arius of Alexandria, in the fourth century AD, denied that Jesus was *consubstantial* (i.e. of the same substance or essence) with God. What on earth could that possibly mean, you are probably asking? Substance? What 'substance'? What exactly do you mean by 'essence'? 'Very little' seems the only reasonable reply. Yet the controversy split Christendom down the middle for a century, and the Emperor Constantine ordered that all copies of Arius's book should be burned. Splitting Christendom by splitting hairs – such has ever been the way of theology.

Do we have one God in three parts, or three Gods in one? The *Catholic Encyclopedia* clears up the matter for us, in a masterpiece of theological close reasoning:

> In the unity of the Godhead there are three Persons, the Father, the Son, and the Holy Spirit, these Three Persons being truly distinct one from another. Thus, in the words of the Athanasian Creed: 'the Father is God, the Son is God, and the Holy Spirit is God, and yet there are not three Gods but one God.'

As if that were not clear enough, the *Encyclopedia* quotes the third-century theologian St Gregory the Miracle Worker:

> There is therefore nothing created, nothing subject to another in the Trinity: nor is there anything that has been added as though it once had not existed, but had entered afterwards: therefore the Father has never been without the Son, nor the Son without the Spirit: and this same Trinity is immutable and unalterable forever.

Whatever miracles may have earned St Gregory his nickname, they were not miracles of honest lucidity. His words convey the characteristically obscurantist flavour of theology, which – unlike science or most other branches of human scholarship – has not moved on in eighteen centuries. Thomas Jefferson, as so often, got it right when he said, 'Ridicule is the only weapon which can be used against unintelligible propositions. Ideas must be distinct before reason can act upon them; and no man ever had a distinct idea of the trinity. It is the mere Abracadabra of the mountebanks calling themselves the priests of Jesus.'

The other thing I cannot help remarking upon is the overweening confidence with which the religious assert minute details for which they neither have, nor could have, any evidence. Perhaps it is the very fact that there is no evidence to support theological opinions, either way, that fosters the characteristic draconian hostility towards those of slightly different opinion, especially, as it happens, in this very field of Trinitarianism.

Jefferson heaped ridicule on the doctrine that, as he put it, 'There are three Gods', in his critique of Calvinism. But it is especially the Roman Catholic branch of Christianity that pushes its recurrent flirtation with polytheism towards runaway inflation. The Trinity is (are?) joined by Mary, 'Queen of Heaven', a goddess in all but name, who surely runs God himself a close second as a target of prayers. The pantheon is further swollen by an army of saints, whose intercessory power makes them, if not demigods, well worth approaching on their own specialist subjects. The Catholic Community Forum helpfully lists 5,120 saints,[18] together with their areas of expertise, which include abdominal pains, abuse victims, anorexia, arms dealers, blacksmiths, broken bones, bomb technicians and bowel disorders, to venture no further than the Bs. And we mustn't forget the four Choirs of Angelic Hosts, arrayed in nine orders: Seraphim, Cherubim, Thrones, Dominions, Virtues, Powers, Principalities, Archangels (heads of all hosts), and just plain old Angels, including our closest friends, the ever-watchful

Guardian Angels. What impresses me about Catholic mythology is partly its tasteless kitsch but mostly the airy nonchalance with which these people make up the details as they go along. It is just shamelessly invented.

Pope John Paul II created more saints than all his predecessors of the past several centuries put together, and he had a special affinity with the Virgin Mary. His polytheistic hankerings were dramatically demonstrated in 1981 when he suffered an assassination attempt in Rome, and attributed his survival to intervention by Our Lady of Fatima: 'A maternal hand guided the bullet.' One cannot help wondering why she didn't guide it to miss him altogether. Others might think the team of surgeons who operated on him for six hours deserved at least a share of the credit; but perhaps their hands, too, were maternally guided. The relevant point is that it wasn't just Our Lady who, in the Pope's opinion, guided the bullet, but specifically Our Lady *of Fatima.* Presumably Our Lady of Lourdes, Our Lady of Guadalupe, Our Lady of Medjugorje, Our Lady of Akita, Our Lady of Zeitoun, Our Lady of Garabandal and Our Lady of Knock were busy on other errands at the time.

How did the Greeks, the Romans and the Vikings cope with such polytheological conundrums? Was Venus just another name for Aphrodite, or were they two distinct goddesses of love? Was Thor with his hammer a manifestation of Wotan, or a separate god? Who cares? Life is too short to bother with the distinction between one figment of the imagination and many. Having gestured towards polytheism to cover myself against a charge of neglect, I shall say no more about it. For brevity I shall refer to all deities, whether poly- or monotheistic, as simply 'God'. I am also conscious that the Abrahamic God is (to put it mildly) aggressively male, and this too I shall accept as a convention in my use of pronouns. More sophisticated theologians proclaim the sexlessness of God, while some feminist theologians seek to redress historic injustices by designating her female. But what, after all, is the difference between a non-existent female and a non-existent male? I suppose that, in the ditzily unreal intersection of theology and feminism, existence might indeed be a less salient attribute than gender.

I am aware that critics of religion can be attacked for failing to credit the fertile diversity of traditions and world-views that have been called religious. Anthropologically informed works, from Sir James Frazer's *Golden Bough* to Pascal Boyer's *Religion Explained* or Scott Atran's *In Gods We Trust*, fascinatingly document the bizarre phenomenology of superstition and ritual. Read such books and marvel at the richness of human gullibility.

But that is not the way of this book. I decry supernaturalism in all its forms, and the most effective way to proceed will be to concentrate on the form most likely to be familiar to my readers – the form that impinges most threateningly on all our societies. Most of my readers will have been reared in one or another of today's three 'great' monotheistic religions (four if you count Mormonism), all of which trace themselves back to the mythological patriarch Abraham, and it will be convenient to keep this family of traditions in mind throughout the rest of the book.

This is as good a moment as any to forestall an inevitable retort to the book, one that would otherwise – as sure as night follows day – turn up in a review: 'The God that Dawkins doesn't believe in is a God that I don't believe in either. I don't believe in an old man in the sky with a long white beard.' That old man is an irrelevant distraction and his beard is as tedious as it is long. Indeed, the distraction is worse than irrelevant. Its very silliness is calculated to distract attention from the fact that what the speaker really believes is not a whole lot less silly. I know you don't believe in an old bearded man sitting on a cloud, so let's not waste any more time on that. I am not attacking any particular version of God or gods. I am attacking God, all gods, anything and everything supernatural, wherever and whenever they have been or will be invented.

MONOTHEISM

The great unmentionable evil at the center of our culture is monotheism. From a barbaric Bronze Age text known as the Old Testament, three anti-human religions have evolved – Judaism, Christianity, and Islam. These are sky-god religions. They are, literally, patriarchal – God is the Omnipotent Father – hence the loathing of women for 2,000 years in those countries afflicted by the sky-god and his earthly male delegates.

GORE VIDAL

The oldest of the three Abrahamic religions, and the clear ancestor of the other two, is Judaism: originally a tribal cult of a single fiercely unpleasant God, morbidly obsessed with sexual restrictions, with the smell of charred flesh, with his own superiority over rival gods and with the exclusiveness of his chosen desert tribe. During the Roman occupation of Palestine, Christianity was founded by Paul of Tarsus as a less ruthlessly monotheistic sect of Judaism and a less exclusive one, which looked outwards from the Jews to the rest of the world. Several centuries later, Muhammad and his followers reverted to the uncompromising monotheism of the Jewish original, but not its exclusiveness, and founded Islam upon a new holy book, the Koran or Qur'an, adding a powerful ideology of military conquest to spread the faith. Christianity, too, was spread by the sword, wielded first by Roman hands after the Emperor Constantine raised it from eccentric cult to official religion, then by the Crusaders, and later by the *conquistadores* and other European invaders and colonists, with missionary accompaniment. For most of my purposes, all three Abrahamic religions can be treated as indistinguishable. Unless otherwise stated, I shall have Christianity mostly in mind, but only because it is the version with which I happen to be most familiar. For my purposes the differences matter less than the similarities. And I shall not be concerned at all with other religions such as Buddhism or Confucianism. Indeed, there is something to be said for treating these not as religions at all but as ethical systems or philosophies of life.

The simple definition of the God Hypothesis with which I began has to be substantially fleshed out if it is to accommodate the Abrahamic God. He not only created the universe; he is a *personal* God dwelling within it, or perhaps outside it (whatever that might mean), possessing the unpleasantly human qualities to which I have alluded.

Personal qualities, whether pleasant or unpleasant, form no part of the deist god of Voltaire and Thomas Paine. Compared with the Old Testament's psychotic delinquent, the deist God of the eighteenth-century Enlightenment is an altogether grander being: worthy of his cosmic creation, loftily unconcerned with human affairs, sublimely aloof from our private thoughts and hopes, caring nothing for our messy sins or mumbled contritions. The deist God is a physicist to end all physics, the alpha and omega of mathematicians, the apotheosis of designers; a hyper-engineer who set up the laws and constants of the universe, fine-tuned them with exquisite precision and foreknowledge, detonated what we would now call the hot big bang, retired and was never heard from again.

In times of stronger faith, deists have been reviled as indistinguishable from atheists. Susan Jacoby, in *Freethinkers: A History of American Secularism*, lists a choice selection of the epithets hurled at poor Tom Paine: 'Judas, reptile, hog, mad dog, souse, louse, archbeast, brute, liar, and of course infidel'. Paine died abandoned (with the honourable exception of Jefferson) by political former friends embarrassed by his anti-Christian views. Nowadays, the ground has shifted so far that deists are more likely to be contrasted with atheists and lumped with theists. They do, after all, believe in a supreme intelligence who created the universe.

SECULARISM, THE FOUNDING FATHERS AND THE RELIGION OF AMERICA

It is conventional to assume that the Founding Fathers of the American Republic were deists. No doubt many of them were, although it has been argued that the greatest of them might have been atheists. Certainly their writings on religion in their own time leave me in no doubt that most of them would have been atheists in ours. But whatever their individual religious views in their own time, the one thing they collectively were is *secularists*, and this is the topic to which I turn in this section, beginning with a – perhaps surprising – quotation from Senator Barry Goldwater in 1981, clearly showing how staunchly that presidential candidate and hero of American conservatism upheld the secular tradition of the Republic's foundation:

> There is no position on which people are so immovable as their religious beliefs. There is no more powerful ally one can claim in a debate than Jesus Christ, or God, or Allah, or whatever one calls this supreme being. But like any powerful weapon, the use of God's name on one's behalf should be used sparingly. The religious factions that are growing throughout our land are not using their religious clout with wisdom. They are trying to force government leaders into following their position 100 percent. If you disagree with these religious groups on a particular moral issue, they complain, they threaten you with a loss of money or votes or both. I'm frankly sick and tired of the political preachers across this country telling me as a citizen that if I want to be a moral person, I must believe in A, B, C, and D. Just who do they think they are? And from where do they presume to claim the right to dictate their moral beliefs to me? And I am even more angry

> as a legislator who must endure the threats of every religious group who thinks it has some God-granted right to control my vote on every roll call in the Senate. I am warning them today: I will fight them every step of the way if they try to dictate their moral convictions to all Americans in the name of conservatism.[19]

The religious views of the Founding Fathers are of great interest to propagandists of today's American right, anxious to push their version of history. Contrary to their view, the fact that the United States was *not* founded as a Christian nation was early stated in the terms of a treaty with Tripoli, drafted in 1796 under George Washington and signed by John Adams in 1797:

> As the Government of the United States of America is not, in any sense, founded on the Christian religion; as it has in itself no character of enmity against the laws, religion, or tranquillity, of Musselmen; and as the said States never have entered into any war or act of hostility against any Mehomitan nation, it is declared by the parties that no pretext arising from religious opinions shall ever produce an interruption of the harmony existing between the two countries.

The opening words of this quotation would cause uproar in today's Washington ascendancy. Yet Ed Buckner has convincingly demonstrated that they caused no dissent at the time,[20] among either politicians or public.

The paradox has often been noted that the United States, founded in secularism, is now the most religiose country in Christendom, while England, with an established church headed by its constitutional monarch, is among the least. I am continually asked why this is, and I do not know. I suppose it is possible that England has wearied of religion after an appalling history of interfaith violence, with Protestants and Catholics alternately gaining the upper hand and systematically murdering the other lot. Another suggestion stems from the observation that America is a nation of immigrants. A colleague points out to me that immigrants, uprooted from the stability and comfort of an extended family in Europe, could well have embraced a church as a kind of kin-substitute on alien soil. It is an interesting idea, worth researching further. There is no doubt that many Americans see their own local church as an important unit of identity, which does indeed have some of the attributes of an extended family.

Yet another hypothesis is that the religiosity of America stems paradoxically from the secularism of its constitution. Precisely because America is legally secular, religion has become free enterprise. Rival churches compete for congregations – not least for the fat tithes that they bring – and the competition is waged with all the aggressive hard-sell techniques of the marketplace. What works for soap flakes works for God, and the result is something approaching religious mania among today's less educated classes. In England, by contrast, religion under the aegis of the established church has become little more than a pleasant social pastime, scarcely recognizable as religious at all. This English tradition is nicely expressed by Giles Fraser, an Anglican vicar who doubles as a philosophy tutor at Oxford, writing in the *Guardian.* Fraser's article is subtitled 'The establishment of the Church of England took God out of religion, but there are risks in a more vigorous approach to faith':

> There was a time when the country vicar was a staple of the English dramatis personae. This tea-drinking, gentle eccentric, with his polished shoes and kindly manners, represented a type of religion that didn't make non-religious people uncomfortable. He wouldn't break into an existential sweat or press you

> against a wall to ask if you were saved, still less launch crusades from the pulpit or plant roadside bombs in the name of some higher power.[21]

(Shades of Betjeman's 'Our Padre', which I quoted at the beginning of Chapter 1.) Fraser goes on to say that 'the nice country vicar in effect inoculated vast swaths of the English against Christianity'. He ends his article by lamenting a more recent trend in the Church of England to take religion seriously again, and his last sentence is a warning: 'the worry is that we may release the genie of English religious fanaticism from the establishment box in which it has been dormant for centuries'.

The genie of religious fanaticism is rampant in present-day America, and the Founding Fathers would have been horrified. Whether or not it is right to embrace the paradox and blame the secular constitution that they devised, the founders most certainly were secularists who believed in keeping religion out of politics, and that is enough to place them firmly on the side of those who object, for example, to ostentatious displays of the Ten Commandments in government-owned public places. But it is tantalizing to speculate that at least some of the Founders might have gone beyond deism. Might they have been agnostics or even out-and-out atheists? The following statement of Jefferson is indistinguishable from what we would now call agnosticism:

> To talk of immaterial existences is to talk of *nothings*. To say that the human soul, angels, god, are immaterial, is to say they are nothings, or that there is no god, no angels, no soul. I cannot reason otherwise . . . without plunging into the fathomless abyss of dreams and phantasms. I am satisfied, and sufficiently occupied with the things which are, without tormenting or troubling myself about those which may indeed be, but of which I have no evidence.

Christopher Hitchens, in his biography *Thomas Jefferson: Author of America*, thinks it likely that Jefferson was an atheist, even in his own time when it was much harder:

> As to whether he was an atheist, we must reserve judgment if only because of the prudence he was compelled to observe during his political life. But as he had written to his nephew, Peter Carr, as early as 1787, one must not be frightened from this inquiry by any fear of its consequences. 'If it ends in a belief that there is no God, you will find incitements to virtue in the comfort and pleasantness you feel in this exercise, and the love of others which it will procure you.'

I find the following advice of Jefferson, again in his letter to Peter Carr, moving:

> Shake off all the fears of servile prejudices, under which weak minds are servilely crouched. Fix reason firmly in her seat, and call on her tribunal for every fact, every opinion. Question with boldness even the existence of a God; because, if there be one, he must more approve of the homage of reason than that of blindfolded fear.

Remarks of Jefferson's such as 'Christianity is the most perverted system that ever shone on man' are compatible with deism but also with atheism. So is James Madison's robust anti-clericalism: 'During almost fifteen centuries has the legal establishment of Christianity been on trial. What has been its fruits? More or less, in all places, pride and indolence in the clergy; ignorance and servility in the laity; in both, superstition, bigotry and persecution.' The same could be said of Benjamin Franklin's 'Lighthouses are more useful than churches.' John Adams seems to have been a deist of a strongly anti-clerical stripe ('The frightful engines of ecclesiastical councils . . .'), and he delivered himself of some splendid tirades against Christianity in particular: 'As I understand the Christian religion, it was, and is, a revelation. But how has it happened that millions of fables, tales, legends, have been blended with both Jewish and Christian revelation that have made them the most bloody religion that ever existed?' And, in another letter, this time to Jefferson, 'I almost shudder at the thought of alluding to the most fatal example of the abuses of grief which the history of mankind has preserved – the Cross. Consider what calamities that engine of grief has produced!'

Whether Jefferson and his colleagues were theists, deists, agnostics or atheists, they were also passionate secularists who believed that the religious opinions of a President, or lack of them, were entirely his own business. All the Founding Fathers, whatever their private religious beliefs, would have been aghast to read the journalist Robert Sherman's report of George Bush Senior's answer when Sherman asked him whether he recognized the equal citizenship and patriotism of Americans who are atheists: 'No, I don't know that atheists should be considered as citizens, nor should they be considered patriots. This is one nation under God.'[22] Assuming Sherman's account to be accurate (unfortunately he didn't use a tape-recorder, and no other newspaper ran the story at the time), try the experiment of replacing 'atheists' with 'Jews' or 'Muslims' or 'Blacks'. That gives the measure of the prejudice and discrimination that American atheists have to endure today. Natalie Angier's 'Confessions of a lonely atheist' is a sad and moving description, in the *New York Times*, of her feelings of isolation as an atheist in today's America.[23] But the isolation of American atheists is an illusion, assiduously cultivated by prejudice. Atheists in America are more numerous than most people realize. As I said in the Preface, American atheists far outnumber religious Jews, yet the Jewish lobby is notoriously one of the most formidably influential in Washington. What might American atheists achieve if they organized themselves properly?*

David Mills, in his admirable book *Atheist Universe*, tells a story which you would dismiss as an unrealistic caricature of police bigotry if it were fiction. A Christian faith-healer ran a 'Miracle Crusade' which came to Mills's home town once a year. Among other things, the faith-healer encouraged diabetics to throw away their insulin, and cancer patients to give up their chemotherapy and pray for a miracle instead. Reasonably enough, Mills decided to organize a peaceful demonstration to warn people. But he made the mistake of going to the police to tell them of his intention and ask for police protection against possible attacks from supporters of the faith-healer. The first police officer to whom he spoke asked, 'Is you gonna protest fir him or 'gin him?' (meaning for or against the faith-healer). When Mills replied, 'Against him,' the policeman said that he himself planned to attend the rally and intended to spit personally in Mills's face as he marched past Mills's demonstration.

Mills decided to try his luck with a second police officer. This one said that if any of the faith-healer's supporters violently confronted Mills, the officer would arrest Mills because he was 'trying to interfere with God's work'. Mills went home and tried telephoning the police station, in the hope of finding more sympathy at a senior level. He was finally connected to a sergeant who said, 'To hell with you, Buddy. No policeman wants to protect a goddamned atheist. I hope somebody bloodies you up good.' Apparently adverbs were in short supply in this police station, along with the milk of human kindness and a sense of duty. Mills relates that he spoke to about seven or eight policemen that day. None of them was helpful, and most of them directly threatened Mills with violence.

---

* Tom Flynn, Editor of *Free Inquiry*, makes the point forcefully ('Secularism's breakthrough moment', *Free Inquiry* 26: 3, 2006, 16–17): 'If atheists are lonely and downtrodden, we have only ourselves to blame. Numerically, we are strong. Let's start punching our weight.'

Anecdotes of such prejudice against atheists abound, but Margaret Downey, founder of the Anti-Discrimination Support Network (ADSN), maintains systematic records of such cases through the Freethought Society of Greater Philadelphia.[24] Her database of incidents, categorized under community, schools, workplace, media, family and government, includes examples of harassment, loss of jobs, shunning by family and even murder.[25] Downey's documented evidence of the hatred and misunderstanding of atheists makes it easy to believe that it is, indeed, virtually impossible for an honest atheist to win a public election in America. There are 435 members of the House of Representatives and 100 members of the Senate. Assuming that the majority of these 535 individuals are an educated sample of the population, it is statistically all but inevitable that a substantial number of them must be atheists. They must have lied, or concealed their true feelings, in order to get elected. Who can blame them, given the electorate they had to convince? It is universally accepted that an admission of atheism would be instant political suicide for any presidential candidate.*

These facts about today's political climate in the United States, and what they imply, would have horrified Jefferson, Washington, Madison, Adams and all their friends. Whether they were atheists, agnostics, deists or Christians, they would have recoiled in horror from the theocrats of early 21st-century Washington. They would have been drawn instead to the secularist founding fathers of post-colonial India, especially the religious Gandhi ('I am a Hindu, I am a Moslem, I am a Jew, I am a Christian, I am a Buddhist!'), and the atheist Nehru:

---

* In March 2007, Representative Pete Stark, US Congressman for the California 13th District, publicly acknowledged his lack of theistic belief. He has since been defeated, leaving Congress in 2012 as the only avowed atheist: http://huffingtonpost.com/2013/09/19/atheists-in-congress_n_3944108.html

> The spectacle of what is called religion, or at any rate organised religion, in India and elsewhere, has filled me with horror and I have frequently condemned it and wished to make a clean sweep of it. Almost always it seemed to stand for blind belief and reaction, dogma and bigotry, superstition, exploitation and the preservation of vested interests.

Nehru's definition of the secular India of Gandhi's dream (would that it had been realized, instead of the partitioning of their country amid an interfaith bloodbath) might almost have been ghosted by Jefferson himself:

> We talk about a secular India . . . Some people think that it means something opposed to religion. That obviously is not correct. What it means is that it is a State which honours all faiths equally and gives them equal opportunities; India has a long history of religious tolerance . . . In a country like India, which has many faiths and religions, no real nationalism can be built up except on the basis of secularity.[26]

The deist God, often associated with the Founding Fathers, is certainly an improvement over the monster of the Bible. Unfortunately it is scarcely more likely that he exists, or ever did. In any of its forms the God Hypothesis is unnecessary.* The God Hypothesis is also very close to being ruled out by the laws of probability. I shall come to that in Chapter 4, after dealing with the alleged proofs of the existence of God in Chapter 3. Meanwhile I turn to agnosticism, and the erroneous notion that the existence or non-existence of God is an untouchable question, forever beyond the reach of science.

---

* 'Sire, I had no need of that hypothesis,' as Laplace said when Napoleon wondered how the famous mathematician had managed to write his book without mentioning God.

THE POVERTY OF AGNOSTICISM

The robust Muscular Christian haranguing us from the pulpit of my old school chapel admitted a sneaking regard for atheists. They at least had the courage of their misguided convictions. What this preacher couldn't stand was agnostics: namby-pamby, mushy pap, weak-tea, weedy, pallid fence-sitters. He was partly right, but for wholly the wrong reason. In the same vein, according to Quentin de la Bédoyère, the Catholic historian Hugh Ross Williamson 'respected the committed religious believer and also the committed atheist. He reserved his contempt for the wishy-washy boneless mediocrities who flapped around in the middle.'[27]

There is nothing wrong with being agnostic in cases where we lack evidence one way or the other. It is the reasonable position. Carl Sagan was proud to be agnostic when asked whether there was life elsewhere in the universe. When he refused to commit himself, his interlocutor pressed him for a 'gut feeling' and he immortally replied: 'But I try not to think with my gut. Really, it's okay to reserve judgment until the evidence is in.'[28] The question of extraterrestrial life is open. Good arguments can be mounted both ways, and we lack the evidence to do more than shade the probabilities one way or the other. Agnosticism, of a kind, is an appropriate stance on many scientific questions, such as what caused the end-Permian extinction, the greatest mass extinction in fossil history. It could have been a meteorite strike like the one that, with greater likelihood on present evidence, caused the later extinction of the dinosaurs. But it could have been any of various other possible causes, or a combination. Agnosticism about the causes of both these mass extinctions is reasonable. How about the question of God? Should we be agnostic about him too? Many have said definitely yes, often with an air of conviction that verges on protesting too much. Are they right?

I'll begin by distinguishing two kinds of agnosticism. TAP, or

Temporary Agnosticism in Practice, is the legitimate fence-sitting where there really is a definite answer, one way or the other, but we so far lack the evidence to reach it (or don't understand the evidence, or haven't time to read the evidence, etc.). TAP would be a reasonable stance towards the Permian extinction. There is a truth out there and one day we hope to know it, though for the moment we don't.

But there is also a deeply inescapable kind of fence-sitting, which I shall call PAP (Permanent Agnosticism in Principle). The fact that the acronym spells a word used by that old school preacher is (almost) accidental. The PAP style of agnosticism is appropriate for questions that can never be answered, no matter how much evidence we gather, because the very idea of evidence is not applicable. The question exists on a different plane, or in a different dimension, beyond the zones where evidence can reach. An example might be that philosophical chestnut, the question whether you see red as I do. Maybe your red is my green, or something completely different from any colour that I can imagine. Philosophers cite this question as one that can never be answered, no matter what new evidence might one day become available. And some scientists and other intellectuals are convinced – too eagerly in my view – that the question of God's existence belongs in the forever inaccessible PAP category. From this, as we shall see, they often make the illogical deduction that the hypothesis of God's existence, and the hypothesis of his non-existence, have exactly equal probability of being right. The view that I shall defend is very different: agnosticism about the existence of God belongs firmly in the temporary or TAP category. Either he exists or he doesn't. It is a scientific question; one day we may know the answer, and meanwhile we can say something pretty strong about the probability.

In the history of ideas, there are examples of questions being answered that had earlier been judged forever out of science's reach. In 1835 the celebrated French philosopher Auguste Comte wrote, of the stars: 'We shall never be able to study, by any method, their chemical composition or their mineralogical structure.' Yet even before Comte had set down these words, Fraunhofer had begun using his spectroscope to analyse the chemical composition of the sun. Now spectroscopists daily confound Comte's agnosticism with their long-distance analyses of the precise chemical composition of even distant stars.[29] Whatever the exact status of Comte's astronomical agnosticism, this cautionary tale suggests, at the very least, that we should hesitate before proclaiming the eternal verity of agnosticism too loudly. Nevertheless, when it comes to God, a great many philosophers and scientists are glad to do so, beginning with the inventor of the word itself, T. H. Huxley.[30]

Huxley explained his coining while rising to a personal attack that it had provoked. The Principal of King's College, London, the Reverend Dr Wace, had poured scorn on Huxley's 'cowardly agnosticism':

> He may prefer to call himself an agnostic; but his real name is an older one – he is an infidel; that is to say, an unbeliever. The word infidel, perhaps, carries an unpleasant significance. Perhaps it is right that it should. It is, and it ought to be, an unpleasant thing for a man to have to say plainly that he does not believe in Jesus Christ.

Huxley was not the man to let that sort of provocation pass him by, and his reply in 1889 was as robustly scathing as we should expect (although never departing from scrupulous good manners: as Darwin's Bulldog, his teeth were sharpened by urbane Victorian irony). Eventually, having dealt Dr Wace his just comeuppance and buried the remains, Huxley returned to the word 'agnostic' and explained how he first came by it. Others, he noted,

> were quite sure they had attained a certain 'gnosis' – had, more or less successfully, solved the problem of

> existence; while I was quite sure I had not, and had a pretty strong conviction that the problem was insoluble. And, with Hume and Kant on my side, I could not think myself presumptuous in holding fast by that opinion . . . So I took thought, and invented what I conceived to be the appropriate title of 'agnostic'.

Later in his speech, Huxley went on to explain that agnostics have no creed, not even a negative one.

> Agnosticism, in fact, is not a creed, but a method, the essence of which lies in the rigorous application of a single principle. . . . Positively the principle may be expressed: In matters of the intellect, follow your reason as far as it will take you, without regard to any other consideration. And negatively: In matters of the intellect, do not pretend that conclusions are certain which are not demonstrated or demonstrable. That I take to be the agnostic faith, which if a man keep whole and undefiled, he shall not be ashamed to look the universe in the face, whatever the future may have in store for him.

To a scientist these are noble words, and one doesn't criticize T. H. Huxley lightly. But Huxley, in his concentration upon the absolute impossibility of proving or disproving God, seems to have been ignoring the shading of *probability*. The fact that we can neither prove nor disprove the existence of something does not put existence and non-existence on an even footing. I don't think Huxley would disagree, and I suspect that when he appeared to do so he was bending over backwards to concede a point, in the interests of securing another one. We have all done this at one time or another.

Contrary to Huxley, I shall suggest that the existence of God is a scientific hypothesis like any other. Even if hard to test in practice, it belongs in the same TAP or temporary agnosticism box as the controversies over the Permian and Cretaceous extinctions. God's existence or non-existence is a scientific fact about the universe, discoverable in principle if not in practice. If he existed and chose to reveal it, God himself could clinch the argument, noisily and unequivocally, in his favour. And even if God's existence is never proved or disproved with certainty one way or the other, available evidence and reasoning may yield an estimate of probability far from 50 per cent.

Let us, then, take the idea of a spectrum of probabilities seriously, and place human judgements about the existence of God along it, between two extremes of opposite certainty. The spectrum is continuous, but it can be represented by the following seven milestones along the way.

1. Strong theist. 100 per cent probability of God. In the words of C. G. Jung, 'I do not believe, I *know*.'
2. Very high probability but short of 100 per cent. *De facto* theist. 'I cannot know for certain, but I strongly believe in God and live my life on the assumption that he is there.'
3. Higher than 50 per cent but not very high. Technically agnostic but leaning towards theism. 'I am very uncertain, but I am inclined to believe in God.'
4. Exactly 50 per cent. Completely impartial agnostic. 'God's existence and non-existence are exactly equiprobable.'
5. Lower than 50 per cent but not very low. Technically agnostic but leaning towards atheism. 'I don't know whether God exists but I'm inclined to be sceptical.'
6. Very low probability, but short of zero. *De facto* atheist. 'I cannot know for certain but I think God is very improbable, and I live my life on the assumption that he is not there.'
7. Strong atheist. 'I know there is no God, with the same conviction as Jung "knows" there is one.'

I'd be surprised to meet many people in category 7, but I include it for symmetry with category 1, which is well populated. It is in the nature of faith that one is capable, like Jung, of holding a belief without adequate reason to do so (Jung also believed that particular books on his shelf spontaneously exploded with a loud bang). Atheists do not have faith; and reason alone could not propel one to total conviction that anything definitely does not exist. Hence category 7 is in practice rather emptier than its opposite number, category 1, which has many devoted inhabitants. I count myself in category 6, but leaning towards 7 – I am agnostic only to the extent that I am agnostic about fairies at the bottom of the garden.

The spectrum of probabilities works well for TAP (temporary agnosticism in practice). It is superficially tempting to place PAP (permanent agnosticism in principle) in the middle of the spectrum, with a 50 per cent probability of God's existence, but this is not correct. PAP agnostics aver that we cannot say anything, one way or the other, on the question of whether or not God exists. The question, for PAP agnostics, is in principle unanswerable, and they should strictly refuse to place themselves anywhere on the spectrum of probabilities. The fact that I cannot know whether your red is the same as my green doesn't make the probability 50 per cent. The proposition on offer is too meaningless to be dignified with a probability. Nevertheless, it is a common error, which we shall meet again, to leap from the premise that the question of God's existence is in principle unanswerable to the conclusion that his existence and his non-existence are equiprobable.

Another way to express that error is in terms of the burden of proof, and in this form it is pleasingly demonstrated by Bertrand Russell's parable of the celestial teapot.[31]

> Many orthodox people speak as though it were the business of sceptics to disprove received dogmas rather than of dogmatists to prove them. This is, of

> course, a mistake. If I were to suggest that between the Earth and Mars there is a china teapot revolving about the sun in an elliptical orbit, nobody would be able to disprove my assertion provided I were careful to add that the teapot is too small to be revealed even by our most powerful telescopes. But if I were to go on to say that, since my assertion cannot be disproved, it is intolerable presumption on the part of human reason to doubt it, I should rightly be thought to be talking nonsense. If, however, the existence of such a teapot were affirmed in ancient books, taught as the sacred truth every Sunday, and instilled into the minds of children at school, hesitation to believe in its existence would become a mark of eccentricity and entitle the doubter to the attentions of the psychiatrist in an enlightened age or of the Inquisitor in an earlier time.

We would not waste time saying so because nobody, so far as I know, worships teapots;* but, if pressed, we would not hesitate to declare our strong belief that there is positively no orbiting teapot. Yet strictly we should all be *teapot agnostics*: we cannot prove, for sure, that there is no celestial teapot. In practice, we move away from teapot agnosticism towards *a-teapotism.*

A friend, who was brought up a Jew and still observes the sabbath and other Jewish customs out of loyalty to his heritage, describes himself as a 'tooth fairy agnostic'. He regards God as no more probable than the tooth fairy. You can't disprove either hypothesis, and both are equally improbable. He is an a-theist to exactly the same large extent that he is an a-fairyist. And agnostic about both, to the same small extent.

---

* Perhaps I spoke too soon. The *Independent on Sunday* of 5 June 2005 carried the following item: 'Malaysian officials say religious sect which built sacred teapot the size of a house has flouted planning regulations.' See also BBC News at http://news.bbc.co.uk/2/hi/asia-pacific/4692039.stm.

Russell's teapot, of course, stands for an infinite number of things whose existence is conceivable and cannot be disproved. That great American lawyer Clarence Darrow said, 'I don't believe in God as I don't believe in Mother Goose.' The journalist Andrew Mueller is of the opinion that pledging yourself to any particular religion 'is no more or less weird than choosing to believe that the world is rhombus-shaped, and borne through the cosmos in the pincers of two enormous green lobsters called Esmerelda and Keith'.[32] A philosophical favourite is the invisible, intangible, inaudible unicorn, disproof of which is attempted yearly by the children at Camp Quest.* A popular deity on the Internet at present – and as undisprovable as Yahweh or any other – is the Flying Spaghetti Monster, who, many claim, has touched them with his noodly appendage.[33] I am delighted to see that the *Gospel of the Flying Spaghetti Monster* has now been published as a book,[34] to great acclaim. I haven't read it myself, but who needs to read a gospel when you just *know* it's true? By the way, it had to happen – a Great Schism has already occurred, resulting in the *Reformed* Church of the Flying Spaghetti Monster.[35]

The point of all these way-out examples is that they are undisprovable, yet nobody thinks the hypothesis of their existence is on an even footing with the hypothesis of their non-existence. Russell's point is that the burden of proof rests with the believers, not the non-believers. Mine is the related point that the odds in favour of the teapot (spaghetti monster / Esmerelda and Keith / unicorn etc.) are not equal to the odds against.

---

* Camp Quest takes the American institution of the summer camp in an entirely admirable direction. Unlike other summer camps that follow a religious or scouting ethos, Camp Quest, founded by Edwin and Helen Kagin in Kentucky, is run by secular humanists, and the children are encouraged to think sceptically for themselves while having a very good time with all the usual outdoor activities (www.camp-quest.org). Other Camp Quests with a similar ethos have now sprung up in Tennessee, Minnesota, Michigan, Ohio and Canada.

The fact that orbiting teapots and tooth fairies are undisprovable is not felt, by any reasonable person, to be the kind of fact that settles any interesting argument. None of us feels an obligation to disprove any of the millions of far-fetched things that a fertile or facetious imagination might dream up. I have found it an amusing strategy, when asked whether I am an atheist, to point out that the questioner is also an atheist when considering Zeus, Apollo, Amon Ra, Mithras, Baal, Thor, Wotan, the Golden Calf and the Flying Spaghetti Monster. I just go one god further.

All of us feel entitled to express extreme scepticism to the point of outright disbelief – except that in the case of unicorns, tooth fairies and the gods of Greece, Rome, Egypt and the Vikings, there is (nowadays) no need to bother. In the case of the Abrahamic God, however, there is a need to bother, because a substantial proportion of the people with whom we share the planet do believe strongly in his existence. Russell's teapot demonstrates that the ubiquity of belief in God, as compared with belief in celestial teapots, does not shift the burden of proof in logic, although it may seem to shift it as a matter of practical politics. That you cannot prove God's non-existence is accepted and trivial, if only in the sense that we can never absolutely prove the non-existence of anything. What matters is not whether God is disprovable (he isn't) but whether his existence is *probable*. That is another matter. Some undisprovable things are sensibly judged far less probable than other undisprovable things. There is no reason to regard God as immune from consideration along the spectrum of probabilities. And there is certainly no reason to suppose that, just because God can be neither proved nor disproved, his probability of existence is 50 per cent. On the contrary, as we shall see.

NOMA

Just as Thomas Huxley bent over backwards to pay lip service to completely impartial agnosticism, right in the middle of my seven-stage spectrum, theists do the same thing from the other direction, and for an equivalent reason. The theologian Alister McGrath makes it the central point of his book *Dawkins' God: Genes, Memes and the Origin of Life*. Indeed, after his admirably fair summary of my scientific works, it seems to be the only point in rebuttal that he has to offer: the undeniable but ignominiously weak point that you cannot disprove the existence of God. On page after page as I read McGrath, I found myself scribbling 'teapot' in the margin. Again invoking T. H. Huxley, McGrath says, 'Fed up with both theists and atheists making hopelessly dogmatic statements on the basis of inadequate empirical evidence, Huxley declared that the God question could not be settled on the basis of the scientific method.'

McGrath goes on to quote Stephen Jay Gould in similar vein: 'To say it for all my colleagues and for the umpteenth millionth time (from college bull sessions to learned treatises): science simply cannot (by its legitimate methods) adjudicate the issue of God's possible superintendence of nature. We neither affirm nor deny it; we simply can't comment on it as scientists.' Despite the confident, almost bullying, tone of Gould's assertion, what, actually, is the justification for it? Why shouldn't we comment on God, as scientists? And why isn't Russell's teapot, or the Flying Spaghetti Monster, equally immune from scientific scepticism? As I shall argue in a moment, a universe with a creative superintendent would be a very different kind of universe from one without. Why is that not a scientific matter?

Gould carried the art of bending over backwards to positively supine lengths in one of his less admired books, *Rocks of Ages*. There he coined the acronym NOMA for the phrase 'non-overlapping magisteria':

> The net, or magisterium, of science covers the empirical realm: what is the universe made of (fact) and why does it work this way (theory). The magisterium of religion extends over questions of

> ultimate meaning and moral value. These two magisteria do not overlap, nor do they encompass all inquiry (consider, for example, the magisterium of art and the meaning of beauty). To cite the old clichés, science gets the age of rocks, and religion the rock of ages; science studies how the heavens go, religion how to go to heaven.

This sounds terrific – right up until you give it a moment's thought. What are these ultimate questions in whose presence religion is an honoured guest and science must respectfully slink away?

Martin Rees, the distinguished Cambridge astronomer whom I have already mentioned, begins his book *Our Cosmic Habitat* by posing two candidate ultimate questions and giving a NOMA-friendly answer. 'The pre-eminent mystery is why anything exists at all. What breathes life into the equations, and actualized them in a real cosmos? Such questions lie beyond science, however: they are the province of philosophers and theologians.' I would prefer to say that if indeed they lie beyond science, they most certainly lie beyond the province of theologians as well (I doubt that philosophers would thank Martin Rees for lumping theologians in with them). I am tempted to go further and wonder in what possible sense theologians can be said to *have* a province. I am still amused when I recall the remark of a former Warden (head) of my Oxford college. A young theologian had applied for a junior research fellowship, and his doctoral thesis on Christian theology provoked the Warden to say, 'I have grave doubts as to whether it's a *subject* at all.'

What expertise can theologians bring to deep cosmological questions that scientists cannot? In another book I recounted the words of an Oxford astronomer who, when I asked him one of those same deep questions, said: 'Ah, now we move beyond the realm of science. This is where I have to hand over to our good friend the chaplain.' I was not quick-witted enough to utter the response that I later wrote: 'But why the chaplain? Why not the gardener or the chef?' Why are scientists so cravenly respectful towards the ambitions of theologians, over questions that theologians are certainly no more qualified to answer than scientists themselves?

It is a tedious cliché (and, unlike many clichés, it isn't even true) that science concerns itself with *how* questions, but only theology is equipped to answer *why* questions. What on Earth *is* a why question? Not every English sentence beginning with the word 'why' is a legitimate question. Why are unicorns hollow? Some questions simply do not deserve an answer. What is the colour of abstraction? What is the smell of hope? The fact that a question can be phrased in a grammatically correct English sentence doesn't make it meaningful, or entitle it to our serious attention. Nor, even if the question is a real one, does the fact that science cannot answer it imply that religion can.

Perhaps there are some genuinely profound and meaningful questions that are forever beyond the reach of science. Maybe quantum theory is already knocking on the door of the unfathomable. But if science cannot answer some ultimate question, what makes anybody think that religion can? I suspect that neither the Cambridge nor the Oxford astronomer really believed that theologians have any expertise that enables them to answer questions that are too deep for science. I suspect that both astronomers were, yet again, bending over backwards to be polite: theologians have nothing worthwhile to say about anything else; let's throw them a sop and let them worry away at a couple of questions that nobody can answer and maybe never will. Unlike my astronomer friends, I don't think we should even throw them a sop. I have yet to see any good reason to suppose that theology (as opposed to biblical history, literature, etc.) is a subject at all.

Similarly, we can all agree that science's entitlement to advise us on moral values is problematic, to say the least. But does Gould really want to cede to *religion* the right to tell us what is good and what is bad? The fact that it has nothing *else* to contribute to human wisdom is no reason to hand religion a free licence to tell us what to do. Which religion, anyway? The one in which we happen to have been brought up? To which chapter, then, of which book of the Bible should we turn – for they are far from unanimous and some of them are odious by any reasonable standards. How many literalists have read enough of the Bible to know that the death penalty is prescribed for adultery, for gathering sticks on the sabbath and for cheeking your parents? If we reject Deuteronomy and Leviticus (as all enlightened moderns do), by what criteria do we then decide which of religion's moral values to *accept*? Or should we pick and choose among all the world's religions until we find one whose moral teaching suits us? If so, again we must ask, by what criterion do we choose? And if we have independent criteria for choosing among religious moralities, why not cut out the middle man and go straight for the moral choice without the religion? I shall return to such questions in Chapter 7.

I simply do not believe that Gould could possibly have meant much of what he wrote in *Rocks of Ages*. As I say, we have all been guilty of bending over backwards to be nice to an unworthy but powerful opponent, and I can only think that this is what Gould was doing. It is conceivable that he really did intend his unequivocally strong statement that science has nothing whatever to say about the question of God's existence: 'We neither affirm nor deny it; we simply can't comment on it as scientists.' This sounds like agnosticism of the permanent and irrevocable kind, full-blown PAP. It implies that science cannot even make *probability* judgements on the question. This remarkably widespread fallacy – many repeat it like a mantra but few of them, I suspect, have thought it through – embodies what I refer to as 'the poverty of agnosticism'. Gould, by the way, was not an impartial agnostic but strongly inclined towards *de facto* atheism. On what basis did he make that judgement, if there is nothing to be said about whether God exists?

The God Hypothesis suggests that the reality we inhabit also contains a supernatural agent who designed the universe and – at least in many versions of the hypothesis – maintains it and even intervenes in it with miracles, which are temporary violations of his own otherwise grandly immutable laws. Richard Swinburne, one of Britain's leading theologians, is surprisingly clear on the matter in his book *Is There a God?*:

> What the theist claims about God is that he does have a power to create, conserve, or annihilate anything, big or small. And he can also make objects move or do anything else . . . He can make the planets move in the way that Kepler discovered that they move, or make gunpowder explode when we set a match to it; or he can make planets move in quite different ways, and chemical substances explode or not explode under quite different conditions from those which now govern their behaviour. God is not limited by the laws of nature; he makes them and he can change or suspend them – if he chooses.

Just too easy, isn't it! Whatever else this is, it is very far from NOMA. And whatever else they may say, those scientists who subscribe to the 'separate magisteria' school of thought should concede that a universe with a supernaturally intelligent creator is a very different kind of universe from one without. The difference between the two hypothetical universes could hardly be more fundamental in principle, even if it is not easy to test in practice. And it undermines the complacently seductive dictum that science must be completely silent about religion's central existence claim. The presence or absence of a creative super-intelligence is unequivocally a scientific question, even if it is not in practice – or not yet – a decided one. So also is the truth or falsehood of every one of the miracle stories that religions rely upon to impress multitudes of the faithful.

Did Jesus have a human father, or was his mother a virgin at the time of his birth? Whether or not there is enough surviving evidence to decide it, this is still a strictly scientific question with a definite answer in principle: yes or no. Did Jesus raise Lazarus from the dead? Did he himself come alive again, three days after being crucified? There is an answer to every such question, whether or not we can discover it in practice, and it is a strictly scientific answer. The methods we should use to settle the matter, in the unlikely event that relevant evidence ever became available, would be purely and entirely scientific methods. To dramatize the point, imagine, by some remarkable set of circumstances, that forensic archaeologists unearthed DNA evidence to show that Jesus really did lack a biological father. Can you imagine religious apologists shrugging their shoulders and saying anything remotely like the following? 'Who cares? Scientific evidence is completely irrelevant to theological questions. Wrong magisterium! We're concerned only with ultimate questions and with moral values. Neither DNA nor any other scientific evidence could ever have any bearing on the matter, one way or the other.'

The very idea is a joke. You can bet your boots that the scientific evidence, if any were to turn up, would be seized upon and trumpeted to the skies. NOMA is popular only because there is no evidence to favour the God Hypothesis. The moment there was the smallest suggestion of any evidence in favour of religious belief, religious apologists would lose no time in throwing NOMA out of the window. Sophisticated theologians aside (and even they are happy to tell miracle stories to the unsophisticated in order to swell congregations), I suspect that alleged miracles provide the strongest reason many believers have for their faith; and miracles, by definition, violate the principles of science.

The Roman Catholic Church on the one hand seems sometimes to aspire to NOMA, but on the other hand lays down the performance of miracles as an essential qualification for elevation to sainthood. The late King of the Belgians is a candidate for sainthood, because of his stand on abortion. Earnest investigations are now going on to discover whether any miraculous cures can be attributed to prayers offered up to him since his death. I am not joking. That is the case, and it is typical of saint stories. I imagine the whole business is an embarrassment to more sophisticated circles within the Church. Why any circles worthy of the name of sophisticated remain within the Church is a mystery at least as deep as those that theologians enjoy.

When faced with miracle stories, Gould would presumably retort along the following lines. The whole point of NOMA is that it is a two-way bargain. The moment religion steps on science's turf and starts to meddle in the real world with miracles, it ceases to be religion in the sense Gould is defending, and his *amicabilis concordia* is broken. Note, however, that the miracle-free religion defended by Gould would not be recognized by most practising theists in the pew or on the prayer mat. It would, indeed, be a grave disappointment to them. To adapt Alice's comment on her sister's book before she fell into Wonderland, what is the use of a God who does no miracles and answers no prayers? Remember Ambrose Bierce's witty definition of the verb 'to pray': 'to ask that the laws of the universe be annulled in behalf of a single petitioner, confessedly unworthy'. There are athletes who believe God helps them win – against opponents who would seem, on the face of it, no less worthy of his favouritism. There are motorists who believe God saves them a parking space – thereby presumably depriving somebody else. This style of theism is embarrassingly popular, and is unlikely to be impressed by anything as (superficially) reasonable as NOMA.

Nevertheless, let us follow Gould and pare our religion down to some sort of non-interventionist minimum: no miracles, no personal communication between God and us in either direction, no monkeying with the laws of physics, no trespassing on the scientific grass. At most, a little deistic input to the initial conditions of the universe so that, in the fullness of time, stars, elements, chemistry and planets develop, and life evolves. Surely that is an adequate separation? Surely

NOMA can survive this more modest and unassuming religion?

Well, you might think so. But I suggest that even a non-interventionist, NOMA God, though less violent and clumsy than an Abrahamic God, is still, when you look at him fair and square, a scientific hypothesis. I return to the point: a universe in which we are alone except for other slowly evolved intelligences is a very different universe from one with an original guiding agent whose intelligent design is responsible for its very existence. I accept that it may not be so easy in practice to distinguish one kind of universe from the other. Nevertheless, there is something utterly special about the hypothesis of ultimate design, and equally special about the only known alternative: gradual evolution in the broad sense. They are close to being irreconcilably different. Like nothing else, evolution really does provide an explanation for the existence of entities whose improbability would otherwise, for practical purposes, rule them out. And the conclusion to the argument, as I shall show in Chapter 4, is close to being terminally fatal to the God Hypothesis.

THE GREAT PRAYER EXPERIMENT

An amusing, if rather pathetic, case study in miracles is the Great Prayer Experiment: does praying for patients help them recover? Prayers are commonly offered for sick people, both privately and in formal places of worship. Darwin's cousin Francis Galton was the first to analyse scientifically whether praying for people is efficacious. He noted that every Sunday, in churches throughout Britain, entire congregations prayed publicly for the health of the royal family. Shouldn't they, therefore, be unusually fit, compared with the rest of us, who are prayed for only by our nearest and dearest?* Galton looked into it, and found no statistical difference. His intention may, in any case,

---

* When my Oxford college elected the Warden whom I quoted earlier, it happened that the Fellows publicly drank his health on three successive evenings. At the third of these dinners, he graciously remarked in his speech of reply: 'I'm feeling better already.'

have been satirical, as also when he prayed over randomized plots of land to see if the plants would grow any faster (they didn't).

More recently, the physicist Russell Stannard (one of Britain's three well-known religious scientists, as we shall see) has thrown his weight behind an initiative, funded by – of course – the Templeton Foundation, to test experimentally the proposition that praying for sick patients improves their health.[36]

Such experiments, if done properly, have to be double blind, and this standard was strictly observed. The patients were assigned, strictly at random, to an experimental group (received prayers) or a control group (received no prayers). Neither the patients, nor their doctors or caregivers, nor the experimenters were allowed to know which patients were being prayed for and which patients were controls. Those who did the experimental praying had to know the names of the individuals for whom they were praying – otherwise, in what sense would they be praying for them rather than for somebody else? But care was taken to tell them only the first name and initial letter of the surname. Apparently that would be enough to enable God to pinpoint the right hospital bed.

The very idea of doing such experiments is open to a generous measure of ridicule, and the project duly received it. As far as I know, Bob Newhart didn't do a sketch about it, but I can distinctly hear his voice:

> What's that you say, Lord? You can't cure me because I'm a member of the control group? . . . Oh I see, my aunt's prayers aren't enough. But Lord, Mr Evans in the next-door bed . . . What was that, Lord? . . . Mr Evans received a thousand prayers per day? But Lord, Mr Evans doesn't know a thousand people . . . Oh, they just referred to him as John E. But Lord, how did you know they didn't mean John Ellsworthy? . . . Oh right, you used your omniscience to work out which John E they meant. But Lord . . .

Valiantly shouldering aside all mockery, the team of researchers soldiered on, spending $2.4 million of Templeton money under the leadership of Dr Herbert Benson, a cardiologist at the Mind/Body Medical Institute near Boston. Dr Benson was earlier quoted in a Templeton press release as 'believing that evidence for the efficacy of intercessory prayer in medicinal settings is mounting'. Reassuringly, then, the research was in good hands, unlikely to be spoiled by sceptical vibrations. Dr Benson and his team monitored 1,802 patients at six hospitals, all of whom received coronary bypass surgery. The patients were divided into three groups. Group 1 received prayers and didn't know it. Group 2 (the control group) received no prayers and didn't know it. Group 3 received prayers and did know it. The comparison between Groups 1 and 2 tests for the efficacy of intercessory prayer. Group 3 tests for possible psychosomatic effects of knowing that one is being prayed for.

Prayers were delivered by the congregations of three churches, one in Minnesota, one in Massachusetts and one in Missouri, all distant from the three hospitals. The praying individuals, as explained, were given only the first name and initial letter of the surname of each patient for whom they were to pray. It is good experimental practice to standardize as far as possible, and they were all, accordingly, told to include in their prayers the phrase 'for a successful surgery with a quick, healthy recovery and no complications'.

The results, reported in the *American Heart Journal* of April 2006, were clear-cut. There was no difference between those patients who were prayed for and those who were not. What a surprise. There was a difference between those who *knew* they had been prayed for and those who did not know one way or the other; but it went in the wrong direction. Those who knew they had been the beneficiaries of prayer suffered significantly more complications than those who did not. Was God doing a bit of smiting, to show his disapproval of the whole barmy enterprise? It seems more probable that those patients who knew they were being prayed for suffered additional stress in consequence: 'performance anxiety', as the experimenters put it. Dr Charles Bethea, one of the researchers, said, 'It may have made them uncertain, wondering am I so sick they had to call in their prayer team?' In today's litigious society, is it too much to hope that those patients suffering heart complications, as a consequence of knowing they were receiving experimental prayers, might put together a class action lawsuit against the Templeton Foundation?

It will be no surprise that this study was opposed by theologians, perhaps anxious about its capacity to bring ridicule upon religion. The Oxford theologian Richard Swinburne, writing after the study failed, objected to it on the grounds that God answers prayers only if they are offered up for good reasons.[37] Praying for somebody rather than somebody else, simply because of the fall of the dice in the design of a double-blind experiment, does not constitute a good reason. God would see through it. That, indeed, was the point of my Bob Newhart satire, and Swinburne is right to make it too. But in other parts of his paper Swinburne himself is beyond satire. Not for the first time, he seeks to justify suffering in a world run by God:

> My suffering provides me with the opportunity to show courage and patience. It provides you with the opportunity to show sympathy and to help alleviate my suffering. And it provides society with the opportunity to choose whether or not to invest a lot of money in trying to find a cure for this or that particular kind of suffering . . . Although a good God regrets our suffering, his greatest concern is surely that each of us shall show patience, sympathy and generosity and, thereby, form a holy character. Some people badly need to be ill for their own sake, and some people badly need to be ill to provide important choices for others. Only in that way can

> some people be encouraged to make serious choices about the sort of person they are to be. For other people, illness is not so valuable.

This grotesque piece of reasoning, so damningly typical of the theological mind, reminds me of an occasion when I was on a television panel with Swinburne, and also with our Oxford colleague Professor Peter Atkins. Swinburne at one point attempted to justify the Holocaust on the grounds that it gave the Jews a wonderful opportunity to be courageous and noble. Peter Atkins splendidly growled, 'May you rot in hell.'*

Another typical piece of theological reasoning occurs further along in Swinburne's article. He rightly suggests that if God wanted to demonstrate his own existence he would find better ways to do it than slightly biasing the recovery statistics of experimental versus control groups of heart patients. If God existed and wanted to convince us of it, he could 'fill the world with super-miracles'. But then Swinburne lets fall his gem: 'There is quite a lot of evidence anyway of God's existence, and too much might not be good for us.' Too much might not be good for us! Read it again. *Too much evidence might not be good for us.* Richard Swinburne is the recently retired holder of one of Britain's most prestigious professorships of theology, and is a Fellow of the British Academy. If it's a theologian you want, they don't come much more distinguished. Perhaps you don't want a theologian.

Swinburne wasn't the only theologian to disown the study after it had failed. The Reverend Raymond J. Lawrence was granted a generous tranche of op-ed space in the *New York Times* to explain why responsible religious leaders 'will breathe

---

* This interchange was edited out of the final broadcast version. That Swinburne's remark is typical of his theology is indicated by his rather similar comment about Hiroshima in *The Existence of God* (2004), page 264: 'Suppose that one less person had been burnt by the Hiroshima atomic bomb. Then there would have been less opportunity for courage and sympathy . . .'

a sigh of relief' that no evidence could be found of intercessory prayer having any effect.[38] Would he have sung a different tune if the Benson study had succeeded in demonstrating the power of prayer? Maybe not, but you can be certain that plenty of other pastors and theologians would. The Reverend Lawrence's piece is chiefly memorable for the following revelation: 'Recently, a colleague told me about a devout, well-educated woman who accused a doctor of malpractice in his treatment of her husband. During her husband's dying days, she charged, the doctor had failed to pray for him.'

Other theologians joined NOMA-inspired sceptics in contending that studying prayer in this way is a waste of money because supernatural influences are by definition beyond the reach of science. But as the Templeton Foundation correctly recognized when it financed the study, the alleged power of intercessory prayer is at least in principle within the reach of science. A double-blind experiment can be done and was done. It could have yielded a positive result. And if it had, can you imagine that a single religious apologist would have dismissed it on the grounds that scientific research has no bearing on religious matters? Of course not.

Needless to say, the negative results of the experiment will not shake the faithful. Bob Barth, the spiritual director of the Missouri prayer ministry which supplied some of the experimental prayers, said: 'A person of faith would say that this study is interesting, but we've been praying a long time and we've seen prayer work, we know it works, and the research on prayer and spirituality is just getting started.' Yeah, right: we know from our *faith* that prayer works, so if evidence fails to show it we'll just soldier on until finally we get the result we want.

THE NEVILLE CHAMBERLAIN SCHOOL OF EVOLUTIONISTS

A possible ulterior motive for those scientists who insist on NOMA – the invulnerability to science of the God Hypothesis

– is a peculiarly American political agenda, provoked by the threat of populist creationism. In parts of the United States, science is under attack from a well-organized, politically well-connected and, above all, well-financed opposition, and the teaching of evolution is in the front-line trench. Scientists could be forgiven for feeling threatened, because most research money comes ultimately from government, and elected representatives have to answer to the ignorant and prejudiced, as well as to the well-informed, among their constituents.

In response to such threats, an evolution defence lobby has sprung up, most notably represented by the National Center for Science Education (NCSE), led by Eugenie Scott, indefatigable activist on behalf of science who has recently produced her own book, *Evolution vs. Creationism.* One of NCSE's main political objectives is to court and mobilize 'sensible' religious opinion: mainstream churchmen and women who have no problem with evolution and may regard it as irrelevant to (or even in some strange way supportive of) their faith. It is to this mainstream of clergy, theologians and non-fundamentalist believers, embarrassed as they are by creationism because it brings religion into disrepute, that the evolution defence lobby tries to appeal. And one way to do this is to bend over backwards in their direction by espousing NOMA – agree that science is completely non-threatening, because it is disconnected from religion's claims.

Another prominent luminary of what we might call the Neville Chamberlain school of evolutionists is the philosopher Michael Ruse. Ruse has been an effective fighter against creationism,[39] both on paper and in court. He claims to be an atheist, but his article in *Playboy* takes the view that

> we who love science must realize that the enemy of our enemies is our friend. Too often evolutionists spend time insulting would-be allies. This is especially true of secular evolutionists. Atheists spend more time running down sympathetic

> Christians than they do countering creationists. When John Paul II wrote a letter endorsing Darwinism, Richard Dawkins's response was simply that the pope was a hypocrite, that he could not be genuine about science and that Dawkins himself simply preferred an honest fundamentalist.

From a purely tactical viewpoint, I can see the superficial appeal of Ruse's comparison with the fight against Hitler: 'Winston Churchill and Franklin Roosevelt did not like Stalin and communism. But in fighting Hitler they realized that they had to work with the Soviet Union. Evolutionists of all kinds must likewise work together to fight creationism.' But I finally come down on the side of my colleague the Chicago geneticist Jerry Coyne, who wrote that Ruse

> fails to grasp the real nature of the conflict. It's not just about evolution versus creationism. To scientists like Dawkins and Wilson [E. O. Wilson, the celebrated Harvard biologist], the *real* war is between rationalism and superstition. Science is but one form of rationalism, while religion is the most common form of superstition. Creationism is just a symptom of what they see as the greater enemy: religion. While religion can exist without creationism, creationism cannot exist without religion.[40]

I do have one thing in common with the creationists. Like me, but unlike the 'Chamberlain school', they will have no truck with NOMA and its separate magisteria. Far from respecting the separateness of science's turf, creationists like nothing better than to trample their dirty hobnails all over it. And they fight dirty, too. Lawyers for creationists, in court cases around the American boondocks, seek out evolutionists who are openly atheists. I know – to my chagrin – that my name has been used in this way. It is an effective tactic because juries selected at random are likely to include individuals brought up to believe that atheists are demons incarnate, on a par with pedophiles or 'terrorists' (today's equivalent of Salem's witches and McCarthy's Commies). Any creationist lawyer who got me on the stand could instantly win over the jury simply by asking me: 'Has your knowledge of evolution influenced you in the direction of becoming an atheist?' I would have to answer yes and, at one stroke, I would have lost the jury. By contrast, the judicially correct answer from the secularist side would be: 'My religious beliefs, or lack of them, are a private matter, neither the business of this court nor connected in any way with my science.' I couldn't honestly say this, for reasons I shall explain in Chapter 4.

The *Guardian* journalist Madeleine Bunting wrote an article entitled 'Why the intelligent design lobby thanks God for Richard Dawkins'.[41] There's no indication that she consulted anybody except Michael Ruse, and her article might as well have been ghost-written by him.* Dan Dennett replied, aptly quoting Uncle Remus:

> I find it amusing that two Brits – Madeleine Bunting and Michael Ruse – have fallen for a version of one of the most famous scams in American folklore (Why the intelligent design lobby thanks God for Richard Dawkins, March 27). When Brer Rabbit gets caught by the fox, he pleads with him: 'Oh, please, please, Brer Fox, whatever you do, don't throw me in that awful briar patch!' – where he ends up safe and sound after the fox does just that. When the American propagandist William Dembski writes tauntingly to Richard Dawkins, telling him to keep

* The same could be said of an article, 'When cosmologies collide', in the *New York Times*, 22 Jan. 2006, by the respected (and usually much better briefed) journalist Judith Shulevitz. General Montgomery's First Rule of War was 'Don't march on Moscow.' Perhaps there should be a First Rule of Science Journalism: 'Interview at least one person other than Michael Ruse.'

> up the good work on behalf of intelligent design, Bunting and Ruse fall for it! 'Oh golly, Brer Fox, your forthright assertion – that evolutionary biology disproves the idea of a creator God – jeopardises the teaching of biology in science class, since teaching that would violate the separation of church and state!' Right. You also ought to soft-pedal physiology, since it declares virgin birth impossible . . .[42]

This whole issue, including an independent invocation of Brer Rabbit in the briar patch, is well discussed by the biologist P. Z. Myers, whose Pharyngula blog can reliably be consulted for trenchant good sense.[43]

I am not suggesting that my colleagues of the appeasement lobby are necessarily dishonest. They may sincerely believe in NOMA, although I can't help wondering how thoroughly they've thought it through and how they reconcile the internal conflicts in their minds. There is no need to pursue the matter for the moment, but anyone seeking to understand the published statements of scientists on religious matters would do well not to forget the political context: the surreal culture wars now rending America. NOMA-style appeasement will surface again in a later chapter. Here, I return to agnosticism and the possibility of chipping away at our ignorance and measurably reducing our uncertainty about the existence or non-existence of God.

LITTLE GREEN MEN

Suppose Bertrand Russell's parable had concerned not a teapot in outer space but *life* in outer space – the subject of Sagan's memorable refusal to think with his gut. Once again we cannot disprove it, and the only strictly rational stance is agnosticism. But the hypothesis is no longer frivolous. We don't immediately scent extreme improbability. We can have an interesting argument based on incomplete evidence, and we can write down the kind of evidence that would decrease our uncertainty. We'd be outraged if our government invested in expensive telescopes for the sole purpose of searching for orbiting teapots. But we can appreciate the case for spending money on SETI, the Search for Extraterrestrial Intelligence, using radio telescopes to scan the skies in the hope of picking up signals from intelligent aliens.

I praised Carl Sagan for disavowing gut feelings about alien life. But one can (and Sagan did) make a sober assessment of what we would need to know in order to estimate the probability. This might start from nothing more than a listing of our points of ignorance, as in the famous Drake Equation which, in Paul Davies's phrase, collects probabilities. It states that to estimate the number of independently evolved civilizations in the universe you must multiply seven terms together. The seven include the number of stars, the number of Earth-like planets per star, and the probability of this, that and the other which I need not list because the only point I am making is that they are all unknown, or estimated with enormous margins of error. When so many terms that are either completely or almost completely unknown are multiplied up, the product – the estimated number of alien civilizations – has such colossal error bars that agnosticism seems a very reasonable, if not the only credible stance.

Some of the terms in the Drake Equation are already less unknown than when he first wrote it down in 1961. At that time, our solar system of planets orbiting a central star was the only one known, together with the local analogies provided by Jupiter's and Saturn's satellite systems. Our best estimate of the number of orbiting systems in the universe was based on theoretical models, coupled with the more informal 'principle of mediocrity': the feeling (born of uncomfortable history lessons from Copernicus, Hubble and others) that there should be nothing particularly unusual about the place where we happen to live. Unfortunately, the principle of mediocrity is in its turn emasculated by the 'anthropic' principle (see Chapter

4): if our solar system really were the only one in the universe, this is precisely where we, as beings who think about such matters, would have to be living. The very fact of our existence could retrospectively determine that we live in an extremely unmediocre place.

But today's estimates of the ubiquity of solar systems are no longer based on the principle of mediocrity; they are informed by direct evidence. The spectroscope, nemesis of Comte's positivism, strikes again. Our telescopes are scarcely powerful enough to see planets around other stars directly. But the position of a star is perturbed by the gravitational pull of its planets as they whirl around it, and spectroscopes can pick up the Doppler shifts in the star's spectrum, at least in cases where the perturbing planet is large. Mostly using this method, at the time of writing we now know of 170 extra-solar planets orbiting 147 stars, but the figure will certainly have increased by the time you read this book. So far, they are bulky 'Jupiters', because only Jupiters are large enough to perturb their stars into the zone of detectability of present-day spectroscopes.

We have at least quantitatively improved our estimate of one previously shrouded term of the Drake Equation. This permits a significant, if still moderate, easing of our agnosticism about the final value yielded by the equation. We must still be agnostic about life on other worlds – but a little bit less agnostic, because we are just that bit less ignorant. Science can chip away at agnosticism, in a way that Huxley bent over backwards to deny for the special case of God. I am arguing that, notwithstanding the polite abstinence of Huxley, Gould and many others, the God question is not in principle and forever outside the remit of science. As with the nature of the stars, *contra* Comte, and as with the likelihood of life in orbit around them, science can make at least probabilistic inroads into the territory of agnosticism.

My definition of the God Hypothesis included the words 'superhuman' and 'supernatural'. To clarify the difference, imagine that a SETI radio telescope actually did pick up a signal from outer space which showed, unequivocally, that we are not alone. It is a non-trivial question, by the way, what kind of signal would convince us of its intelligent origin. A good approach is to turn the question around. What should we intelligently do in order to advertise our presence to extra-terrestrial listeners? Rhythmic pulses wouldn't do it. Jocelyn Bell Burnell, the radio astronomer who first discovered the pulsar in 1967, was moved by the precision of its 1.33-second periodicity to name it, tongue in cheek, the LGM (Little Green Men) signal. She later found a second pulsar, elsewhere in the sky and of different periodicity, which pretty much disposed of the LGM hypothesis. Metronomic rhythms can be generated by many non-intelligent phenomena, from swaying branches to dripping water, from time lags in self-regulating feedback loops to spinning and orbiting celestial bodies. More than a thousand pulsars have now been found in our galaxy, and it is generally accepted that each one is a spinning neutron star emitting radio energy that sweeps around like a lighthouse beam. It is amazing to think of a star rotating on a timescale of seconds (imagine if each of our days lasted 1.33 seconds instead of 24 hours), but just about everything we know of neutron stars is amazing. The point is that the pulsar phenomenon is now understood as a product of simple physics, not intelligence.

Nothing simply rhythmic, then, would announce our intelligent presence to the waiting universe. Prime numbers are often mentioned as the recipe of choice, since it is difficult to think of a purely physical process that could generate them. Whether by detecting prime numbers or by some other means, imagine that SETI does come up with unequivocal evidence of extra-terrestrial intelligence, followed, perhaps, by a massive transmission of knowledge and wisdom, along the science-fiction lines of Fred Hoyle's *A for Andromeda* or Carl Sagan's *Contact.* How should we respond? A pardonable reaction would be something akin to worship, for any civilization capable of broadcasting a signal over such an immense distance is likely to be greatly superior to ours. Even if that civilization is not more advanced than ours at the time of transmission, the enormous distance between us entitles us to calculate that they must be millennia ahead of us by the time the message reaches us (unless they have driven themselves extinct, which is not unlikely).

Whether we ever get to know about them or not, there are very probably alien civilizations that are superhuman, to the point of being god-like in ways that exceed anything a theologian could possibly imagine. Their technical achievements would seem as supernatural to us as ours would seem to a Dark Age peasant transported to the twenty-first century. Imagine his response to a laptop computer, a mobile telephone, a hydrogen bomb or a jumbo jet. As Arthur C. Clarke put it, in his Third Law: 'Any sufficiently advanced technology is indistinguishable from magic.' The miracles wrought by our technology would have seemed to the ancients no less remarkable than the tales of Moses parting the waters, or Jesus walking upon them. The aliens of our SETI signal would be to us like gods, just as missionaries were treated as gods (and exploited the undeserved honour to the hilt) when they turned up in Stone Age cultures bearing guns, telescopes, matches, and almanacs predicting eclipses to the second.

In what sense, then, would the most advanced SETI aliens not *be* gods? In what sense would they be superhuman but not supernatural? In a very important sense, which goes to the heart of this book. The crucial difference between gods and god-like extraterrestrials lies not in their properties but in their provenance. Entities that are complex enough to be intelligent are products of an evolutionary process. No matter how god-like they may seem when we encounter them, they didn't start that way. Science-fiction authors, such as Daniel F. Galouye in *Counterfeit World*, have even suggested (and I cannot think how to disprove it) that we live in a computer simulation, set up by some vastly superior civilization. But the simulators themselves would have to come from somewhere. The laws of probability forbid all notions of their spontaneously appearing without simpler antecedents. They probably owe their existence to a (perhaps unfamiliar) version of Darwinian evolution: some sort of cumulatively ratcheting 'crane' as opposed to 'skyhook', to use Daniel Dennett's terminology.[44] Skyhooks – including all gods – are magic spells. They do no *bona fide* explanatory work and demand more explanation than they provide. Cranes are explanatory devices that actually do explain. Natural selection is the champion crane of all time. It has lifted life from primeval simplicity to the dizzy heights of complexity, beauty and apparent design that dazzle us today. This will be a dominant theme of Chapter 4, 'Why there almost certainly is no God'. But first, before proceeding with my main reason for actively disbelieving in God's existence, I have a responsibility to dispose of the positive arguments for belief that have been offered through history.

CHAPTER 3

Arguments for God's existence

A professorship of theology should have no place in our institution.

THOMAS JEFFERSON

Arguments for the existence of God have been codified for centuries by theologians, and supplemented by others, including purveyors of misconceived 'common sense'.

THOMAS AQUINAS' 'PROOFS'

The five 'proofs' asserted by Thomas Aquinas in the thirteenth century don't prove anything, and are easily – though I hesitate to say so, given his eminence – exposed as vacuous. The first three are just different ways of saying the same thing, and they can be considered together. All involve an infinite regress – the answer to a question raises a prior question, and so on *ad infinitum*.

1 *The Unmoved Mover.* Nothing moves without a prior mover. This leads us to a regress, from which the only escape is God. Something had to make the first move, and that something we call God.

2 *The Uncaused Cause.* Nothing is caused by itself. Every

effect has a prior cause, and again we are pushed back into regress. This has to be terminated by a first cause, which we call God.

3 *The Cosmological Argument.* There must have been a time when no physical things existed. But, since physical things exist now, there must have been something non-physical to bring them into existence, and that something we call God.

All three of these arguments rely upon the idea of a regress and invoke God to terminate it. They make the entirely unwarranted assumption that God himself is immune to the regress. Even if we allow the dubious luxury of arbitrarily conjuring up a terminator to an infinite regress and giving it a name, simply because we need one, there is absolutely no reason to endow that terminator with any of the properties normally ascribed to God: omnipotence, omniscience, goodness, creativity of design, to say nothing of such human attributes as listening to prayers, forgiving sins and reading innermost thoughts. Incidentally, it has not escaped the notice of logicians that omniscience and omnipotence are mutually incompatible. If God is omniscient, he must already know how he is going to intervene to change the course of history using his omnipotence. But that means he can't change his mind about his intervention, which means he is not omnipotent. Karen Owens has captured this witty little paradox in equally engaging verse:

> Can omniscient God, who
> Knows the future, find
> The omnipotence to
> Change His future mind?

To return to the infinite regress and the futility of invoking God to terminate it, it is more parsimonious to conjure up, say, a 'big bang singularity', or some other physical concept as yet unknown. Calling it God is at best unhelpful and at worst perniciously misleading. Edward Lear's Nonsense Recipe for Crumboblious Cutlets invites us to 'Procure some strips of beef, and having cut them into the smallest possible pieces, proceed to cut them still smaller, eight or perhaps nine times.' Some regresses do reach a natural terminator. Scientists used to wonder what would happen if you could dissect, say, gold into the smallest possible pieces. Why shouldn't you cut one of those pieces in half and produce an even smaller smidgen of gold? The regress in this case is decisively terminated by the atom. The smallest possible piece of gold is a nucleus consisting of exactly seventy-nine protons and a slightly larger number of neutrons, attended by a swarm of seventy-nine electrons. If you 'cut' gold any further than the level of the single atom, whatever else you get it is not gold. The atom provides a natural terminator to the Crumboblious Cutlets type of regress. It is by no means clear that God provides a natural terminator to the regresses of Aquinas. That's putting it mildly, as we shall see later. Let's move on down Aquinas' list.

4 *The Argument from Degree.* We notice that things in the world differ. There are degrees of, say, goodness or perfection. But we judge these degrees only by comparison with a maximum. Humans can be both good and bad, so the maximum goodness cannot rest in us. Therefore there must be some other maximum to set the standard for perfection, and we call that maximum God.

That's an argument? You might as well say, people vary in smelliness but we can make the comparison only by reference to a perfect maximum of conceivable smelliness. Therefore there must exist a pre-eminently peerless stinker, and we call him God. Or substitute any dimension of comparison you like, and derive an equivalently fatuous conclusion.

5 *The Teleological Argument*, or *Argument from Design.* Things in the world, especially living things, look as though they have been designed. Nothing that we know looks designed unless it is designed. Therefore there must have been a designer, and we call him God.* Aquinas himself used the analogy of an arrow moving towards a target, but a modern heat-seeking anti-aircraft missile would have suited his purpose better.

The argument from design is the only one still in regular use today, and it still sounds to many like the ultimate knockdown argument. The young Darwin was impressed by it when, as a Cambridge undergraduate, he read it in William Paley's *Natural Theology*. Unfortunately for Paley, the mature Darwin blew it out of the water. There has probably never been a more devastating rout of popular belief by clever reasoning than Charles Darwin's destruction of the argument from design. It was so unexpected. Thanks to Darwin, it is no longer true to say that nothing that we know looks designed unless it is designed. Evolution by natural selection produces an excellent simulacrum of design, mounting prodigious heights of complexity and elegance. And among these eminences of pseudo-design are nervous systems which – among their more modest accomplishments – manifest goal-seeking behaviour that, even in a tiny insect, resembles a sophisticated heat-seeking missile more than a simple arrow on target. I shall return to the argument from design in Chapter 4.

THE ONTOLOGICAL ARGUMENT AND OTHER *A PRIORI* ARGUMENTS

Arguments for God's existence fall into two main categories, the *a priori* and the *a posteriori*. Thomas Aquinas' five are

* I cannot help being reminded of the immortal syllogism that was smuggled into a Euclidean proof by a schoolfriend, when we were studying geometry together: 'Triangle ABC looks isosceles. Therefore . . .'

a posteriori arguments, relying upon inspection of the world. The most famous of the *a priori* arguments, those that rely upon pure armchair ratiocination, is the *ontological argument*, proposed by St Anselm of Canterbury in 1078 and restated in different forms by numerous philosophers ever since. An odd aspect of Anselm's argument is that it was originally addressed not to humans but to God himself, in the form of a prayer (you'd think that any entity capable of listening to a prayer would need no convincing of his own existence).

It is possible to conceive, Anselm said, of a being than which nothing greater can be conceived. Even an atheist can conceive of such a superlative being, though he would deny its existence in the real world. But, goes the argument, a being that doesn't exist in the real world is, by that very fact, less than perfect. Therefore we have a contradiction and, hey presto, God exists!

Let me translate this infantile argument into the appropriate language, which is the language of the playground:

> 'Bet you I can prove God exists.'
>
> 'Bet you can't.'
>
> 'Right then, imagine the most perfect perfect *perfect* thing possible.'
>
> 'Okay, now what?'
>
> 'Now, is that perfect perfect *perfect* thing real? Does it exist?'
>
> 'No, it's only in my mind.'
>
> 'But if it was real it would be even more perfect, because a really really perfect thing would have to be better than a silly old imaginary thing. So I've proved that God exists. Nur Nurny Nur Nur. All atheists are fools.'

I had my childish wiseacre choose the word 'fools' advisedly. Anselm himself quoted the first verse of Psalm 14, 'The fool hath said in his heart, There is no God,' and he had the cheek to use the name 'fool' (Latin *insipiens*) for his hypothetical atheist:

> Hence, even the fool is convinced that something exists in the understanding, at least, than which nothing greater can be conceived. For, when he hears of this, he understands it. And whatever is understood, exists in the understanding. And assuredly that, than which nothing greater can be conceived, cannot exist in the understanding alone. For, suppose it exists in the understanding alone: then it can be conceived to exist in reality; which is greater.

The very idea that grand conclusions could follow from such logomachist trickery offends me aesthetically, so I must take care to refrain from bandying words like 'fool'. Bertrand Russell (no fool) interestingly said, 'It is easier to feel convinced that [the ontological argument] must be fallacious than it is to find out precisely where the fallacy lies.' Russell himself, as a young man, was briefly convinced by it:

> I remember the precise moment, one day in 1894, as I was walking along Trinity Lane, when I saw in a flash (or thought I saw) that the ontological argument is valid. I had gone out to buy a tin of tobacco; on my way back, I suddenly threw it up in the air, and exclaimed as I caught it: 'Great Scott, the ontological argument is sound.'

Why, I wonder, didn't he say something like: 'Great Scott, the ontological argument seems to be plausible. But isn't it too good to be true that a grand truth about the cosmos should follow from a mere word game? I'd better set to work to resolve what is perhaps a paradox like those of Zeno.' The Greeks had a hard time seeing through Zeno's 'proof' that Achilles would never catch the tortoise.* But they had the sense not to conclude that therefore Achilles really would fail to catch the tortoise. Instead, they called it a paradox and waited for later generations of mathematicians to explain it. Russell himself, of course, was as well qualified as anyone to understand why no tobacco tins should be thrown up in celebration of Achilles' failure to catch the tortoise. Why didn't he exercise the same caution over St Anselm? I suspect that he was an exaggeratedly fair-minded atheist, over-eager to be disillusioned if logic seemed to require it.† Or perhaps the answer lies in something Russell himself wrote in 1946, long after he had rumbled the ontological argument:

> The real question is: Is there anything we can think of which, by the mere fact that we can think of it, is shown to exist outside our thought? Every philosopher would *like* to say yes, because a philosopher's job is to find out things about the world by thinking rather than observing. If yes is the right

---

* Zeno's paradox is too well known for the details to be promoted out of a footnote. Achilles can run ten times as fast as the tortoise, so he gives the animal, say, 100 yards' start. Achilles runs 100 yards, and the tortoise is now 10 yards ahead. Achilles runs the 10 yards and the tortoise is now 1 yard ahead. Achilles runs the 1 yard, and the tortoise is still a tenth of a yard ahead . . . and so on *ad infinitum*, so Achilles never catches the tortoise.

† We might be seeing something similar today in the over-publicized tergiversation of the philosopher Antony Flew, who announced in his old age that he had been converted to belief in some sort of deity (triggering a frenzy of eager repetition all around the Internet). On the other hand, Russell was a great philosopher. Russell won the Nobel Prize. Maybe Flew's alleged conversion will be rewarded with the Templeton Prize. A first step in that direction is his ignominious decision to accept, in 2006, the 'Phillip E. Johnson Award for Liberty and Truth'. The first holder of the Phillip E. Johnson Award was Phillip E. Johnson, the lawyer credited with founding the Intelligent Design 'wedge strategy'. Flew will be the second holder. The awarding university is BIOLA, the Bible Institute of Los Angeles. One can't help wondering whether Flew realizes that he is being used. See Victor Stenger, 'Flew's flawed science', *Free Inquiry* 25: 2, 2005, 17–18; www.secularhumanism.org/index.php?section=library&page=stenger_25_2.

> answer, there is a bridge from pure thought to things. If not, not.

My own feeling, to the contrary, would have been an automatic, deep suspicion of any line of reasoning that reached such a significant conclusion without feeding in a single piece of data from the real world. Perhaps that indicates no more than that I am a scientist rather than a philosopher. Philosophers down the centuries have indeed taken the ontological argument seriously, both for and against. The atheist philosopher J. L. Mackie gives a particularly clear discussion in *The Miracle of Theism.* I mean it as a compliment when I say that you could almost define a philosopher as someone who won't take common sense for an answer.

The most definitive refutations of the ontological argument are usually attributed to the philosophers David Hume (1711–76) and Immanuel Kant (1724–1804). Kant identified the trick card up Anselm's sleeve as his slippery assumption that 'existence' is more 'perfect' than non-existence. The American philosopher Norman Malcolm put it like this: 'The doctrine that existence is a perfection is remarkably queer. It makes sense and is true to say that my future house will be a better one if it is insulated than if it is not insulated; but what could it mean to say that it will be a better house if it exists than if it does not?'[45] The Australian philosopher Douglas Gasking devised an ironic parody of Anselm's argument, which he did not record, but which has been reconstructed by William Grey of the University of Queensland as follows (this formulation was wrongly attributed in previous printings).

1 The creation of the world is the most marvellous achievement imaginable.

2 The merit of an achievement is the product of (a) its intrinsic quality, and (b) the ability of its creator.

3 The greater the disability (or handicap) of the creator, the more impressive the achievement.

4 The most formidable handicap for a creator would be non-existence.

5 Therefore if we suppose that the universe is the product of an existent creator we can conceive a greater being – namely, one who created everything while not existing.

6 An existing God therefore would not be a being greater than which a greater cannot be conceived because an even more formidable and incredible creator would be a God which did not exist.

Ergo:

7 God does not exist.

Needless to say, Gasking didn't really prove that God does not exist. By the same token, Anselm didn't prove that he does. The only difference is, Gasking was being funny on purpose. As he realized, the existence or non-existence of God is too big a question to be decided by 'dialectical prestidigitation'. And I don't think the slippery use of existence as an indicator of perfection is the worst of the argument's problems. I've forgotten the details, but I once piqued a gathering of theologians and philosophers by adapting the ontological argument to prove that pigs can fly. They felt the need to resort to Modal Logic to prove that I was wrong.

The ontological argument, like all *a priori* arguments for the existence of God, reminds me of the old man in Aldous Huxley's *Point Counter Point* who discovered a mathematical proof of the existence of God:

> You know the formula, *m* over nought equals infinity, *m* being any positive number? Well, why not reduce the equation to a simpler form by multiplying both sides by nought. In which case you have *m* equals infinity times nought. That is to say that a

> positive number is the product of zero and infinity. Doesn't that demonstrate the creation of the universe by an infinite power out of nothing? Doesn't it?

Unfortunately, the famous story of Diderot, the encyclopedist of the Enlightenment, and Euler, the Swiss mathematician, is open to doubt. According to legend, Catherine the Great staged a debate between the two of them in which the pious Euler threw down the challenge to the atheistic Diderot: 'Monsieur, $(a + b^n)/n = x$, therefore God exists. Reply!' The point of the myth is that Diderot was no mathematician and therefore had to withdraw in confusion. However, as B. H. Brown pointed out in the *American Mathematical Monthly* (1942), Diderot was actually rather a good mathematician, and would have been unlikely to fall for what might be called the Argument from Blinding with Science (in this case mathematics). David Mills, in *Atheist Universe*, transcribes a radio interview of himself by a religious spokesman, who invoked the Law of Conservation of Mass-Energy in a weirdly ineffectual attempt to blind with science: 'Since we're all composed of matter and energy, doesn't that scientific principle lend credibility to a belief in eternal life?' Mills replied more patiently and politely than I would have, for what the interviewer was saying, translated into English, was no more than: 'When we die, none of the atoms of our body (and none of the energy) are lost. Therefore we are immortal.'

Even I, with my long experience, have never encountered wishful thinking as silly as that. I have, however, met many of the wonderful 'proofs' collected at http://www.godlessgeeks.com/LINKS/GodProof.htm, a richly comic numbered list of 'Over Three Hundred Proofs of God's Existence'. Here's a hilarious half-dozen, beginning with Proof Number 36.

36 ***Argument from Incomplete Devastation:*** A plane crashed killing 143 passengers and crew. But one child survived with only third-degree burns. Therefore God exists.

37 *Argument from Possible Worlds:* If things had been different, then things would be different. That would be bad. Therefore God exists.

38 *Argument from Sheer Will:* I do believe in God! I do believe in God! I do I do I do. I do believe in God! Therefore God exists.

39 *Argument from Non-belief:* The majority of the world's population are non-believers in Christianity. This is just what Satan intended. Therefore God exists.

40 *Argument from Post-Death Experience:* Person X died an atheist. He now realizes his mistake. Therefore God exists.

41 *Argument from Emotional Blackmail:* God loves you. How could you be so heartless as not to believe in him? Therefore God exists.

THE ARGUMENT FROM BEAUTY

Another character in the Aldous Huxley novel just mentioned proved the existence of God by playing Beethoven's string quartet no. 15 in A minor ('*heiliger Dankgesang*') on a gramophone. Unconvincing as that sounds, it does represent a popular strand of argument. I have given up counting the number of times I receive the more or less truculent challenge: 'How do you account for Shakespeare, then?' (Substitute Schubert, Michelangelo, etc. to taste.) The argument will be so familiar, I needn't document it further. But the logic behind it is never spelled out, and the more you think about it the more vacuous you realize it to be. Obviously Beethoven's late quartets are sublime. So are Shakespeare's sonnets. They are sublime if God is there and they are sublime if he isn't. They do not prove the existence of God; they prove the existence of Beethoven and of Shakespeare. A great conductor is credited with saying: 'If you have Mozart to listen to, why would you need God?'

I once was the guest of the week on a British radio show called *Desert Island Discs*. You have to choose the eight records you would take with you if marooned on a desert island. Among my choices was '*Mache dich mein Herze rein*' from Bach's *St Matthew Passion*. The interviewer was unable to understand how I could choose religious music without being religious. You might as well say, how can you enjoy *Wuthering Heights* when you know perfectly well that Cathy and Heathcliff never really existed?

But there is an additional point that I might have made, and which needs to be made whenever religion is given credit for, say, the Sistine Chapel or Raphael's *Annunciation*. Even great artists have to earn a living, and they will take commissions where they are to be had. I have no reason to doubt that Raphael and Michelangelo were Christians – it was pretty much the only option in their time – but the fact is almost incidental. Its enormous wealth had made the Church the dominant patron of the arts. If history had worked out differently, and Michelangelo had been commissioned to paint a ceiling for a giant Museum of Science, mightn't he have produced something at least as inspirational as the Sistine Chapel? How sad that we shall never hear Beethoven's *Mesozoic Symphony*, or Mozart's opera *The Expanding Universe*. And what a shame that we are deprived of Haydn's *Evolution Oratorio* – but that does not stop us from enjoying his *Creation*. To approach the argument from the other side, what if, as my wife chillingly suggests to me, Shakespeare had been obliged to work to commissions from the Church? We'd surely have lost *Hamlet*, *King Lear* and *Macbeth*. And what would we have gained in return? Such stuff as dreams are made on? Dream on.

If there is a logical argument linking the existence of great art to the existence of God, it is not spelled out by its proponents. It is simply assumed to be self-evident, which it most certainly is not. Maybe it is to be seen as yet another version of the argument from design: Schubert's musical brain is a wonder of improbability, even more so than the vertebrate eye. Or, more ignobly, perhaps it's a sort of jealousy of genius. How dare another human being make such beautiful music/poetry/art, when I can't? It must be God that did it.

THE ARGUMENT FROM PERSONAL 'EXPERIENCE'

One of the cleverer and more mature of my undergraduate contemporaries, who was deeply religious, went camping in the Scottish isles. In the middle of the night he and his girlfriend were woken in their tent by the voice of the devil – Satan himself; there could be no possible doubt: the voice was in every sense diabolical. My friend would never forget this horrifying experience, and it was one of the factors that later drove him to be ordained. My youthful self was impressed by his story, and I recounted it to a gathering of zoologists relaxing in the Rose and Crown Inn, Oxford. Two of them happened to be experienced ornithologists, and they roared with laughter. 'Manx Shearwater!' they shouted in delighted chorus. One of them added that the diabolical shrieks and cackles of this species have earned it, in various parts of the world and various languages, the local nickname 'Devil Bird'.

Many people believe in God because they believe they have seen a vision of him – or of an angel or a virgin in blue – with their own eyes. Or he speaks to them inside their heads. This argument from personal experience is the one that is most convincing to those who claim to have had one. But it is the least convincing to anyone else, and anyone knowledgeable about psychology.

You say you have experienced God directly? Well, some people have experienced a pink elephant, but that probably doesn't impress you. Peter Sutcliffe, the Yorkshire Ripper, distinctly heard the voice of Jesus telling him to kill women, and he was locked up for life. George W. Bush says that God told him to invade Iraq (a pity God didn't vouchsafe him a revelation that there were no weapons of mass destruction).

Individuals in asylums think they are Napoleon or Charlie Chaplin, or that the entire world is conspiring against them, or that they can broadcast their thoughts into other people's heads. We humour them but don't take their internally revealed beliefs seriously, mostly because not many people share them. Religious experiences are different only in that the people who claim them are numerous. Sam Harris was not being overly cynical when he wrote, in *The End of Faith*:

> We have names for people who have many beliefs for which there is no rational justification. When their beliefs are extremely common we call them 'religious'; otherwise, they are likely to be called 'mad', 'psychotic' or 'delusional' . . . Clearly there is sanity in numbers. And yet, it is merely an accident of history that it is considered normal in our society to believe that the Creator of the universe can hear your thoughts, while it is demonstrative of mental illness to believe that he is communicating with you by having the rain tap in Morse code on your bedroom window. And so, while religious people are not generally mad, their core beliefs absolutely are.

I shall return to the subject of hallucinations in Chapter 10.

The human brain runs first-class simulation software. Our eyes don't present to our brains a faithful photograph of what is out there, or an accurate movie of what is going on through time. Our brains construct a continuously updated model: updated by coded pulses chattering along the optic nerve, but constructed nevertheless. Optical illusions are vivid reminders of this.[46] A major class of illusions, of which the Necker Cube is an example, arise because the sense data that the brain receives are compatible with two alternative models of reality. The brain, having no basis for choosing between them, alternates, and we experience a series of flips from one internal model to the other. The picture we are looking at appears, almost literally, to flip over and become something else.

The simulation software in the brain is especially adept at constructing faces and voices. I have on my windowsill a plastic mask of Einstein. When seen from the front, it looks like a solid face, not surprisingly. What is surprising is that, when seen from behind – the hollow side – it also looks like a solid face, and our perception of it is very odd indeed. As the viewer moves around, the face seems to follow – and not in the weak, unconvincing sense that the Mona Lisa's eyes are said to follow you. The hollow mask really *really* looks as though it is moving. People who haven't previously seen the illusion gasp with amazement. Even stranger, if the mask is mounted on a slowly rotating turntable, it appears to turn in the correct direction when you are looking at the solid side, but in the *opposite* direction when the hollow side comes into view. The result is that, when you watch the transition from one side to the other, the coming side appears to 'eat' the going side. It is a stunning illusion, well worth going to some trouble to see. Sometimes you can get surprisingly close to the hollow face and still not see that it is 'really' hollow. When you do see it, again there is a sudden flip, which may be reversible.

Why does it happen? There is no trick in the construction of the mask. Any hollow mask will do it. The trickery is all in the brain of the beholder. The internal simulating software receives data indicating the presence of a face, perhaps nothing more than a pair of eyes, a nose and a mouth in approximately the right places. Having received these sketchy clues, the brain does the rest. The face simulation software kicks into action and it constructs a fully solid model of a face, even though the reality presented to the eyes is a hollow mask. The illusion of rotation in the wrong direction comes about because (it's quite hard, but if you think it through carefully you will confirm it) reverse rotation is the only way to make sense of the optical data when a hollow mask rotates while being perceived to be a solid mask.[47] It is like the illusion of a rotating radar dish that you sometimes see at airports. Until the brain flips to the correct model of the radar dish, an incorrect model is seen rotating in the wrong direction but in a weirdly cock-eyed way.

I say all this just to demonstrate the formidable power of the brain's simulation software. It is well capable of constructing 'visions' and 'visitations' of the utmost veridical power. To simulate a ghost or an angel or a Virgin Mary would be child's play to software of this sophistication. And the same thing works for hearing. When we hear a sound, it is not faithfully transported up the auditory nerve and relayed to the brain as if by a high-fidelity Bang & Olufsen. As with vision, the brain constructs a sound model, based upon continuously updated auditory nerve data. That is why we hear a trumpet blast as a single note, rather than as the composite of pure-tone harmonics that gives it its brassy snarl. A clarinet playing the same note sounds 'woody', and an oboe sounds 'reedy', because of different balances of harmonics. If you carefully manipulate a sound synthesizer to bring in the separate harmonics one by one, the brain hears them as a combination of pure tones for a short while, until its simulation software 'gets it', and from then on we experience only a single note of pure trumpet or oboe or whatever it is. The vowels and consonants of speech are constructed in the brain in the same kind of way, and so, at another level, are higher-order phonemes and words.

Once, as a child, I heard a ghost: a male voice murmuring, as if in recitation or prayer. I could almost, but not quite, make out the words, which seemed to have a serious, solemn timbre. I had been told stories of priest holes in ancient houses, and I was a little frightened. But I got out of bed and crept up on the source of the sound. As I got closer, it grew louder, and then suddenly it 'flipped' inside my head. I was now close enough to discern what it really was. The wind, gusting through the keyhole, was creating sounds which the simulation software in my brain had used to construct a model of male speech, solemnly intoned. Had I been a more impressionable child, it is possible that I would have 'heard' not just unintelligible speech but particular words and even sentences. And had I been both impressionable and religiously brought up, I wonder what words the wind might have spoken.

On another occasion, when I was about the same age, I saw a giant round face gazing, with unspeakable malevolence, out through the window of an otherwise ordinary house in a seaside village. In trepidation, I approached until I was close enough to see what it really was: just a vaguely face-like pattern created by the chance fall of the curtains. The face itself, and its evil mien, had been constructed in my fearful child's brain. On 11 September 2001, pious people thought they saw the face of Satan in the smoke rising from the Twin Towers: a superstition backed by a photograph which was published on the Internet and widely circulated.

Constructing models is something the human brain is very good at. When we are asleep it is called dreaming; when we are awake we call it imagination or, when it is exceptionally vivid, hallucination. As Chapter 10 will show, children who have 'imaginary friends' sometimes see them clearly, exactly as if they were real. If we are gullible, we don't recognize hallucination or lucid dreaming for what it is and we claim to have seen or heard a ghost; or an angel; or God; or – especially if we happen to be young, female and Catholic – the Virgin Mary. Such visions and manifestations are certainly not good grounds for believing that ghosts or angels, gods or virgins, are actually there.

On the face of it mass visions, such as the report that seventy thousand pilgrims at Fatima in Portugal in 1917 saw the sun 'tear itself from the heavens and come crashing down upon the multitude',[48] are harder to write off. It is not easy to explain how seventy thousand people could share the same hallucination. But it is even harder to accept that it really happened without the rest of the world, outside Fatima, seeing it too – and not just seeing it, but feeling it as the catastrophic destruction of the solar system, including acceleration forces sufficient to hurl everybody into space. David Hume's pithy test for a miracle comes irresistibly to mind: 'No testimony is sufficient to establish a miracle, unless the testimony be of such a kind, that its falsehood would be more miraculous than the fact which it endeavours to establish.'

It may seem improbable that seventy thousand people could simultaneously be deluded, or could simultaneously collude in a mass lie. Or that history is mistaken in recording that seventy thousand people claimed to see the sun dance. Or that they all simultaneously saw a mirage (they had been persuaded to stare at the sun, which can't have done much for their eyesight). But any of those apparent improbabilities is far more probable than the alternative: that the Earth was suddenly yanked sideways in its orbit, and the solar system destroyed, with nobody outside Fatima noticing. I mean, Portugal is not that isolated.*

That is really all that needs to be said about personal 'experiences' of gods or other religious phenomena. If you've had such an experience, you may well find yourself believing firmly that it was real. But don't expect the rest of us to take your word for it, especially if we have the slightest familiarity with the brain and its powerful workings.

THE ARGUMENT FROM SCRIPTURE

There are still some people who are persuaded by scriptural evidence to believe in God. A common argument, attributed among others to C. S. Lewis (who should have known better), states that, since Jesus claimed to be the Son of God, he must have been either right or else insane or a liar: 'Mad, Bad or God'. Or, with artless alliteration, 'Lunatic, Liar or Lord'. The historical evidence that Jesus claimed any sort of divine status is minimal. But even if that evidence were good, the trilemma on offer would be ludicrously inadequate. A fourth possibility, almost too obvious to need mentioning, is that Jesus was honestly mistaken. Plenty of people are. In any case, as I said, there is no good historical evidence that he ever thought he was divine.

---

* Although admittedly my wife's parents once stayed in a Paris hotel called the *Hôtel de l'Univers et du Portugal.*

The fact that something is written down is persuasive to people not used to asking questions like: 'Who wrote it, and when?' 'How did they know what to write?' 'Did they, in their time, really mean what we, in our time, understand them to be saying?' 'Were they unbiased observers, or did they have an agenda that coloured their writing?' Ever since the nineteenth century, scholarly theologians have made an overwhelming case that the gospels are not reliable accounts of what happened in the history of the real world. All were written long after the death of Jesus, and also after the epistles of Paul, which mention almost none of the alleged facts of Jesus' life. All were then copied and recopied, through many different 'Chinese Whispers generations' (see Chapter 5) by fallible scribes who, in any case, had their own religious agendas.

A good example of the colouring by religious agendas is the whole heart-warming legend of Jesus' birth in Bethlehem, followed by Herod's massacre of the innocents. When the gospels were written, many years after Jesus' death, nobody knew where he was born. But an Old Testament prophecy (Micah 5: 2) had led Jews to expect that the long-awaited Messiah would be born in Bethlehem. In the light of this prophecy, John's gospel specifically remarks that his followers were surprised that he was *not* born in Bethlehem: 'Others said, This is the Christ. But some said, Shall Christ come out of Galilee? Hath not the scripture said, That Christ cometh of the seed of David, and out of the town of Bethlehem, where David was?'

Matthew and Luke handle the problem differently, by deciding that Jesus *must* have been born in Bethlehem after all. But they get him there by different routes. Matthew has Mary and Joseph in Bethlehem all along, moving to Nazareth only long after the birth of Jesus, on their return from Egypt where they fled from King Herod and the massacre of the innocents. Luke, by contrast, acknowledges that Mary and Joseph lived in Nazareth before Jesus was born. So how to get them to Bethlehem at the crucial moment, in order to fulfil the prophecy? Luke says that, in the time when Cyrenius (Quirinius) was governor of Syria, Caesar Augustus decreed a census for taxation purposes, and everybody had to go 'to his own city'. Joseph was 'of the house and lineage of David' and therefore he had to go to 'the city of David, which is called Bethlehem'. That must have seemed like a good solution. Except that historically it is complete nonsense, as A. N. Wilson in *Jesus* and Robin Lane Fox in *The Unauthorized Version* (among others) have pointed out. David, if he existed, lived nearly a thousand years before Mary and Joseph. Why on earth would the Romans have required Joseph to go to the city where a remote ancestor had lived a millennium earlier? It is as though I were required to specify, say, Ashby-de-la-Zouch as my home town on a census form, if it happened that I could trace my ancestry back to the Seigneur de Dakeyne, who came over with William the Conqueror and settled there.

Moreover, Luke screws up his dating by tactlessly mentioning events that historians are capable of independently checking. There was indeed a census under Governor Quirinius – a local census, not one decreed by Caesar Augustus for the Empire as a whole – but it happened too late: in AD 6, long after Herod's death. Lane Fox concludes that 'Luke's story is historically impossible and internally incoherent', but he sympathizes with Luke's plight and his desire to fulfil the prophecy of Micah.

In the December 2004 issue of *Free Inquiry*, Tom Flynn, the Editor of that excellent magazine, assembled a collection of articles documenting the contradictions and gaping holes in the well-loved Christmas story. Flynn himself lists the many contradictions between Matthew and Luke, the only two evangelists who treat the birth of Jesus at all.[49] Robert Gillooly shows how all the essential features of the Jesus legend, including the star in the east, the virgin birth, the veneration of the baby by kings, the miracles, the execution, the resurrection and the ascension are borrowed – every last one of them – from other religions already in existence in the Mediterranean and

Near East region. Flynn suggests that Matthew's desire to fulfil messianic prophecies (descent from David, birth in Bethlehem) for the benefit of Jewish readers came into headlong collision with Luke's desire to adapt Christianity for the Gentiles, and hence to press the familiar hot buttons of pagan Hellenistic religions (virgin birth, worship by kings, etc.). The resulting contradictions are glaring, but consistently overlooked by the faithful.

Sophisticated Christians do not need Ira Gershwin to convince them that 'The things that you're li'ble / To read in the Bible / It ain't necessarily so'. But there are many unsophisticated Christians out there who think it absolutely is necessarily so – who take the Bible very seriously indeed as a literal and accurate record of history and hence as evidence supporting their religious beliefs. Do these people never open the book that they believe is the literal truth? Why don't they notice those glaring contradictions? Shouldn't a literalist worry about the fact that Matthew traces Joseph's descent from King David via twenty-eight intermediate generations, while Luke has forty-one generations? Worse, there is almost no overlap in the names on the two lists! In any case, if Jesus really was born of a virgin, Joseph's ancestry is irrelevant and cannot be used to fulfil, on Jesus' behalf, the Old Testament prophecy that the Messiah should be descended from David.

The American biblical scholar Bart Ehrman, in a book whose subtitle is *The Story Behind Who Changed the New Testament and Why*, unfolds the huge uncertainty befogging the New Testament texts.* In the introduction to the book, Professor Ehrman movingly charts his personal educational journey from Bible-believing fundamentalist to thoughtful sceptic, a journey driven by his dawning realization of the massive fallibility of the

---

* I give the subtitle because that is all I am confident of. The main title of my copy of the book, published by Continuum of London, is *Whose Word Is It?* I can find nothing in this edition to say whether it is the same book as the American publication by Harper San Francisco, which I haven't seen, whose main title is *Misquoting Jesus*. I presume they are the same book, but why do publishers do this kind of thing?

scriptures. Significantly, as he moved up the hierarchy of American universities, from rock bottom at the 'Moody Bible Institute', through Wheaton College (a little bit higher on the scale, but still the alma mater of Billy Graham) to Princeton Theological Seminary, he was at every step warned that he would have trouble maintaining his fundamentalist Christianity in the face of dangerous progressivism. So it proved; and we, his readers, are the beneficiaries. Other refreshingly iconoclastic books of biblical criticism are Robin Lane Fox's *The Unauthorized Version*, already mentioned, and Jacques Berlinerblau's *The Secular Bible: Why Nonbelievers Must Take Religion Seriously*.

The four gospels that made it into the official canon were chosen, more or less arbitrarily, out of a larger sample of at least a dozen including the Gospels of Thomas, Peter, Nicodemus, Philip, Bartholomew and Mary Magdalen.[50] Some of these gospels, the known Apocrypha of the time, were the additional gospels that Thomas Jefferson was referring to in his letter to his nephew:

> I forgot to observe, when speaking of the New Testament, that you should read all the histories of Christ, as well of those whom a council of ecclesiastics have decided for us, to be Pseudo-evangelists, as those they named Evangelists. Because these Pseudo-evangelists pretended to inspiration, as much as the others, and you are to judge their pretensions by your own reason, and not by the reason of those ecclesiastics.

The gospels that didn't make it were omitted by those ecclesiastics perhaps because they included stories that were even more embarrassingly implausible than those in the four canonical ones. The Infant Gospel of Thomas, for example, has numerous anecdotes about the child Jesus abusing his magical powers in the manner of a mischievous fairy, impishly transforming his playmates into goats, or turning mud into sparrows, or giving his father a hand with the carpentry by miraculously lengthening a piece of wood.* It will be said that nobody believes crude miracle stories such as those in the Gospel of Thomas anyway. But there is no more and no less reason to believe the four canonical gospels. All have the status of legends, as factually dubious as the stories of King Arthur and his Knights of the Round Table.

Most of what the four canonical gospels share is derived from a common source, either Mark's gospel or a lost work of which Mark is the earliest extant descendant. Nobody knows who the four evangelists were, but they almost certainly never met Jesus personally. Much of what they wrote was in no sense an honest attempt at history but was simply rehashed from the Old Testament, because the gospel-makers were devoutly convinced that the life of Jesus must fulfil Old Testament prophecies. It is even possible to mount a serious, though not widely supported, historical case that Jesus never lived at all, as has been done by, among others, Professor G. A. Wells of the University of London in a number of books, including *Did Jesus Exist?*.

Although Jesus probably existed, reputable biblical scholars do not in general regard the New Testament (and obviously not the Old Testament) as a reliable record of what actually happened in history, and I shall not consider the Bible further as evidence

---

* A. N. Wilson, in his biography of Jesus, casts doubt on the story that Joseph was a carpenter at all. The Greek word *tekton* does indeed mean carpenter, but it was translated from the Aramaic word *naggar*, which could mean craftsman or learned man. This is one of several constructive mistranslations that bedevil the Bible, the most famous being the mistranslation of Isaiah's Hebrew for young woman (*almah*) into the Greek for virgin (*parthenos*). An easy mistake to make (think of the English words 'maid' and 'maiden' to see how it might have happened), this one translator's slip was to be wildly inflated and give rise to the whole preposterous legend of Jesus' mother being a virgin! The only competitor for the title of champion constructive mistranslation of all time also concerns virgins. Ibn Warraq has hilariously argued that in the famous promise of seventy-two virgins to every Muslim martyr, 'virgins' is a mistranslation of 'white raisins of crystal clarity'. Now, if only that had been more widely known, how many innocent victims of suicide missions might have been saved? (Ibn Warraq, 'Virgins? What virgins?', *Free Inquiry* 26: 1, 2006, 45–6.)

for any kind of deity. In the farsighted words of Thomas Jefferson, writing to his predecessor, John Adams, 'The day will come when the mystical generation of Jesus, by the Supreme Being as his father, in the womb of a virgin, will be classed with the fable of the generation of Minerva in the brain of Jupiter.'

Dan Brown's novel *The Da Vinci Code*, and the film made from it, are arousing huge controversy in church circles. Christians are encouraged to boycott the film and picket cinemas that show it. It is indeed fabricated from start to finish: invented, made-up fiction. In that respect, it is exactly like the gospels. The only difference between *The Da Vinci Code* and the gospels is that the gospels are ancient fiction while *The Da Vinci Code* is modern fiction.

THE ARGUMENT FROM ADMIRED RELIGIOUS SCIENTISTS

> *The immense majority of intellectually eminent men disbelieve in Christian religion, but they conceal the fact in public, because they are afraid of losing their incomes.*
>
> BERTRAND RUSSELL

'Newton was religious. Who are you to set yourself up as superior to Newton, Galileo, Kepler, etc. etc. etc.? If God was good enough for the likes of them, just who do you think you are?' Not that it makes much difference to such an already bad argument, some apologists even add the name of Darwin, about whom persistent, but demonstrably false, rumours of a deathbed conversion continually come around like a bad smell,*

* Even I have been honoured by prophecies of deathbed conversion. Indeed, they recur with monotonous regularity (see e.g. Steer 2003), each repetition trailing dewy fresh clouds of illusion that it is witty, and the first. I should probably take the precaution of installing a tape-recorder to protect my posthumous reputation. Lalla Ward adds, 'Why mess around with deathbeds? If you're going to sell out, do it in good time to win the Templeton Prize and blame it on senility.'

ever since they were deliberately started by a certain 'Lady Hope', who spun a touching yarn of Darwin resting against the pillows in the evening light, leafing through the New Testament and confessing that evolution was all wrong. In this section I shall concentrate mostly on scientists, because – for reasons that are perhaps not too hard to imagine – those who trot out the names of admired individuals as religious exemplars very commonly choose scientists.

Newton did indeed claim to be religious. So did almost everybody until – significantly I think – the nineteenth century, when there was less social and judicial pressure than in earlier centuries to profess religion, and more scientific support for abandoning it. There have been exceptions, of course, in both directions. Even before Darwin, not everybody was a believer, as James Haught shows in his *2000 Years of Disbelief: Famous People with the Courage to Doubt.* And some distinguished scientists went on believing after Darwin. We have no reason to doubt Michael Faraday's sincerity as a Christian even after the time when he must have known of Darwin's work. He was a member of the Sandemanian sect, which believed (past tense because they are now virtually extinct) in a literal interpretation of the Bible, ritually washed the feet of newly inducted members and drew lots to determine God's will. Faraday became an Elder in 1860, the year after *The Origin of Species* was published, and he died a Sandemanian in 1867. The experimentalist Faraday's theorist counterpart, James Clerk Maxwell, was an equally devout Christian. So was that other pillar of nineteenth-century British physics, William Thomson, Lord Kelvin, who tried to demonstrate that evolution was ruled out for lack of time. That great thermodynamicist's erroneous datings assumed that the sun was some kind of fire, burning fuel which would have to run out in tens of millions of years, not thousands of millions. Kelvin obviously could not be expected to know about nuclear energy. Pleasingly, at the British Association meeting of 1903, it fell to Sir George Darwin, Charles's second son, to vindicate his un-knighted father by invoking the Curies' discovery of radium, and confound the earlier estimate of the still living Lord Kelvin.

Great scientists who profess religion become harder to find through the twentieth century, but they are not particularly rare. I suspect that most of the more recent ones are religious only in the Einsteinian sense which, I argued in Chapter 1, is a misuse of the word. Nevertheless, there are some genuine specimens of good scientists who are sincerely religious in the full, traditional sense. Among contemporary British scientists, the same three names crop up with the likeable familiarity of senior partners in a firm of Dickensian lawyers: Peacocke, Stannard and Polkinghorne. All three have either won the Templeton Prize or are on the Templeton Board of Trustees. After amicable discussions with all of them, both in public and in private, I remain baffled, not so much by their belief in a cosmic lawgiver of some kind, as by their belief in the details of the Christian religion: resurrection, forgiveness of sins and all.

There are some corresponding examples in the United States, for example Francis Collins, administrative head of the American branch of the official Human Genome Project.* But, as in Britain, they stand out for their rarity and are a subject of amused bafflement to their peers in the academic community. In 1996, in the gardens of his old college at Cambridge, Clare, I interviewed my friend Jim Watson, founding genius of the Human Genome Project, for a BBC television documentary that I was making on Gregor Mendel, founding genius of genetics itself. Mendel, of course, was a religious man, an Augustinian monk; but that was in the nineteenth century, when becoming a monk was the easiest way for the young Mendel to pursue his science. For him, it was the equivalent of a research grant. I asked Watson whether he knew many religious scientists today. He replied: 'Virtually none. Occasionally I meet them, and I'm a bit embarrassed [laughs]

---

* Not to be confused with the unofficial human genome project, led by that brilliant (and non-religious) 'buccaneer' of science, Craig Venter.

because, you know, I can't believe anyone accepts truth by revelation.'

Francis Crick, Watson's co-founder of the whole molecular genetics revolution, resigned his fellowship at Churchill College, Cambridge, because of the college's decision to build a chapel (at the behest of a benefactor). In my interview with Watson at Clare, I conscientiously put it to him that, unlike him and Crick, some people see no conflict between science and religion, because they claim science is about how things work and religion is about what it is all for. Watson retorted: 'Well I don't think we're *for* anything. We're just products of evolution. You can say, "Gee, your life must be pretty bleak if you don't think there's a purpose." But I'm anticipating having a good lunch.' We did have a good lunch, too.

The efforts of apologists to find genuinely distinguished modern scientists who are religious have an air of desperation, generating the unmistakably hollow sound of bottoms of barrels being scraped. The only website I could find that claimed to list 'Nobel Prize-winning Scientific Christians' came up with six, out of a total of several hundred scientific Nobelists. Of these six, it turned out that four were not Nobel Prize-winners at all; and at least one, to my certain knowledge, is a non-believer who attends church for purely social reasons. A more systematic study by Benjamin Beit-Hallahmi 'found that among Nobel Prize laureates in the sciences, as well as those in literature, there was a remarkable degree of irreligiosity, as compared to the populations they came from'.[51]

A study in the leading journal *Nature* by Larson and Witham in 1998 showed that of those American scientists considered eminent enough by their peers to have been elected to the National Academy of Sciences (equivalent to being a Fellow of the Royal Society in Britain) only about 7 per cent believe in a personal God.[52] This overwhelming preponderance of atheists is almost the exact opposite of the profile of the American population at large, of whom more than 90 per cent are believers in some sort of supernatural being. The figure for less eminent scientists, not elected to the National Academy, is intermediate. As with the more distinguished sample, religious believers are in a minority, but a less dramatic minority of about 40 per cent. It is completely as I would expect that American scientists are less religious than the American public generally, and that the most distinguished scientists are the least religious of all. What is remarkable is the polar opposition between the religiosity of the American public at large and the atheism of the intellectual elite.[53]

It is faintly amusing that the leading creationist website, 'Answers in Genesis', cites the Larson and Witham study, not in evidence that there might be something wrong with religion, but as a weapon in their internal battle against those rival religious apologists who claim that evolution is compatible with religion. Under the headline 'National Academy of Science is Godless to the Core',[54] 'Answers in Genesis' is pleased to quote the concluding paragraph of Larson and Witham's letter to the editor of *Nature*:

> As we compiled our findings, the NAS [National Academy of Sciences] issued a booklet encouraging the teaching of evolution in public schools, an ongoing source of friction between the scientific community and some conservative Christians in the United States. The booklet assures readers, 'Whether God exists or not is a question about which science is neutral.' NAS president Bruce Alberts said: 'There are many very outstanding members of this academy who are very religious people, people who believe in evolution, many of them biologists.' Our survey suggests otherwise.

Alberts, one feels, embraced 'NOMA' for the reasons I discussed in 'The Neville Chamberlain school of evolutionists' (see Chapter 2). 'Answers in Genesis' has a very different agenda.

The equivalent of the US National Academy of Sciences in

Britain (and the Commonwealth, including Canada, Australia, New Zealand, India, Pakistan, anglophone Africa, etc.) is the Royal Society. As this book goes to press, my colleagues R. Elisabeth Cornwell and Michael Stirrat are writing up their comparable, but more thorough, research on the religious opinions of the Fellows of the Royal Society (FRS). The authors' conclusions will be published in full later, but they have kindly allowed me to quote preliminary results here. They used a standard technique for scaling opinion, the Likert-type seven-point scale. All 1,074 Fellows of the Royal Society who possess an email address (the great majority) were polled, and about 23 per cent responded (a good figure for this kind of study). They were offered various propositions, for example: 'I believe in a personal God, that is one who takes an interest in individuals, hears and answers prayers, is concerned with sin and transgressions, and passes judgement.' For each such proposition, they were invited to choose a number from 1 (strong disagreement) to 7 (strong agreement). It is a little hard to compare the results directly with the Larson and Witham study, because Larson and Witham offered their academicians only a three-point scale, not a seven-point scale, but the overall trend is the same. The overwhelming majority of FRS, like the overwhelming majority of US Academicians, are atheists. Only 3.3 per cent of the Fellows agreed strongly with the statement that a personal god exists (i.e. chose 7 on the scale), while 78.8 per cent strongly disagreed (i.e. chose 1 on the scale). If you define 'believers' as those who chose 6 or 7, and if you define 'unbelievers' as those who chose 1 or 2, there were a massive 213 unbelievers and a mere 12 believers. Like Larson and Witham, and as also noted by Beit-Hallahmi and Argyle, Cornwell and Stirrat found a small but significant tendency for biological scientists to be even more atheistic than physical scientists. For the details, and all the rest of their very interesting conclusions, please refer to their own paper when it is published.[55]

Moving on from the elite scientists of the National Academy

and the Royal Society, is there any evidence that, in the population at large, atheists are likely to be drawn from among the better educated and more intelligent? Several research studies have been published on the statistical relationship between religiosity and educational level, or religiosity and IQ. Michael Shermer, in *How We Believe: The Search for God in an Age of Science*, describes a large survey of randomly chosen Americans that he and his colleague Frank Sulloway carried out. Among their many interesting results was the discovery that religiosity is indeed negatively correlated with education (more highly educated people are less likely to be religious). Religiosity is also negatively correlated with interest in science and (strongly) with political liberalism. None of this is surprising, nor is the fact that there is a positive correlation between religiosity and parents' religiosity. Sociologists studying British children have found that only about one in twelve break away from their parents' religious beliefs.

As you might expect, different researchers measure things in different ways, so it is hard to compare different studies. Meta-analysis is the technique whereby an investigator looks at all the research papers that have been published on a topic, and counts up the number of papers that have concluded one thing, versus the number that have concluded something else. On the subject of religion and IQ, the only meta-analysis known to me was published by Paul Bell in *Mensa Magazine* in 2002 (Mensa is the society of individuals with a high IQ, and their journal not surprisingly includes articles on the one thing that draws them together).[56] Bell concluded: 'Of 43 studies carried out since 1927 on the relationship between religious belief and one's intelligence and/or educational level, all but four found an inverse connection. That is, the higher one's intelligence or education level, the less one is likely to be religious or hold "beliefs" of any kind.'

A meta-analysis is almost bound to be less specific than any one of the studies that contributed to it. It would be nice to have more studies along these lines, as well as more studies of the members of elite bodies such as other national academies, and winners of major prizes and medals such as the Nobel, the Crafoord, the Fields, the Kyoto, the Cosmos and others. I hope that future editions of this book will include such data. A reasonable conclusion from existing studies is that religious apologists might be wise to keep quieter than they habitually do on the subject of admired role models, at least where scientists are concerned.

PASCAL'S WAGER

The great French mathematician Blaise Pascal reckoned that, however long the odds against God's existence might be, there is an even larger asymmetry in the penalty for guessing wrong. You'd better believe in God, because if you are right you stand to gain eternal bliss and if you are wrong it won't make any difference anyway. On the other hand, if you don't believe in God and you turn out to be wrong you get eternal damnation, whereas if you are right it makes no difference. On the face of it the decision is a no-brainer. Believe in God.

There is something distinctly odd about the argument, however. Believing is not something you can decide to do as a matter of policy. At least, it is not something I can decide to do as an act of will. I can decide to go to church and I can decide to recite the Nicene Creed, and I can decide to swear on a stack of bibles that I believe every word inside them. But none of that can make me actually believe it if I don't. Pascal's Wager could only ever be an argument for *feigning* belief in God. And the God that you claim to believe in had better not be of the omniscient kind or he'd see through the deception. The ludicrous idea that believing is something you can *decide* to do is deliciously mocked by Douglas Adams in *Dirk Gently's Holistic Detective Agency*, where we meet the robotic Electric Monk, a labour-saving device that you buy 'to do your believing for you'. The *de luxe* model is advertised as 'Capable of believing things they wouldn't believe in Salt Lake City'.

But why, in any case, do we so readily accept the idea that the one thing you must do if you want to please God is *believe* in him? What's so special about believing? Isn't it just as likely that God would reward kindness, or generosity, or humility? Or sincerity? What if God is a scientist who regards honest seeking after truth as the supreme virtue? Indeed, wouldn't the designer of the universe *have* to be a scientist? Bertrand Russell was asked what he would say if he died and found himself confronted by God, demanding to know why Russell had not believed in him. 'Not enough evidence, God, not enough evidence,' was Russell's (I almost said immortal) reply. Mightn't God respect Russell for his courageous scepticism (let alone for the courageous pacifism that landed him in prison in the First World War) far more than he would respect Pascal for his cowardly bet-hedging? And, while we cannot know which way God would jump, we don't need to *know* in order to refute Pascal's Wager. We are talking about a bet, remember, and Pascal wasn't claiming that his wager enjoyed anything but very long odds. Would you *bet* on God's valuing dishonestly faked belief (or even honest belief) over honest scepticism?

Then again, suppose the god who confronts you when you die turns out to be Baal, and suppose Baal is just as jealous as his old rival Yahweh was said to be. Mightn't Pascal have been better off wagering on no god at all rather than on the wrong god? Indeed, doesn't the sheer number of potential gods and goddesses on whom one might bet vitiate Pascal's whole logic? Pascal was probably joking when he promoted his wager, just as I am joking in my dismissal of it. But I have encountered people, for example in the question session after a lecture, who have seriously advanced Pascal's Wager as an argument in favour of believing in God, so it was right to give it a brief airing here.

Is it possible, finally, to argue for a sort of anti-Pascal wager? Suppose we grant that there is indeed some small chance that God exists. Nevertheless, it could be said that you will lead a better, fuller life if you bet on his not existing, than if you bet on his existing and therefore squander your precious time on worshipping him, sacrificing to him, fighting and dying for him, etc. I won't pursue the question here, but readers might like to bear it in mind when we come to later chapters on the evil consequences that can flow from religious belief and observance.

BAYESIAN ARGUMENTS

I think the oddest case I have seen attempted for the existence of God is the Bayesian argument recently put forward by Stephen Unwin in *The Probability of God*. I hesitated before including this argument, which is both weaker and less hallowed by antiquity than others. Unwin's book, however, received considerable journalistic attention when it was published in 2003, and it does give the opportunity to bring some explanatory threads together. I have some sympathy with his aims because, as argued in Chapter 2, I believe the existence of God as a scientific hypothesis is, at least in principle, investigable. Also, Unwin's quixotic attempt to put a number on the probability is quite agreeably funny.

The book's subtitle, *A Simple Calculation that Proves the Ultimate Truth*, has all the hallmarks of a late addition by the publisher, because such overweening confidence is not to be found in Unwin's text. The book is better seen as a 'How To' manual, a sort of *Bayes' Theorem for Dummies*, using the existence of God as a semi-facetious case study. Unwin could equally well have used a hypothetical murder as his test case to demonstrate Bayes' Theorem. The detective marshals the evidence. The fingerprints on the revolver point to Mrs Peacock. Quantify that suspicion by slapping a numerical likelihood on her. However, Professor Plum had a motive to frame her. Reduce the suspicion of Mrs Peacock by a corresponding numerical value. The forensic evidence suggests a 70 per cent likelihood that the revolver was fired accurately from a long distance, which argues for a culprit with military training. Quantify our raised suspicion of Colonel Mustard. The Reverend Green has the most plausible motive for murder.* Increase our numerical assessment of his likelihood. But the long blond hair on the victim's jacket could only belong to Miss Scarlet . . . and so on. A mix of more or less subjectively judged likelihoods churns around in the detective's mind, pulling him in different directions. Bayes' Theorem is supposed to help him to a conclusion. It is a mathematical engine for combining many estimated likelihoods and coming up with a final verdict, which bears its own quantitative estimate of likelihood. But of course that final estimate can only be as good as the original numbers fed in. These are usually subjectively judged, with all the doubts that inevitably flow from that. The GIGO principle (Garbage In, Garbage Out) is applicable here – and, in the case of Unwin's God example, applicable is too mild a word.

Unwin is a risk management consultant who carries a torch for Bayesian inference, as against rival statistical methods. He illustrates Bayes' Theorem by taking on, not a murder, but the biggest test case of all, the existence of God. The plan is to start with complete uncertainty, which he chooses to quantify by assigning the existence and non-existence of God a 50 per cent starting likelihood each. Then he lists six facts that might bear on the matter, puts a numerical weighting on each, feeds the six numbers into the engine of Bayes' Theorem and sees what number pops out. The trouble is that (to repeat) the six weightings are not measured quantities but simply Stephen Unwin's own personal judgements, turned into numbers for the sake of the exercise. The six facts are:

1 We have a sense of goodness.

2 People do evil things (Hitler, Stalin, Saddam Hussein).

---

* The Reverend Green is the character's name in the versions of *Cluedo* sold in Britain (where the game originated), Australia, New Zealand, India and all other English-speaking areas except North America, where he suddenly becomes Mr Green. What is that all about?

3 Nature does evil things (earthquakes, tsunamis, hurricanes).

4 There might be minor miracles (I lost my keys and found them again).

5 There might be major miracles (Jesus might have risen from the dead).

6 People have religious experiences.

For what it is worth (nothing, in my opinion), at the end of a ding-dong Bayesian race in which God surges ahead in the betting, then drops way back, then claws his way up to the 50 per cent mark from which he started, he finally ends up enjoying, in Unwin's estimation, a 67 per cent likelihood of existing. Unwin then decides that his Bayesian verdict of 67 per cent isn't high enough, so he takes the bizarre step of boosting it to 95 per cent by an emergency injection of 'faith'. It sounds like a joke, but that really is how he proceeds. I wish I could say how he justifies it, but there really is nothing to say. I have met this kind of absurdity elsewhere, when I have challenged religious but otherwise intelligent scientists to justify their belief, given their admission that there is no evidence: 'I admit that there's no evidence. There's a *reason* why it's called faith' (this last sentence uttered with almost truculent conviction, and no hint of apology or defensiveness).

Surprisingly, Unwin's list of six statements does not include the argument from design, nor any of Aquinas' five 'proofs', nor any of the various ontological arguments. He has no truck with them: they don't contribute even a minor fillip to his numerical estimate of God's likelihood. He discusses them and, as a good statistician, dismisses them as empty. I think this is to his credit, although his reason for discounting the design argument is different from mine. But the arguments that he does admit through his Bayesian door are, it seems to me, just as weak. That is only to say that the subjective likelihood weightings I would give to them are different from his, and *who cares* about subjective judgements anyway? He thinks the fact that we have a sense of right and wrong counts strongly in God's favour, whereas I don't see that it should really shift him, in either direction, from his initial prior expectation. Chapters 6 and 7 will show that there is no good case to be made for our possession of a sense of right and wrong having any clear connection with the existence of a supernatural deity. As in the case of our ability to appreciate a Beethoven quartet, our sense of goodness (though not necessarily our inducement to follow it) would be the way it is with a God and without a God.

On the other hand, Unwin thinks the existence of evil, especially natural catastrophes such as earthquakes and tsunamis, counts strongly *against* the likelihood that God exists. Here, Unwin's judgement is opposite to mine but goes along with many uncomfortable theologians. 'Theodicy' (the vindication of divine providence in the face of the existence of evil) keeps theologians awake at night. The authoritative *Oxford Companion to Philosophy* gives the problem of evil as 'the most powerful objection to traditional theism'. But it is an argument only against the existence of a good God. Goodness is no part of the *definition* of the God Hypothesis, merely a desirable add-on.

Admittedly, people of a theological bent are often chronically incapable of distinguishing what is true from what they'd like to be true. But, for a more sophisticated believer in some kind of supernatural intelligence, it is childishly easy to overcome the problem of evil. Simply postulate a nasty god – such as the one who stalks every page of the Old Testament. Or, if you don't like that, invent a separate evil god, call him Satan, and blame his cosmic battle against the good god for the evil in the world. Or – a more sophisticated solution – postulate a god with grander things to do than fuss about human distress. Or a god who is not indifferent to suffering but regards it as the price that has to be paid for free will in an orderly, lawful cosmos. Theologians can be found buying into all these rationalizations.

For these reasons, if I were redoing Unwin's Bayesian exercise, neither the problem of evil nor moral considerations in general would shift me far, one way or the other, from the null hypothesis (Unwin's 50 per cent). But I don't want to argue the point because, in any case, I can't get excited about personal opinions, whether Unwin's or mine.

There is a much more powerful argument, which does not depend upon subjective judgement, and it is the argument from improbability. It really does transport us dramatically away from 50 per cent agnosticism, far towards the extreme of theism in the view of many theists, far towards the extreme of atheism in my view. I have alluded to it several times already. The whole argument turns on the familiar question 'Who made God?', which most thinking people discover for themselves. A designer God cannot be used to explain organized complexity because any God capable of designing anything would have to be complex enough to demand the same kind of explanation in his own right. God presents an infinite regress from which he cannot help us to escape. This argument, as I shall show in the next chapter, demonstrates that God, though not technically disprovable, is very very improbable indeed.

CHAPTER 4

Why there almost certainly is no God

The priests of the different religious sects . . . dread the advance of science as witches do the approach of daylight, and scowl on the fatal harbinger announcing the subdivision of the duperies on which they live.

THOMAS JEFFERSON

THE ULTIMATE BOEING 747

The argument from improbability is the big one. In the traditional guise of the argument from design, it is easily today's most popular argument offered in favour of the existence of God and it is seen, by an amazingly large number of theists, as completely and utterly convincing. It is indeed a very strong and, I suspect, unanswerable argument – but in precisely the opposite direction from the theist's intention. The argument from improbability, properly deployed, comes close to proving that God does *not* exist. My name for the statistical demonstration that God almost certainly does not exist is the Ultimate Boeing 747 gambit.

The name comes from Fred Hoyle's amusing image of the Boeing 747 and the scrapyard. I am not sure whether Hoyle ever wrote it down himself, but it was attributed to him by his close colleague Chandra Wickramasinghe and is presumably authentic.[57] Hoyle said that the probability of life originating on Earth is no greater than the chance that a hurricane, sweeping through a scrapyard, would have the luck to assemble a Boeing 747. Others have borrowed the metaphor to refer to the later evolution of complex living bodies, where it has a spurious plausibility. The odds against assembling a fully functioning horse, beetle or ostrich by randomly shuffling its parts are up there in 747 territory. This, in a nutshell, is the creationist's favourite argument – an argument that could be made only by somebody who doesn't understand the first thing about natural selection: somebody who thinks natural selection is a theory of chance whereas – in the relevant sense of chance – it is the opposite.

The creationist misappropriation of the argument from improbability always takes the same general form, and it doesn't make any difference if the creationist chooses to masquerade in the politically expedient fancy dress of 'intelligent design' (ID).* Some observed phenomenon – often a living creature or one of its more complex organs, but it could be anything from a molecule up to the universe itself – is correctly extolled as statistically improbable. Sometimes the language of information theory is used: the Darwinian is challenged to explain the source of all the information in living matter, in the technical sense of information content as a measure of improbability or 'surprise value'. Or the argument may invoke the economist's hackneyed motto: there's no such thing as a free lunch – and Darwinism is accused of trying to get something for nothing. In fact, as I shall show in this chapter, Darwinian natural selection is the only known solution to the otherwise unanswerable riddle of where the information comes from. It turns out to be the God Hypothesis that tries to get something for nothing. God tries to have his free lunch and be it too. However statistically improbable the entity you seek to explain by invoking a designer, the designer himself has got to be at least as improbable. God is the Ultimate Boeing 747.

---

* Intelligent design has been unkindly described as creationism in a cheap tuxedo.

The argument from improbability states that complex things could not have come about by chance. But many people *define* 'come about by chance' as a synonym for 'come about in the absence of deliberate design'. Not surprisingly, therefore, they think improbability is evidence of design. Darwinian natural selection shows how wrong this is with respect to biological improbability. And although Darwinism may not be directly relevant to the inanimate world – cosmology, for example – it raises our consciousness in areas outside its original territory of biology.

A deep understanding of Darwinism teaches us to be wary of the easy assumption that design is the only alternative to chance, and teaches us to seek out graded ramps of slowly increasing complexity. Before Darwin, philosophers such as Hume understood that the improbability of life did not mean it had to be designed, but they couldn't imagine the alternative. After Darwin, we all should feel, deep in our bones, suspicious of the very idea of design. The illusion of design is a trap that has caught us before, and Darwin should have immunized us by raising our consciousness. Would that he had succeeded with all of us.

NATURAL SELECTION AS A CONSCIOUSNESS-RAISER

In a science-fiction starship, the astronauts were homesick: 'Just to think that it's springtime back on Earth!' You may not immediately see what's wrong with this, so deeply ingrained is the unconscious northern hemisphere chauvinism in those of us who live there, and even some who don't. 'Unconscious' is exactly right. That is where consciousness-raising comes in. It is for a deeper reason than gimmicky fun that, in Australia and New Zealand, you can buy maps of the world with the South Pole on top. What splendid consciousness-raisers those maps would be, pinned to the walls of our northern hemisphere classrooms. Day after day, the children would be reminded that 'north' is an arbitrary polarity which has no monopoly on 'up'. The map would intrigue them as well as raise their consciousness. They'd go home and tell their parents – and, by the way, giving children something with which to surprise their parents is one of the greatest gifts a teacher can bestow.

It was the feminists who raised my consciousness of the power of consciousness-raising. 'Herstory' is obviously ridiculous, if only because the 'his' in 'history' has no etymological connection with the masculine pronoun. It is as etymologically silly as the sacking, in 1999, of a Washington official whose use of 'niggardly' was held to give racial offence. But even daft examples like 'niggardly' or 'herstory' succeed in raising consciousness. Once we have smoothed our philological hackles and stopped laughing, herstory shows us history from a different point of view. Gendered pronouns notoriously are the front line of such consciousness-raising. He or she must ask himself or herself whether his or her sense of style could ever allow himself or herself to write like this. But if we can just get over the clunking infelicity of the language, it raises our consciousness to the sensitivities of half the human race. Man, mankind, the Rights of Man, all men are created equal, one man one vote – English too often seems to exclude woman.* When I was young, it never occurred to me that women might feel slighted by a phrase like 'the future of man'. During the intervening decades, we have all had our consciousness raised. Even those who still use 'man' instead of 'human' do so with an air of self-conscious apology – or truculence, taking a stand for traditional language, even deliberately to rile feminists. All participants in the *Zeitgeist* have had their consciousness raised, even those who choose to respond negatively by digging in their heels and redoubling the offence.

---

* Classical Latin and Greek were better equipped. Latin *homo* (Greek *anthropo-*) means human, as opposed to *vir* (*andro-*) which means man, and *femina* (*gyne-*) which means woman. Thus anthropology pertains to all humanity, where andrology and gynecology are sexually exclusive branches of medicine.

Feminism shows us the power of consciousness-raising, and I want to borrow the technique for natural selection. Natural selection not only explains the whole of life; it also raises our consciousness to the power of science to explain how organized complexity can emerge from simple beginnings without any deliberate guidance. A full understanding of natural selection encourages us to move boldly into other fields. It arouses our suspicion, in those other fields, of the kind of false alternatives that once, in pre-Darwinian days, beguiled biology. Who, before Darwin, could have guessed that something so apparently *designed* as a dragonfly's wing or an eagle's eye was really the end product of a long sequence of non-random but purely natural causes?

Douglas Adams's moving and funny account of his own conversion to radical atheism – he insisted on the 'radical' in case anybody should mistake him for an agnostic – is testimony to the power of Darwinism as a consciousness-raiser. I hope I shall be forgiven the self-indulgence that will become apparent in the following quotation. My excuse is that Douglas's conversion by my earlier books – which did not set out to convert anyone – inspired me to dedicate to his memory this book – which does! In an interview, reprinted posthumously in *The Salmon of Doubt*, he was asked by a journalist how he became an atheist. He began his reply by explaining how he became an agnostic, and then proceeded:

> And I thought and thought and thought. But I just didn't have enough to go on, so I didn't really come to any resolution. I was extremely doubtful about the idea of god, but I just didn't know enough about anything to have a good working model of any other explanation for, well, life, the universe, and everything to put in its place. But I kept at it, and I kept reading and I kept thinking. Sometime around my early thirties I stumbled upon evolutionary biology, particularly in the form of Richard

> Dawkins's books *The Selfish Gene* and then *The Blind Watchmaker*, and suddenly (on, I think the second reading of *The Selfish Gene*) it all fell into place. It was a concept of such stunning simplicity, but it gave rise, naturally, to all of the infinite and baffling complexity of life. The awe it inspired in me made the awe that people talk about in respect of religious experience seem, frankly, silly beside it. I'd take the awe of understanding over the awe of ignorance any day.[58]

The concept of stunning simplicity that he was talking about was, of course, nothing to do with me. It was Darwin's theory of evolution by natural selection – the ultimate scientific consciousness-raiser. Douglas, I miss you. You are my cleverest, funniest, most open-minded, wittiest, tallest, and possibly only convert. I hope this book might have made you laugh – though not as much as you made me.

That scientifically savvy philosopher Daniel Dennett pointed out that evolution counters one of the oldest ideas we have: 'the idea that it takes a big fancy smart thing to make a lesser thing. I call that the trickle-down theory of creation. You'll never see a spear making a spear maker. You'll never see a horse shoe making a blacksmith. You'll never see a pot making a potter.'[59] Darwin's discovery of a workable process that does that very counter-intuitive thing is what makes his contribution to human thought so revolutionary, and so loaded with the power to raise consciousness.

It is surprising how necessary such consciousness-raising is, even in the minds of excellent scientists in fields other than biology. Fred Hoyle was a brilliant physicist and cosmologist, but his Boeing 747 misunderstanding, and other mistakes in biology such as his attempt to dismiss the fossil *Archaeopteryx* as a hoax, suggest that he needed to have his consciousness raised by some good exposure to the world of natural selection. At an intellectual level, I suppose he understood natural selection. But perhaps you need to be steeped in natural selection, immersed in it, swim about in it, before you can truly appreciate its power.

Other sciences raise our consciousness in different ways. Fred Hoyle's own science of astronomy puts us in our place, metaphorically as well as literally, scaling down our vanity to fit the tiny stage on which we play out our lives – our speck of debris from the cosmic explosion. Geology reminds us of our brief existence both as individuals and as a species. It raised John Ruskin's consciousness and provoked his memorable heart cry of 1851: 'If only the Geologists would let me alone, I could do very well, but those dreadful hammers! I hear the clink of them at the end of every cadence of the Bible verses.' Evolution does the same thing for our sense of time – not surprisingly, since it works on the geological timescale. But Darwinian evolution, specifically natural selection, does something more. It shatters the illusion of design within the domain of biology, and teaches us to be suspicious of any kind of design hypothesis in physics and cosmology as well. I think the physicist Leonard Susskind had this in mind when he wrote, 'I'm not an historian but I'll venture an opinion: Modern cosmology really began with Darwin and Wallace. Unlike anyone before them, they provided explanations of our existence that completely rejected supernatural agents . . . Darwin and Wallace set a standard not only for the life sciences but for cosmology as well.'[60] Other physical scientists who are far above needing any such consciousness-raising are Victor Stenger, whose book *Has Science Found God?* (the answer is no) I strongly recommend,* and Peter Atkins, whose *Creation Revisited* is my favourite work of scientific prose poetry.

I am continually astonished by those theists who, far from having their consciousness raised in the way that I propose,

---

* See also his 2007 book *God, the Failed Hypothesis: How Science Shows that God Does Not Exist.*

seem to rejoice in natural selection as 'God's way of achieving his creation'. They note that evolution by natural selection would be a very easy and neat way to achieve a world full of life. God wouldn't need to do anything at all! Peter Atkins, in the book just mentioned, takes this line of thought to a sensibly godless conclusion when he postulates a hypothetically lazy God who tries to get away with as little as possible in order to make a universe containing life. Atkins's lazy God is even lazier than the deist God of the eighteenth-century Enlightenment: *deus otiosus* – literally God at leisure, unoccupied, unemployed, superfluous, useless. Step by step, Atkins succeeds in reducing the amount of work the lazy God has to do until he finally ends up doing nothing at all: he might as well not bother to exist. My memory vividly hears Woody Allen's perceptive whine: 'If it turns out that there is a God, I don't think that he's evil. But the worst that you can say about him is that basically he's an under-achiever.'

IRREDUCIBLE COMPLEXITY

It is impossible to exaggerate the magnitude of the problem that Darwin and Wallace solved. I could mention the anatomy, cellular structure, biochemistry and behaviour of literally any living organism by example. But the most striking feats of apparent design are those picked out – for obvious reasons – by creationist authors, and it is with gentle irony that I derive mine from a creationist book. *Life – How Did It Get Here?*, with no named author but published by the Watchtower Bible and Tract Society in sixteen languages and eleven million copies, is obviously a firm favourite because no fewer than six of those eleven million copies have been sent to me as unsolicited gifts by well-wishers from around the world.

Picking a page at random from this anonymous and lavishly distributed work, we find the sponge known as Venus' Flower Basket (*Euplectella*), accompanied by a quotation from Sir David Attenborough, no less: 'When you look at a complex sponge skeleton such as that made of silica spicules which is known as Venus' Flower Basket, the imagination is baffled. How could quasi-independent microscopic cells collaborate to secrete a million glassy splinters and construct such an intricate and beautiful lattice? We do not know.' The Watchtower authors lose no time in adding their own punchline: 'But one thing we do know: Chance is not the likely designer.' No indeed, chance is not the likely designer. That is one thing on which we can all agree. The statistical improbability of phenomena such as *Euplectella*'s skeleton is the central problem that any theory of life must solve. The greater the statistical improbability, the less plausible is chance as a solution: that is what improbable means. But the candidate solutions to the riddle of improbability are not, as is falsely implied, design and chance. They are design and natural selection. Chance is not a solution, given the high levels of improbability we see in living organisms, and no sane biologist ever suggested that it was. Design is not a real solution either, as we shall see later; but for the moment I want to continue demonstrating the problem that any theory of life must solve: the problem of how to escape from chance.

Turning Watchtower's page, we find the wonderful plant known as Dutchman's Pipe (*Aristolochia trilobata*), all of whose parts seem elegantly designed to trap insects, cover them with pollen and send them on their way to another Dutchman's Pipe. The intricate elegance of the flower moves Watchtower to ask: 'Did all of this happen by chance? Or did it happen by intelligent design?' Once again, no of *course* it didn't happen by chance. Once again, intelligent design is not the proper alternative to chance. Natural selection is not only a parsimonious, plausible and elegant solution; it is the only workable alternative to chance that has ever been suggested. Intelligent design suffers from exactly the same objection as chance. It is simply not a plausible solution to the riddle of statistical improbability. And the higher the improbability, the more implausible intelligent design becomes. Seen clearly, intelligent design will turn out to be a redoubling of the problem. Once again, this is because the designer himself (/herself/itself) immediately raises the bigger problem of his own origin. Any entity capable of intelligently designing something as improbable as a Dutchman's Pipe (or a universe) would have to be even more improbable than a Dutchman's Pipe. Far from terminating the vicious regress, God aggravates it with a vengeance.

Turn another Watchtower page for an eloquent account of the giant redwood (*Sequoiadendron giganteum*), a tree for which I have a special affection because I have one in my garden – a mere baby, scarcely more than a century old, but still the tallest tree in the neighbourhood. 'A puny man, standing at a sequoia's base, can only gaze upward in silent awe at its massive grandeur. Does it make sense to believe that the shaping of this majestic giant and of the tiny seed that packages it was not by design?' Yet again, if you think the only alternative to design is chance then, no, it does not make sense. But again the authors omit all mention of the real alternative, natural selection, either because they genuinely don't understand it or because they don't want to.

The process by which plants, whether tiny pimpernels or massive wellingtonias, acquire the energy to build themselves is photosynthesis. Watchtower again: ' "There are about seventy separate chemical reactions involved in photosynthesis," one biologist said. "It is truly a miraculous event." Green plants have been called nature's "factories" – beautiful, quiet, nonpolluting, producing oxygen, recycling water and feeding the world. Did they just happen by chance? Is that truly believable?' No, it is not believable; but the repetition of example after example gets us nowhere. Creationist 'logic' is always the same. Some natural phenomenon is too statistically improbable, too complex, too beautiful, too awe-inspiring to have come into existence by chance. Design is the only alternative to chance that the authors can imagine. Therefore a designer must have done it. And science's answer to this faulty logic is also always the same. Design is not the only alternative to chance. Natural selection is a better alternative. Indeed, design is not a real alternative at all because it raises an even bigger problem than it solves: who designed the designer? Chance and design both fail as solutions to the problem of statistical improbability, because one of them is the problem, and the other one regresses to it. Natural selection is a real solution. It is the only workable solution that has ever been suggested. And it is not only a workable solution, it is a solution of stunning elegance and power.

What is it that makes natural selection succeed as a solution to the problem of improbability, where chance and design both fail at the starting gate? The answer is that natural selection is a cumulative process, which breaks the problem of improbability up into small pieces. Each of the small pieces is slightly improbable, but not prohibitively so. When large numbers of these slightly improbable events are stacked up in series, the end product of the accumulation is very very improbable indeed, improbable enough to be far beyond the reach of chance. It is these end products that form the subjects of the creationist's wearisomely recycled argument. The creationist completely misses the point, because he (women should for once not mind being excluded by the pronoun) insists on treating the genesis of statistical improbability as a single, one-off event. He doesn't understand the power of *accumulation*.

In *Climbing Mount Improbable*, I expressed the point in a parable. One side of the mountain is a sheer cliff, impossible to climb, but on the other side is a gentle slope to the summit. On the summit sits a complex device such as an eye or a bacterial flagellar motor. The absurd notion that such complexity could spontaneously self-assemble is symbolized by leaping from the foot of the cliff to the top in one bound. Evolution, by contrast, goes around the back of the mountain and creeps up the gentle slope to the summit: easy! The principle of climbing the gentle slope as opposed to leaping up the precipice is so simple, one is tempted to marvel that it took so long for a Darwin to arrive on the scene and discover it. By the time he did, nearly two centuries had elapsed since Newton's *annus mirabilis*, although his achievement seems, on the face of it, harder than Darwin's.

Another favourite metaphor for extreme improbability is the combination lock on a bank vault. Theoretically, a bank robber could get lucky and hit upon the right combination of numbers by chance. In practice, the bank's combination lock is designed with enough improbability to make this tantamount to impossible – almost as unlikely as Fred Hoyle's Boeing 747. But imagine a badly designed combination lock that gave out little hints progressively – the equivalent of the 'getting warmer' of children playing Hunt the Slipper. Suppose that when each one of the dials approaches its correct setting, the vault door opens another chink, and a dribble of money trickles out. The burglar would home in on the jackpot in no time.

Creationists who attempt to deploy the argument from improbability in their favour always assume that biological adaptation is a question of the jackpot or nothing. Another name for the 'jackpot or nothing' fallacy is 'irreducible complexity' (IC). Either the eye sees or it doesn't. Either the wing flies or it doesn't. There are assumed to be no useful intermediates. But this is simply wrong. Such intermediates abound in practice – which is exactly what we should expect in theory. The combination lock of life is a 'getting warmer, getting cooler, getting warmer' Hunt the Slipper device. Real life seeks the gentle slopes at the back of Mount Improbable, while creationists are blind to all but the daunting precipice at the front.

Darwin devoted an entire chapter of the *Origin of Species* to 'Difficulties on the theory of descent with modification', and it is fair to say that this brief chapter anticipated and disposed of every single one of the alleged difficulties that have since been proposed, right up to the present day. The most formidable difficulties are Darwin's 'organs of extreme perfection and complication', sometimes erroneously described as 'irreducibly complex'. Darwin singled out the eye as posing a particularly challenging problem: 'To suppose that the eye with all its inimitable contrivances for adjusting the focus to different distances, for admitting different amounts of light, and for the correction of spherical and chromatic aberration, could have been formed by natural selection, seems, I freely confess, absurd in the highest degree.' Creationists gleefully quote this sentence again and again. Needless to say, they never quote what follows. Darwin's fulsomely free confession turned out to be a rhetorical device. He was drawing his opponents towards him so that his punch, when it came, struck the harder. The punch, of course, was Darwin's effortless explanation of exactly how the eye evolved by gradual degrees. Darwin may not have used the phrase 'irreducible complexity', or 'the smooth gradient up Mount Improbable', but he clearly understood the principle of both.

'What is the use of half an eye?' and 'What is the use of half a wing?' are both instances of the argument from 'irreducible complexity'. A functioning unit is said to be irreducibly complex if the removal of one of its parts causes the whole to cease functioning. This has been assumed to be self-evident for both eyes and wings. But as soon as we give these assumptions a moment's thought, we immediately see the fallacy. A cataract patient with the lens of her eye surgically removed can't see clear images without glasses, but can see enough not to bump into a tree or fall over a cliff. Half a wing is indeed not as good as a whole wing, but it is certainly better than no wing at all. Half a wing could save your life by easing your fall from a tree of a certain height. And 51 per cent of a wing could save you if you fall from a slightly taller tree. Whatever fraction of a wing you have, there is a fall from which it will save your life where a slightly smaller winglet would not. The thought experiment of trees of different height, from which one might fall, is just one way to see, in theory, that there must be a smooth gradient of advantage all the way from 1 per cent of a wing to 100 per cent. The forests are replete with gliding or parachuting animals illustrating, in practice, every step of the way up that particular slope of Mount Improbable.

By analogy with the trees of different height, it is easy to imagine situations in which half an eye would save the life of an animal where 49 per cent of an eye would not. Smooth gradients are provided by variations in lighting conditions, variations in the distance at which you catch sight of your prey – or your predators. And, as with wings and flight surfaces, plausible intermediates are not only easy to imagine: they are abundant all around the animal kingdom. A flatworm has an eye that, by any sensible measure, is less than half a human eye. *Nautilus* (and perhaps its extinct ammonite cousins who dominated Paleozoic and Mesozoic seas) has an eye that is intermediate in quality between flatworm and human. Unlike the flatworm eye, which can detect light and shade but see no image, the *Nautilus* 'pinhole camera' eye makes a real image; but it is a blurred and dim image compared to ours. It would be spurious precision to put numbers on the improvement, but nobody could sanely deny that these invertebrate eyes, and many others, are all better than no eye at all, and all lie on a continuous and shallow slope up Mount Improbable, with our eyes near a peak – not the highest peak but a high one. In *Climbing Mount Improbable*, I devoted a whole chapter each to the eye and the wing, demonstrating how easy it was for them to evolve by slow (or even, maybe, not all that slow) gradual degrees, and I will leave the subject here.

So, we have seen that eyes and wings are certainly not irreducibly complex; but what is more interesting than these particular examples is the general lesson we should draw. The fact that so many people have been dead wrong over these obvious cases should serve to warn us of other examples that are less obvious, such as the cellular and biochemical cases now being touted by those creationists who shelter under the politically expedient euphemism of 'intelligent design theorists'.

We have a cautionary tale here, and it is telling us this: do not just declare things to be irreducibly complex; the chances are that you haven't looked carefully enough at the details, or thought carefully enough about them. On the other hand, we on the science side must not be too dogmatically confident. Maybe there is something out there in nature that really does preclude, by its *genuinely* irreducible complexity, the smooth gradient of Mount Improbable. The creationists are right that, if genuinely irreducible complexity could be properly demonstrated, it would wreck Darwin's theory. Darwin himself said as much: 'If it could be demonstrated that any complex organ existed which could not possibly have been formed by numerous, successive, slight modifications, my theory would absolutely break down. But I can find no such case.' Darwin could find no such case, and nor has anybody since Darwin's time, despite strenuous, indeed desperate, efforts. Many candidates for this holy grail of creationism have been proposed. None has stood up to analysis.

In any case, even though genuinely irreducible complexity would wreck Darwin's theory if it were ever found, who is to say that it wouldn't wreck the intelligent design theory as well? Indeed, it already *has* wrecked the intelligent design theory, for, as I keep saying and will say again, however little we know about God, the one thing we can be sure of is that he would have to be very very complex and presumably irreducibly so!

THE WORSHIP OF GAPS

Searching for particular examples of irreducible complexity is a fundamentally unscientific way to proceed: a special case of arguing from present ignorance. It appeals to the same faulty logic as 'the God of the Gaps' strategy condemned by the theologian Dietrich Bonhoeffer. Creationists eagerly seek a gap in present-day knowledge or understanding. If an apparent gap is found, it is *assumed* that God, by default, must fill it. What worries thoughtful theologians such as Bonhoeffer is that gaps shrink as science advances, and God is threatened with eventually having nothing to do and nowhere to hide. What worries scientists is something else. It is an essential part of the scientific enterprise to admit ignorance, even to exult in ignorance as a challenge to future conquests. As my friend Matt Ridley has written, 'Most scientists are bored by what they have already discovered. It is ignorance that drives them on.' Mystics exult in mystery and want it to stay mysterious. Scientists exult in mystery for a different reason: it gives them something to do. More generally, as I shall repeat in Chapter 8, one of the truly bad effects of religion is that it teaches us that it is a virtue to be satisfied with not understanding.

Admissions of ignorance and temporary mystification are vital to good science. It is therefore unfortunate, to say the least, that the main strategy of creation propagandists is the negative one of seeking out gaps in scientific knowledge and claiming to fill them with 'intelligent design' by default. The following is hypothetical but entirely typical. A creationist speaking: 'The elbow joint of the lesser spotted weasel frog is irreducibly complex. No part of it would do any good at all until the whole was assembled. Bet you can't think of a way in which the weasel frog's elbow could have evolved by slow gradual degrees.' If the scientist fails to give an immediate and comprehensive answer, the creationist draws a *default* conclusion: 'Right then, the alternative theory, "intelligent design", wins by default.' Notice the biased logic: if theory A fails in some particular, theory B must be right. Needless to say, the argument is not applied the other way around. We are encouraged to leap to the default theory without even looking to see whether it fails in the very same particular as the theory it is alleged to replace. Intelligent design – ID – is granted a Get Out Of Jail Free card, a charmed immunity to the rigorous demands made of evolution.

But my present point is that the creationist ploy undermines the scientist's natural – indeed necessary – rejoicing in (temporary) uncertainty. For purely political reasons, today's scientist might hesitate before saying: 'Hm, interesting point. I wonder how the weasel frog's ancestors *did* evolve their elbow joint. I'm not a specialist in weasel frogs, I'll have to go to the University Library and take a look. Might make an interesting project for a graduate student.' The moment a scientist said something like that – and long before the student began the project – the default conclusion would become a headline in a creationist pamphlet: 'Weasel frog could only have been designed by God.'

There is, then, an unfortunate hook-up between science's methodological need to seek out areas of ignorance in order to target research, and ID's need to seek out areas of ignorance in order to claim victory by default. It is precisely the fact that ID has no evidence of its own, but thrives like a weed in gaps left by scientific knowledge, that sits uneasily with science's need to identify and proclaim the very same gaps as a prelude to researching them. In this respect, science finds itself in alliance with sophisticated theologians like Bonhoeffer, united against the common enemies of naïve, populist theology and the gap theology of intelligent design.

The creationists' love affair with 'gaps' in the fossil record symbolizes their whole gap theology. I once introduced a chapter on the so-called Cambrian Explosion with the sentence, 'It is as though the fossils were planted there without any evolutionary history.' Again, this was a rhetorical overture, intended to whet the reader's appetite for the full explanation that was to follow. Sad hindsight tells me now how predictable it was that my patient explanation would be excised and my overture itself gleefully quoted out of context. Creationists adore 'gaps' in the fossil record, just as they adore gaps generally.

Many evolutionary transitions are elegantly documented by more or less continuous series of gradually changing intermediate fossils. Some are not, and these are the famous 'gaps'. Michael Shermer has wittily pointed out that if a new fossil discovery neatly bisects a 'gap', the creationist will declare that there are now twice as many gaps! But in any case, note yet again the unwarranted use of a default. If there are no fossils to document a postulated evolutionary transition, the default assumption is that there was no evolutionary transition, therefore God must have intervened.

It is utterly illogical to demand complete documentation of every step of any narrative, whether in evolution or any other science. You might as well demand, before convicting somebody of murder, a complete cinematic record of the murderer's every step leading up to the crime, with no missing frames. Only a tiny fraction of corpses fossilize, and we are lucky to have as many intermediate fossils as we do. We could easily have had no fossils at all, and still the evidence for evolution from other sources, such as molecular genetics and geographical distribution, would be overwhelmingly strong. On the other hand, evolution makes the strong prediction that if a *single* fossil turned up in the *wrong* geological stratum, the theory would be blown out of the water. When challenged by a zealous Popperian to say how evolution could ever be falsified, J. B. S. Haldane famously growled: 'Fossil rabbits in the Precambrian.' No such anachronistic fossils have ever been authentically found, despite discredited creationist legends of human skulls in the Coal Measures and human footprints interspersed with dinosaurs'.

Gaps, by default in the mind of the creationist, are filled by God. The same applies to all apparent precipices on the massif of Mount Improbable, where the graded slope is not immediately obvious or is otherwise overlooked. Areas where there is a lack of data, or a lack of understanding, are automatically assumed to belong, by default, to God. The speedy resort to a dramatic proclamation of 'irreducible complexity' represents a failure of the imagination. Some biological organ, if not an eye then a bacterial flagellar motor or a biochemical pathway, is *decreed* without further argument to be irreducibly complex. No attempt is made to *demonstrate* irreducible complexity. Notwithstanding the cautionary tales of eyes, wings and many other things, each new candidate for the dubious accolade is assumed to be transparently, self-evidently irreducibly complex, its status asserted by fiat. But think about it. Since irreducible complexity is being deployed as an argument for design, it should no more be asserted by fiat than design itself. You might as well simply assert that the weasel frog (bombardier beetle, etc.) demonstrates design, without further argument or justification. That is no way to do science.

The logic turns out to be no more convincing than this: 'I [insert own name] am personally unable to think of any way in which [insert biological phenomenon] could have been built up step by step. Therefore it is irreducibly complex. That means it is designed.' Put it like that, and you immediately see that it is vulnerable to some scientist coming along and finding an intermediate; or at least imagining a plausible intermediate. Even if no scientists do come up with an explanation, it is plain bad logic to assume that 'design' will fare any better. The reasoning that underlies 'intelligent design' theory is lazy and defeatist – classic 'God of the Gaps' reasoning. I have previously dubbed it the Argument from Personal Incredulity.

Imagine that you are watching a really great magic trick. The celebrated conjuring duo Penn and Teller have a routine in which they simultaneously appear to shoot each other with pistols, and each appears to catch the bullet in his teeth. Elaborate precautions are taken to scratch identifying marks on the bullets before they are put in the guns, the whole procedure is witnessed at close range by volunteers from the audience who have experience of firearms, and apparently all possibilities for trickery are eliminated. Teller's marked bullet ends up in Penn's mouth and Penn's marked bullet ends up in Teller's. I [Richard Dawkins] am utterly unable to think of any way in which this could be a trick. The Argument from Personal Incredulity screams from the depths of my prescientific brain centres, and almost compels me to say, 'It must be a miracle. There is no scientific explanation. It's got to be supernatural.' But the still small voice of scientific education speaks a different message. Penn and Teller are world-class illusionists. There is a perfectly good explanation. It is just that I am too naïve, or too unobservant, or too unimaginative, to think of it. That is the proper response to a conjuring trick. It is also the proper response to a biological phenomenon that appears to be irreducibly complex. Those people who leap from personal bafflement at a natural phenomenon straight to a hasty invocation of the supernatural are no better than the fools who see a conjuror bending a spoon and leap to the conclusion that it is 'paranormal'.

In his book *Seven Clues to the Origin of Life*, the Scottish chemist A. G. Cairns-Smith makes an additional point, using the analogy of an arch. A free-standing arch of rough-hewn stones and no mortar can be a stable structure, but it is irreducibly complex: it collapses if any one stone is removed. How, then, was it built in the first place? One way is to pile a solid heap of stones, then carefully remove stones one by one. More generally, there are many structures that are irreducible in the sense that they cannot survive the subtraction of any part, but which were built with the aid of scaffolding that was subsequently subtracted and is no longer visible. Once the structure is completed, the scaffolding can be removed safely and the structure remains standing. In evolution, too, the organ or structure you are looking at may have had scaffolding in an ancestor which has since been removed.

'Irreducible complexity' is not a new idea, but the phrase itself was invented by the creationist Michael Behe in 1996.[61] He is credited (if credited is the word) with moving creationism into a new area of biology: biochemistry and cell biology, which he saw as perhaps a happier hunting ground for gaps than eyes or wings. His best approach to a good example (still a bad one) was the bacterial flagellar motor.

The flagellar motor of bacteria is a prodigy of nature. It drives the only known example, outside human technology, of a freely rotating axle. Wheels for big animals would, I suspect, be genuine examples of irreducible complexity, and this is probably why they don't exist. How would the nerves and blood vessels get across the bearing?* The flagellum is a thread-like

* There is an example in fiction. The children's writer Philip Pullman, in *His Dark Materials*, imagines a species of animals, the 'mulefa', that co-exist with trees that produce perfectly round seedpods with a hole in the centre. These pods the mulefa adopt as wheels. The wheels, not being part of the body, have no nerves or blood vessels to get twisted around the 'axle' (a strong claw of horn or bone). Pullman perceptively notes an additional point: the system works only because the planet is paved with natural basalt ribbons, which serve as 'roads'. Wheels are no good over rough country.

propeller, with which the bacterium burrows its way through the water. I say 'burrows' rather than 'swims' because, on the bacterial scale of existence, a liquid such as water would not feel as a liquid feels to us. It would feel more like treacle, or jelly, or even sand, and the bacterium would seem to burrow or screw its way through the water rather than swim. Unlike the so-called flagellum of larger organisms like protozoans, the bacterial flagellum doesn't just wave about like a whip, or row like an oar. It has a true, freely rotating axle which turns continuously inside a bearing, driven by a remarkable little molecular motor. At the molecular level, the motor uses essentially the same principle as muscle, but in free rotation rather than in intermittent contraction.* It has been happily described as a tiny outboard motor (although by engineering standards – and unusually for a biological mechanism – it is a spectacularly inefficient one).

Without a word of justification, explanation or amplification, Behe simply *proclaims* the bacterial flagellar motor to be irreducibly complex. Since he offers no argument in favour of his assertion, we may begin by suspecting a failure of his imagination. He further alleges that specialist biological literature has ignored the problem. The falsehood of this allegation was massively and (to Behe) embarrassingly documented in the court of Judge John E. Jones in Pennsylvania in 2005, where Behe was testifying as an expert witness on behalf of a group of creationists who had tried to impose 'intelligent design' creationism on the science curriculum of a local public school – a move of 'breathtaking inanity', to quote Judge Jones

---

* Fascinatingly, the muscle principle is deployed in yet a third mode in some insects such as flies, bees and bugs, in which the flight muscle is intrinsically oscillatory, like a reciprocating engine. Whereas other insects such as locusts send nervous instructions for each wing stroke (as a bird does), bees send an instruction to switch on (or switch off) the oscillatory motor. Bacteria have a mechanism which is neither a simple contractor (like a bird's flight muscle) nor a reciprocator (like a bee's flight muscle), but a true rotator: in that respect it is like an electric motor or a Wankel engine.

(phrase and man surely destined for lasting fame). This wasn't the only embarrassment Behe suffered at the hearing, as we shall see.

The key to demonstrating irreducible complexity is to show that none of the parts could have been useful on its own. They all needed to be in place before any of them could do any good (Behe's favourite analogy is a mousetrap). In fact, molecular biologists have no difficulty in finding parts functioning outside the whole, both for the flagellar motor and for Behe's other alleged examples of irreducible complexity. The point is well put by Kenneth Miller of Brown University, for my money the most persuasive nemesis of 'intelligent design', not least because he is a devout Christian. I frequently recommend Miller's book, *Finding Darwin's God*, to religious people who write to me having been bamboozled by Behe.

In the case of the bacterial rotary engine, Miller calls our attention to a mechanism called the Type Three Secretory System or TTSS.[62] The TTSS is not used for rotatory movement. It is one of several systems used by parasitic bacteria for pumping toxic substances through their cell walls to poison their host organism. On our human scale, we might think of pouring or squirting a liquid through a hole; but, once again, on the bacterial scale things look different. Each molecule of secreted substance is a large protein with a definite, three-dimensional structure on the same scale as the TTSS's own: more like a solid sculpture than a liquid. Each molecule is individually propelled through a carefully shaped mechanism, like an automated slot machine dispensing, say, toys or bottles, rather than a simple hole through which a substance might 'flow'. The goods-dispenser itself is made of a rather small number of protein molecules, each one comparable in size and complexity to the molecules being dispensed through it. Interestingly, these bacterial slot machines are often similar across bacteria that are not closely related. The genes for making them have probably been 'copied and pasted' from other bacteria: something that bacteria are remarkably adept at doing, and a fascinating topic in its own right, but I must press on.

The protein molecules that form the structure of the TTSS are very similar to components of the flagellar motor. To the evolutionist it is clear that TTSS components were commandeered for a new, but not wholly unrelated, function when the flagellar motor evolved. Given that the TTSS is tugging molecules through itself, it is not surprising that it uses a rudimentary version of the principle used by the flagellar motor, which tugs the molecules of the axle round and round. Evidently, crucial components of the flagellar motor were already in place and working before the flagellar motor evolved. Commandeering existing mechanisms is an obvious way in which an apparently irreducibly complex piece of apparatus could climb Mount Improbable.

A lot more work needs to be done, of course, and I'm sure it will be. Such work would never be done if scientists were satisfied with a lazy default such as 'intelligent design theory' would encourage. Here is the message that an imaginary 'intelligent design theorist' might broadcast to scientists: 'If you don't understand how something works, never mind: just give up and say God did it. You don't know how the nerve impulse works? Good! You don't understand how memories are laid down in the brain? Excellent! Is photosynthesis a bafflingly complex process? Wonderful! Please don't go to work on the problem, just give up, and appeal to God. Dear scientist, don't *work* on your mysteries. Bring us your mysteries, for we can use them. Don't squander precious ignorance by researching it away. We need those glorious gaps as a last refuge for God.' St Augustine said it quite openly: 'There is another form of temptation, even more fraught with danger. This is the disease of curiosity. It is this which drives us to try and discover the secrets of nature, those secrets which are beyond our understanding, which can avail us nothing and which man should not wish to learn' (quoted in Freeman 2002).

Another of Behe's favourite alleged examples of 'irreducible complexity' is the immune system. Let Judge Jones himself take up the story:

> In fact, on cross-examination, Professor Behe was questioned concerning his 1996 claim that science would never find an evolutionary explanation for the immune system. He was presented with fifty-eight peer-reviewed publications, nine books, and several immunology textbook chapters about the evolution of the immune system; however, he simply insisted that this was still not sufficient evidence of evolution, and that it was not 'good enough.'

Behe, under cross-examination by Eric Rothschild, chief counsel for the plaintiffs, was forced to admit that he hadn't read most of those fifty-eight peer-reviewed papers. Hardly surprising, for immunology is hard work. Less forgivable is that Behe dismissed such research as 'unfruitful'. It certainly is unfruitful if your aim is to make propaganda among gullible laypeople and politicians, rather than to discover important truths about the real world. After listening to Behe, Rothschild eloquently summed up what every honest person in that courtroom must have felt:

> Thankfully, there are scientists who do search for answers to the question of the origin of the immune system . . . It's our defense against debilitating and fatal diseases. The scientists who wrote those books and articles toil in obscurity, without book royalties or speaking engagements. Their efforts help us combat and cure serious medical conditions. By contrast, Professor Behe and the entire intelligent design movement are doing nothing to advance scientific or medical knowledge and are telling future generations of scientists, don't bother.[63]

As the American geneticist Jerry Coyne put it in his review of Behe's book: 'If the history of science shows us anything, it is that we get nowhere by labelling our ignorance "God". Or, in the words of an eloquent blogger, commenting on an article on intelligent design in the *Guardian* by Coyne and me,

> Why is God considered an explanation for anything? It's not – it's a failure to explain, a shrug of the shoulders, an 'I dunno' dressed up in spirituality and ritual. If someone credits something to God, generally what it means is that they haven't a clue, so they're attributing it to an unreachable, unknowable sky-fairy. Ask for an explanation of where that bloke came from, and odds are you'll get a vague, pseudo-philosophical reply about having always existed, or being outside nature. Which, of course, explains nothing.[64]

Darwinism raises our consciousness in other ways. Evolved organs, elegant and efficient as they often are, also demonstrate revealing flaws – exactly as you'd expect if they have an evolutionary history, and exactly as you would not expect if they were designed. I have discussed examples in other books: the recurrent laryngeal nerve, for one, which betrays its evolutionary history in a massive and wasteful detour on its way to its destination. Many of our human ailments, from lower back pain to hernias, prolapsed uteruses and our susceptibility to sinus infections, result directly from the fact that we now walk upright with a body that was shaped over hundreds of millions of years to walk on all fours. Our consciousness is also raised by the cruelty and wastefulness of natural selection. Predators seem beautifully 'designed' to catch prey animals, while the prey animals seem equally beautifully 'designed' to escape them. Whose side is God on?[65]

THE ANTHROPIC PRINCIPLE: PLANETARY VERSION

Gap theologians who may have given up on eyes and wings, flagellar motors and immune systems, often pin their remaining hopes on the origin of life. The root of evolution in non-biological chemistry somehow seems to present a bigger gap than any particular transition during subsequent evolution. And in one sense it is a bigger gap. That one sense is quite specific, and it offers no comfort to the religious apologist. The origin of life only had to happen once. We therefore can allow it to have been an extremely improbable event, many orders of magnitude more improbable than most people realize, as I shall show. Subsequent evolutionary steps are duplicated, in more or less similar ways, throughout millions and millions of species independently, and continually and repeatedly throughout geological time. Therefore, to explain the evolution of complex life, we cannot resort to the same kind of statistical reasoning as we are able to apply to the origin of life. The events that constitute run-of-the-mill evolution, as distinct from its singular origin (and perhaps a few special cases), cannot have been very improbable.

This distinction may seem puzzling, and I must explain it further, using the so-called anthropic principle. The anthropic principle was named by the mathematician Brandon Carter in 1974 and expanded by the physicists John Barrow and Frank Tipler in their book on the subject.[66] The anthropic argument is usually applied to the cosmos, and I'll come to that. But I'll introduce the idea on a smaller, planetary scale. We exist here on Earth. Therefore Earth must be the kind of planet that is capable of generating and supporting us, however unusual, even unique, that kind of planet might be. For example, our kind of life cannot survive without liquid water. Indeed, exobiologists searching for evidence of extraterrestrial life are scanning the heavens, in practice, for signs of water. Around a typical star like our sun, there is a so-called Goldilocks zone – not too hot and not too cold, but just right – for planets with liquid water. A thin band of orbits lies between those that are too far from the star, where water freezes, and too close, where it boils.

Presumably, too, a life-friendly orbit has to be nearly circular. A fiercely elliptical orbit, like that of the newly discovered tenth planet informally known as Xena, would at best allow the planet to whizz briefly through the Goldilocks zone once every few (Earth) decades or centuries. Xena itself doesn't get into the Goldilocks zone at all, even at its closest approach to the sun, which it reaches once every 560 Earth years. The temperature of Halley's Comet varies between about 47°C at perihelion and minus 270°C at aphelion. Earth's orbit, like those of all the planets, is technically an ellipse (it is closest to the sun in January and furthest away in July*); but a circle is a special case of an ellipse, and Earth's orbit is so close to circular that it never strays out of the Goldilocks zone. Earth's situation in the solar system is propitious in other ways that singled it out for the evolution of life. The massive gravitational vacuum cleaner of Jupiter is well placed to intercept asteroids that might otherwise threaten us with lethal collision. Earth's single relatively large moon serves to stabilize our axis of rotation,[67] and helps to foster life in various other ways. Our sun is unusual in not being a binary, locked in mutual orbit with a companion star. It is possible for binary stars to have planets, but their orbits are likely to be too chaotically variable to encourage the evolution of life.

Two main explanations have been offered for our planet's peculiar friendliness to life. The design theory says that God made the world, placed it in the Goldilocks zone, and deliberately set up all the details for our benefit. The anthropic approach is very different, and it has a faintly Darwinian feel.

---

* If you find that surprising, you may be suffering from northern hemisphere chauvinism, as described on page 139.

The great majority of planets in the universe are not in the Goldilocks zones of their respective stars, and not suitable for life. None of that majority has life. However small the minority of planets with just the right conditions for life may be, we necessarily have to be on one of that minority, because here we are thinking about it.

It is a strange fact, incidentally, that religious apologists love the anthropic principle. For some reason that makes no sense at all, they think it supports their case. Precisely the opposite is true. The anthropic principle, like natural selection, is an *alternative* to the design hypothesis. It provides a rational, design-free explanation for the fact that we find ourselves in a situation propitious to our existence. I think the confusion arises in the religious mind because the anthropic principle is only ever mentioned in the context of the problem that it solves, namely the fact that we live in a life-friendly place. What the religious mind then fails to grasp is that two candidate solutions are offered to the problem. God is one. The anthropic principle is the other. They are *alternatives.*

Liquid water is a necessary condition for life as we know it, but it is far from sufficient. Life still has to originate in the water, and the origin of life may have been a highly improbable occurrence. Darwinian evolution proceeds merrily once life has originated. But how does life get started? The origin of life was the chemical event, or series of events, whereby the vital conditions for natural selection first came about. The major ingredient was heredity, either DNA or (more probably) something that copies like DNA but less accurately, perhaps the related molecule RNA. Once the vital ingredient – some kind of genetic molecule – is in place, true Darwinian natural selection can follow, and complex life emerges as the eventual consequence. But the spontaneous arising by chance of the first hereditary molecule strikes many as improbable. Maybe it is – very very improbable, and I shall dwell on this, for it is central to this section of the book.

The origin of life is a flourishing, if speculative, subject for research. The expertise required for it is chemistry and it is not mine. I watch from the sidelines with engaged curiosity, and I shall not be surprised if, within the next few years, chemists report that they have successfully midwifed a new origin of life in the laboratory. Nevertheless it hasn't happened yet, and it is still possible to maintain that the probability of its happening is, and always was, exceedingly low – although it did happen once!

Just as we did with the Goldilocks orbits, we can make the point that, however improbable the origin of life might be, we know it happened on Earth because we are here. Again as with temperature, there are two hypotheses to explain what happened – the design hypothesis and the scientific or 'anthropic' hypothesis. The design approach postulates a God who wrought a deliberate miracle, struck the prebiotic soup with divine fire and launched DNA, or something equivalent, on its momentous career.

Again, as with Goldilocks, the anthropic alternative to the design hypothesis is statistical. Scientists invoke the magic of large numbers. It has been estimated that there are between 1 billion and 30 billion planets in our galaxy, and about 100 billion galaxies in the universe. Knocking a few noughts off for reasons of ordinary prudence, a billion billion is a conservative estimate of the number of available planets in the universe. Now, suppose the origin of life, the spontaneous arising of something equivalent to DNA, really was a quite staggeringly improbable event. Suppose it was so improbable as to occur on only one in a billion planets. A grant-giving body would laugh at any chemist who admitted that the chance of his proposed research succeeding was only one in a hundred. But here we are talking about odds of one in a billion. And yet . . . even with such absurdly long odds, life will still have arisen on a billion planets – of which Earth, of course, is one.[68]

This conclusion is so surprising, I'll say it again. If the odds of life originating spontaneously on a planet were a billion to one against, nevertheless that stupefyingly improbable event would still happen on a billion planets. The chance of finding any one of those billion life-bearing planets recalls the proverbial needle in a haystack. But we don't have to go out of our way to find a needle because (back to the anthropic principle) any beings capable of looking must necessarily be sitting on one of those prodigiously rare needles before they even start the search.

Any probability statement is made in the context of a certain level of ignorance. If we know nothing about a planet, we may postulate the odds of life's arising on it as, say, one in a billion. But if we now import some new assumptions into our estimate, things change. A particular planet may have some peculiar properties, perhaps a special profile of element abundances in its rocks, which shift the odds in favour of life's emerging. Some planets, in other words, are more 'Earth-like' than others. Earth itself, of course, is especially Earth-like! This should give encouragement to our chemists trying to recreate the event in the lab, for it could shorten the odds against their success. But my earlier calculation demonstrated that even a chemical model with odds of success as low as one in a billion would *still* predict that life would arise on a billion planets in the universe. And the beauty of the anthropic principle is that it tells us, against all intuition, that a chemical model need only predict that life will arise on *one* planet in a billion billion to give us a good and entirely satisfying explanation for the presence of life here. I do not for a moment believe the origin of life was anywhere near so improbable in practice. I think it is definitely worth spending money on trying to duplicate the event in the lab and – by the same token, on SETI, because I think it is likely that there is intelligent life elsewhere.

Even accepting the most pessimistic estimate of the probability that life might spontaneously originate, this statistical argument completely demolishes any suggestion that we should postulate design to fill the gap. Of all the apparent gaps in the evolutionary story, the origin of life gap can seem unbridgeable to brains calibrated to assess likelihood and risk on an everyday scale: the scale on which grant-giving bodies assess research proposals submitted by chemists. Yet even so big a gap as this is easily filled by statistically informed science, while the very same statistical science rules out a divine creator on the 'Ultimate 747' grounds we met earlier.

But now, to return to the interesting point that launched this section. Suppose somebody tried to explain the general phenomenon of biological adaptation along the same lines as we have just applied to the origin of life: appealing to an immense number of available planets. The observed fact is that every species, and every organ that has ever been looked at within every species, is good at what it does. The wings of birds, bees and bats are good at flying. Eyes are good at seeing. Leaves are good at photosynthesizing. We live on a planet where we are surrounded by perhaps ten million species, each one of which independently displays a powerful illusion of apparent design. Each species is well fitted to its particular way of life. Could we get away with the 'huge numbers of planets' argument to explain all these separate illusions of design? No, we could not, repeat *not*. Don't even think about it. This is important, for it goes to the heart of the most serious misunderstanding of Darwinism.

It doesn't matter how many planets we have to play with, lucky chance could never be enough to explain the lush diversity of living complexity on Earth in the same way as we used it to explain the existence of life here in the first place. The evolution of life is a completely different case from the origin of life because, to repeat, the origin of life was (or could have been) a unique event which had to happen only once. The adaptive fit of species to their separate environments, on the other hand, is millionfold, and ongoing.

It is clear that here on Earth we are dealing with a generalized *process* for optimizing biological species, a process that works all over the planet, on all continents and islands, and at all times. We can safely predict that, if we wait another ten million years, a whole new set of species will be as well adapted to their ways of life as today's species are to theirs. This is a recurrent, predictable, multiple phenomenon, not a piece of statistical luck recognized with hindsight. And, thanks to Darwin, we know how it is brought about: by natural selection.

The anthropic principle is impotent to explain the multifarious details of living creatures. We really need Darwin's powerful crane to account for the diversity of life on Earth, and especially the persuasive illusion of design. The origin of life, by contrast, lies outside the reach of that crane, because natural selection cannot proceed without it. Here the anthropic principle comes into its own. We can deal with the unique origin of life by postulating a very large number of planetary opportunities. Once that initial stroke of luck has been granted – and the anthropic principle most decisively grants it to us – natural selection takes over: and natural selection is emphatically not a matter of luck.

Nevertheless, it may be that the origin of life is not the only major gap in the evolutionary story that is bridged by sheer luck, anthropically justified. For example, my colleague Mark Ridley in *Mendel's Demon* (gratuitously and confusingly retitled *The Cooperative Gene* by his American publishers) has suggested that the origin of the eucaryotic cell (our kind of cell, with a nucleus and various other complicated features such as mitochondria, which are not present in bacteria) was an even more momentous, difficult and statistically improbable step than the origin of life. The origin of consciousness might be another major gap whose bridging was of the same order of improbability. One-off events like this might be explained by the anthropic principle, along the following lines. There are billions of planets that have developed life at the level of bacteria, but only a fraction of these life forms ever made it across the gap to something like the eucaryotic cell. And of these, a yet smaller fraction managed to cross the later Rubicon to consciousness. If both of these are one-off events, we are not dealing with a ubiquitous and all-pervading *process*, as we are with ordinary, run-of-the-mill biological adaptation. The anthropic principle states that, since we are alive, eucaryotic and conscious, our planet has to be one of the intensely rare planets that has bridged all three gaps.

Natural selection works because it is a cumulative one-way street to improvement. It needs some luck to get started, and the 'billions of planets' anthropic principle grants it that luck. Maybe a few later gaps in the evolutionary story also need major infusions of luck, with anthropic justification. But whatever else we may say, *design* certainly does not work as an explanation for life, because design is ultimately not cumulative and it therefore raises bigger questions than it answers – it takes us straight back along the Ultimate 747 infinite regress.

We live on a planet that is friendly to our kind of life, and we have seen two reasons why this is so. One is that life has evolved to flourish in the conditions provided by the planet. This is because of natural selection. The other reason is the anthropic one. There are billions of planets in the universe, and, however small the minority of evolution-friendly planets may be, our planet necessarily has to be one of them. Now it is time to take the anthropic principle back to an earlier stage, from biology back to cosmology.

THE ANTHROPIC PRINCIPLE: COSMOLOGICAL VERSION

We live not only on a friendly planet but also in a friendly universe. It follows from the fact of our existence that the laws of physics must be friendly enough to allow life to arise. It is no accident that when we look at the night sky we see stars, for stars are a necessary prerequisite for the existence of most of the chemical elements, and without chemistry there could be no life. Physicists have calculated that, if the laws and constants of physics had been even slightly different, the universe would have developed in such a way that life would have been impossible. Different physicists put it in different ways, but the conclusion is always much the same.* Martin Rees, in *Just Six Numbers*, lists six fundamental constants, which are believed to hold all around the universe. Each of these six numbers is finely tuned in the sense that, if it were slightly different, the universe would be comprehensively different and presumably unfriendly to life.†

An example of Rees's six numbers is the magnitude of the so-called 'strong' force, the force that binds the components of an atomic nucleus: the nuclear force that has to be overcome when one 'splits' the atom. It is measured as E, the proportion of the mass of a hydrogen nucleus that is converted to energy when hydrogen fuses to form helium. The value of this number in our universe is 0.007, and it looks as though it had to be very close to this value in order for any chemistry (which is a prerequisite for life) to exist. Chemistry as we know it consists of the combination and recombination of the ninety or so naturally occurring elements of the periodic table. Hydrogen is the simplest and commonest of the elements. All the other elements in the universe are made ultimately from hydrogen by nuclear fusion. Nuclear fusion is a difficult process which occurs in the intensely hot conditions of the interiors of stars (and in hydrogen bombs). Relatively small stars, such as our sun, can make only light elements such as helium, the second

---

* The physicist Victor Stenger (in e.g. *God, the Failed Hypothesis*) dissents from this consensus, and is unpersuaded that the physical laws and constants are particularly friendly to life. Nevertheless, I shall bend over backwards to accept the 'friendly universe' consensus, in order to show that, in any case, it cannot be used to support theism.

† I say 'presumably', partly because we don't know how different alien forms of life might be, and partly because it is possible that we make a mistake if we consider only the consequences of changing one constant at a time. Could there be other *combinations* of values of the six numbers which would turn out to be friendly to life, in ways that we do not discover if we consider them only one at a time? Nevertheless, I shall proceed, for simplicity, as though we really do have a big problem to explain in the apparent fine-tuning of the fundamental constants.

lightest in the periodic table after hydrogen. It takes larger and hotter stars to develop the high temperatures needed to forge most of the heavier elements, in a cascade of nuclear fusion processes whose details were worked out by Fred Hoyle and two colleagues (an achievement for which, mysteriously, Hoyle was not given a share of the Nobel Prize received by the others). These big stars may explode as supernovas, scattering their materials, including the elements of the periodic table, in dust clouds. These dust clouds eventually condense to form new stars and planets, including our own. This is why Earth is rich in elements over and above the ubiquitous hydrogen: elements without which chemistry, and life, would be impossible.

The relevant point here is that the value of the strong force crucially determines how far up the periodic table the nuclear fusion cascade goes. If the strong force were too small, say 0.006 instead of 0.007, the universe would contain nothing but hydrogen, and no interesting chemistry could result. If it were too large, say 0.008, all the hydrogen would have fused to make heavier elements. A chemistry without hydrogen could not generate life as we know it. For one thing, there would be no water. The Goldilocks value – 0.007 – is just right for yielding the richness of elements that we need for an interesting and life-supporting chemistry.

I won't go through the rest of Rees's six numbers. The bottom line for each of them is the same. The actual number sits in a Goldilocks band of values outside which life would not have been possible. How should we respond to this? Yet again, we have the theist's answer on the one hand, and the anthropic answer on the other. The theist says that God, when setting up the universe, tuned the fundamental constants of the universe so that each one lay in its Goldilocks zone for the production of life. It is as though God had six knobs that he could twiddle, and he carefully tuned each knob to its Goldilocks value. As ever, the theist's answer is deeply unsatisfying, because it leaves the existence of God unexplained. A God capable of calculating the Goldilocks values for the six numbers would have to be at least as improbable as the finely tuned combination of numbers itself, and that's very improbable indeed. This is exactly the premise of the whole discussion we are having. It follows that the theist's answer has utterly failed to make any headway towards solving the problem at hand. I see no alternative but to dismiss it, while at the same time marvelling at the number of people who can't see the problem and seem genuinely satisfied by the 'Divine Knob-Twiddler' argument.

Maybe the psychological reason for this amazing blindness has something to do with the fact that many people have not had their consciousness raised, as biologists have, by natural selection and its power to tame improbability. J. Anderson Thomson, from his perspective as an evolutionary psychiatrist, points me to an additional reason, the psychological bias that we all have towards personifying inanimate objects as agents. As Thomson says, we are more inclined to mistake a shadow for a burglar than a burglar for a shadow. A false positive might be a waste of time. A false negative could be fatal. In a letter to me, he suggested that, in our ancestral past, our greatest challenge in our environment came from each other. 'The legacy of that is the default assumption, often fear, of human intention. We have a great deal of difficulty seeing anything other than *human* causation.' We naturally generalized that to divine intention. I shall return to the seductiveness of 'agents' in Chapter 5.

Biologists, with their raised consciousness of the power of natural selection to explain the rise of improbable things, are unlikely to be satisfied with any theory that evades the problem of improbability altogether. And the theistic response to the riddle of improbability is an evasion of stupendous proportions. It is more than a restatement of the problem, it is a grotesque amplification of it. Let's turn, then, to the anthropic alternative. The anthropic answer, in its most general form, is that we could only be discussing the question in the kind of universe that was capable of producing us. Our existence therefore determines that the fundamental constants of physics had to be in their respective Goldilocks zones. Different physicists espouse different kinds of anthropic solutions to the riddle of our existence.

Hard-nosed physicists say that the six knobs were never free to vary in the first place. When we finally reach the long-hoped-for Theory of Everything, we shall see that the six key numbers depend upon each other, or on something else as yet unknown, in ways that we today cannot imagine. The six numbers may turn out to be no freer to vary than is the ratio of a circle's circumference to its diameter. It will turn out that there is only one way for a universe to be. Far from God being needed to twiddle six knobs, there are no knobs to twiddle.

Other physicists (Martin Rees himself would be an example) find this unsatisfying, and I think I agree with them. It is indeed perfectly plausible that there is only one way for a universe to be. But why did that one way have to be such a set-up for our eventual evolution? Why did it have to be the kind of universe which seems almost as if, in the words of the theoretical physicist Freeman Dyson, it 'must have known we were coming'? The philosopher John Leslie uses the analogy of a man sentenced to death by firing squad. It is just possible that all ten men of the firing squad will miss their victim. With hindsight, the survivor who finds himself in a position to reflect upon his luck can cheerfully say, 'Well, obviously they all missed, or I wouldn't be here thinking about it.' But he could still, forgivably, wonder why they all missed, and toy with the hypothesis that they were bribed, or drunk.

This objection can be answered by the suggestion, which Martin Rees himself supports, that there are many universes, co-existing like bubbles of foam, in a 'multiverse' (or 'megaverse', as Leonard Susskind prefers to call it).* The laws and constants of any one universe, such as our observable

---

* Susskind (2006) gives a splendid advocacy of the anthropic principle in the megaverse. He says the idea is hated by most physicists. I can't understand why. I think it is beautiful – perhaps because my consciousness has been raised by Darwin.

universe, are by-laws. The multiverse as a whole has a plethora of alternative sets of by-laws. The anthropic principle kicks in to explain that we have to be in one of those universes (presumably a minority) whose by-laws happened to be propitious to our eventual evolution and hence contemplation of the problem.

An intriguing version of the multiverse theory arises out of considerations of the ultimate fate of our universe. Depending upon the values of numbers such as Martin Rees's six constants, our universe may be destined to expand indefinitely, or it may stabilize at an equilibrium, or the expansion may reverse itself and go into contraction, culminating in the so-called 'big crunch'. Some big crunch models have the universe then bouncing back into expansion, and so on indefinitely with, say, a 20-billion-year cycle time. The standard model of our universe says that time itself began in the big bang, along with space, some 13 billion years ago. The serial big crunch model would amend that statement: our time and space did indeed begin in our big bang, but this was just the latest in a long series of big bangs, each one initiated by the big crunch that terminated the previous universe in the series. Nobody understands what goes on in singularities such as the big bang, so it is conceivable that the laws and constants are reset to new values, each time. If bang–expansion–contraction–crunch cycles have been going on for ever like a cosmic accordion, we have a serial, rather than a parallel, version of the multiverse. Once again, the anthropic principle does its explanatory duty. Of all the universes in the series, only a minority have their 'dials' tuned to biogenic conditions. And, of course, the present universe has to be one of that minority, because we are in it. As it turns out, this serial version of the multiverse must now be judged less likely than it once was, because recent evidence is starting to steer us away from the big crunch model. It now looks as though our own universe is destined to expand for ever.

Another theoretical physicist, Lee Smolin, has developed a tantalizingly Darwinian variant on the multiverse theory, including both serial and parallel elements. Smolin's idea, expounded in *The Life of the Cosmos*, hinges on the theory that daughter universes are born of parent universes, not in a fully fledged big crunch but more locally in black holes. Smolin adds a form of heredity: the fundamental constants of a daughter universe are slightly 'mutated' versions of the constants of its parent. Heredity is the essential ingredient of Darwinian natural selection, and the rest of Smolin's theory follows naturally. Those universes that have what it takes to 'survive' and 'reproduce' come to predominate in the multiverse. 'What it takes' includes lasting long enough to 'reproduce'. Because the act of reproduction takes place in black holes, successful universes must have what it takes to make black holes. This ability entails various other properties. For example, the tendency for matter to condense into clouds and then stars is a prerequisite to making black holes. Stars also, as we have seen, are the precursors to the development of interesting chemistry, and hence life. So, Smolin suggests, there has been a Darwinian natural selection of universes in the multiverse, directly favouring the evolution of black hole fecundity and indirectly favouring the production of life. Not all physicists are enthusiastic about Smolin's idea, although the Nobel Prize-winning physicist Murray Gell-Mann is quoted as saying: 'Smolin? Is he that young guy with those crazy ideas? He may not be wrong.'[69] A mischievous biologist might wonder whether some other physicists are in need of Darwinian consciousness-raising.

It is tempting to think (and many have succumbed) that to postulate a plethora of universes is a profligate luxury which should not be allowed. If we are going to permit the extravagance of a multiverse, so the argument runs, we might as well be hung for a sheep as a lamb and allow a God. Aren't they both equally unparsimonious ad hoc hypotheses, and equally unsatisfactory? People who think that have not had their consciousness raised by natural selection. The key difference between the genuinely extravagant God hypothesis and the apparently extravagant multiverse hypothesis is one of statistical improbability. The multiverse, for all that it is extravagant, is simple. God, or any intelligent, decision-taking, calculating agent, would have to be highly improbable in the very same statistical sense as the entities he is supposed to explain. The multiverse may seem extravagant in sheer *number* of universes. But if each one of those universes is simple in its fundamental laws, we are still not postulating anything highly improbable. The very opposite has to be said of any kind of intelligence.

Some physicists are known to be religious (Russell Stannard and the Reverend John Polkinghorne are the two British examples I have mentioned). Predictably, they seize upon the improbability of the physical constants all being tuned in their more or less narrow Goldilocks zones, and suggest that there must be a cosmic intelligence who deliberately did the tuning. I have already dismissed all such suggestions as raising bigger problems than they solve. But what attempts have theists made to reply? How do they cope with the argument that any God capable of designing a universe, carefully and foresightfully tuned to lead to our evolution, must be a supremely complex and improbable entity who needs an even bigger explanation than the one he is supposed to provide?

The theologian Richard Swinburne, as we have learned to expect, thinks he has an answer to this problem, and he expounds it in his book *Is There a God?*. He begins by showing that his heart is in the right place by convincingly demonstrating why we should always prefer the simplest hypothesis that fits the facts. Science explains complex things in terms of the interactions of simpler things, ultimately the interactions of fundamental particles. I (and I dare say you) think it a beautifully simple idea that all things are made of fundamental particles which, although exceedingly numerous, are drawn from a small, finite set of *types* of particle. If we are sceptical, it is likely to be because we think the idea too simple. But for Swinburne it is not simple at all, quite the reverse.

Given that the number of particles of any one type, say

electrons, is large, Swinburne thinks it too much of a coincidence that so many should have the same properties. One electron, he could stomach. But billions and billions of electrons, *all with the same properties*, that is what really excites his incredulity. For him it would be simpler, more natural, less demanding of explanation, if all electrons were different from each other. Worse, no one electron should naturally retain its properties for more than an instant at a time; each should change capriciously, haphazardly and fleetingly from moment to moment. That is Swinburne's view of the simple, native state of affairs. Anything more uniform (what you or I would call more simple) requires a special explanation. 'It is only because electrons and bits of copper and all other material objects have the same powers in the twentieth century as they did in the nineteenth century that things are as they are now.'

Enter God. God comes to the rescue by deliberately and continuously sustaining the properties of all those billions of electrons and bits of copper, and neutralizing their otherwise ingrained inclination to wild and erratic fluctuation. That is why when you've seen one electron you've seen them all; that is why bits of copper all behave like bits of copper, and that is why each electron and each bit of copper stays the same as itself from microsecond to microsecond and from century to century. It is because God constantly keeps a finger on each and every particle, curbing its reckless excesses and whipping it into line with its colleagues to keep them all the same.

But how can Swinburne possibly maintain that this hypothesis of God simultaneously keeping a gazillion fingers on wayward electrons is a *simple* hypothesis? It is, of course, precisely the opposite of simple. Swinburne pulls off the trick to his own satisfaction by a breathtaking piece of intellectual *chutzpah*. He asserts, without justification, that God is only a *single* substance. What brilliant economy of explanatory causes, compared with all those gigazillions of independent electrons all just happening to be the same!

> Theism claims that every other object which exists is caused to exist and kept in existence by just one substance, God. And it claims that every property which every substance has is due to God causing or permitting it to exist. It is a hallmark of a simple explanation to postulate few causes. There could in this respect be no simpler explanation than one which postulated only one cause. Theism is simpler than polytheism. And theism postulates for its one cause, a person [with] infinite power (God can do anything logically possible), infinite knowledge (God knows everything logically possible to know), and infinite freedom.

Swinburne generously concedes that God cannot accomplish feats that are *logically* impossible, and one feels grateful for this forbearance. Having said that, there is no limit to the explanatory purposes to which God's infinite power is put. Is science having a little difficulty explaining X? No problem. Don't give X another glance. God's infinite power is effortlessly wheeled in to explain X (along with everything else), and it is always a supremely *simple* explanation because, after all, there is only one God. What could be simpler than that?

Well, actually, almost everything. A God capable of continuously monitoring and controlling the individual status of every particle in the universe *cannot* be simple. His existence is going to need a mammoth explanation in its own right. Worse (from the point of view of simplicity), other corners of God's giant consciousness are simultaneously preoccupied with the doings and emotions and prayers of every single human being – and whatever intelligent aliens there might be on other planets in this and 100 billion other galaxies. He even, according to Swinburne, has to decide continuously *not* to intervene miraculously to save us when we get cancer. That would never do, for, 'If God answered most prayers for a relative to recover from cancer, then cancer would no longer be a problem for humans to solve.'

And *then* what would we find to do with our time?

Not all theologians go as far as Swinburne. Nevertheless, the remarkable suggestion that the God Hypothesis is *simple* can be found in other modern theological writings. Keith Ward, then Regius Professor of Divinity at Oxford, was very clear on the matter in his 1996 book *God, Chance and Necessity*:

> As a matter of fact, the theist would claim that God is a very elegant, economical and fruitful explanation for the existence of the universe. It is economical because it attributes the existence and nature of absolutely everything in the universe to just one being, an ultimate cause which assigns a reason for the existence of everything, including itself. It is elegant because from one key idea – the idea of the most perfect possible being – the whole nature of God and the existence of the universe can be intelligibly explicated.

Like Swinburne, Ward mistakes what it means to explain something, and he also seems not to understand what it means to say of something that it is simple. I am not clear whether Ward really thinks God is simple, or whether the above passage represented a temporary 'for the sake of argument' exercise. Sir John Polkinghorne, in *Science and Christian Belief*, quotes Ward's earlier criticism of the thought of Thomas Aquinas: 'Its basic error is in supposing that God is logically simple – simple not just in the sense that his being is indivisible, but in the much stronger sense that what is true of any part of God is true of the whole. It is quite coherent, however, to suppose that God, while indivisible, is internally complex.' Ward gets it right here. Indeed, the biologist Julian Huxley, in 1912, defined complexity in terms of 'heterogeneity of parts', by which he meant a particular kind of functional indivisibility.[70]

Elsewhere, Ward gives evidence of the difficulty the theological mind has in grasping where the complexity of life comes from. He quotes another theologian-scientist, the biochemist Arthur Peacocke (the third member of my trio of British religious scientists), as postulating the existence in living matter of a 'propensity for increased complexity'. Ward characterizes this as 'some inherent weighting of evolutionary change which favours complexity'. He goes on to suggest that such a bias 'might be some weighting of the mutational process, to ensure that more complex mutations occurred'. Ward is sceptical of this, as well he should be. The evolutionary drive towards complexity comes, in those lineages where it comes at all, not from any inherent propensity for increased complexity, and not from biased mutation. It comes from natural selection: the process which, as far as we know, is the only process ultimately capable of generating complexity out of simplicity. The theory of natural selection is genuinely simple. So is the origin from which it starts. That which it explains, on the other hand, is complex almost beyond telling: more complex than anything we can imagine, save a God capable of designing it.

An Interlude at Cambridge

At a recent Cambridge conference on science and religion, where I put forward the argument I am here calling the Ultimate 747 argument, I encountered what, to say the least, was a cordial failure to achieve a meeting of minds on the question of God's simplicity. The experience was a revealing one, and I'd like to share it.

First I should confess (that is probably the right word) that the conference was sponsored by the Templeton Foundation. The audience was a small number of hand-picked science journalists from Britain and America. I was the token atheist among the eighteen invited speakers. One of the journalists, John Horgan, reported that they had each been paid the handsome sum of $15,000 to attend the conference, on top of all expenses. This surprised me. My long experience of academic conferences included no instances where the audience (as opposed to the speakers) was paid to attend. If I had known, my suspicions would immediately have been aroused. Was Templeton using his money to suborn science journalists and subvert their scientific integrity? John Horgan later wondered the same thing and wrote an article about his whole experience.[71] In it he revealed, to my chagrin, that my advertised involvement as a speaker had helped him and others to overcome their doubts:

> The British biologist Richard Dawkins, whose participation in the meeting helped convince me and other fellows of its legitimacy, was the only speaker who denounced religious beliefs as incompatible with science, irrational, and harmful. The other speakers – three agnostics, one Jew, a deist, and 12 Christians (a Muslim philosopher canceled at the last minute) – offered a perspective clearly skewed in favor of religion and Christianity.

Horgan's article is itself endearingly ambivalent. Despite his misgivings, there were aspects of the experience that he clearly valued (and so did I, as will become apparent below). Horgan wrote:

> My conversations with the faithful deepened my appreciation of why some intelligent, well-educated people embrace religion. One reporter discussed the experience of speaking in tongues, and another described having an intimate relationship with Jesus. My convictions did not change, but others' did. At least one fellow said that his faith was wavering as a result of Dawkins's dissection of religion. And if the Templeton Foundation can help bring about even such a tiny step toward my vision of a world without religion, how bad can it be?

Horgan's article was given a second airing by the literary

agent John Brockman on his 'Edge' website (often described as an on-line scientific *salon*) where it elicited varying responses, including one from the theoretical physicist Freeman Dyson. I responded to Dyson, quoting from his acceptance speech when he won the Templeton Prize. Whether he liked it or not, by accepting the Templeton Prize Dyson had sent a powerful signal to the world. It would be taken as an endorsement of religion by one of the world's most distinguished physicists.

> 'I am content to be one of the multitude of Christians who do not care much about the doctrine of the Trinity or the historical truth of the gospels.'

But isn't that exactly what any atheistic scientist *would* say, if he wanted to sound Christian? I gave further quotations from Dyson's acceptance speech, satirically interspersing them with imagined questions (in italics) to a Templeton official:

> *Oh, you want something a bit more profound, as well? How about . . .*
>
> 'I do not make any clear distinction between mind and God. God is what mind becomes when it has passed beyond the scale of our comprehension.'
>
> *Have I said enough yet, and can I get back to doing physics now? Oh, not enough yet? OK then, how about this:*
>
> 'Even in the gruesome history of the twentieth century, I see some evidence of progress in religion. The two individuals who epitomized the evils of our century, Adolf Hitler and Joseph Stalin, were both avowed atheists.'*
>
> *Can I go now?*

---

* This calumny is dealt with in Chapter 7.

Dyson could easily refute the implication of these quotations from his Templeton acceptance speech, if only he would explain clearly what evidence he finds to believe in God, in something more than just the Einsteinian sense which, as I explained in Chapter 1, we can all trivially subscribe to. If I understand Horgan's point, it is that Templeton's money corrupts science. I am sure Freeman Dyson is way above being corrupted. But his acceptance speech is still unfortunate if it seems to set an example to others. The Templeton Prize is two orders of magnitude larger than the inducements offered to the journalists at Cambridge, having been explicitly set up to be larger than the Nobel Prize. In Faustian vein, my friend the philosopher Daniel Dennett once joked to me, 'Richard, if ever you fall on hard times . . .'

For better or worse, I attended two days at the Cambridge conference, giving a talk of my own and taking part in the discussion of several other talks. I challenged the theologians to answer the point that a God capable of designing a universe, or anything else, would have to be complex and statistically improbable. The strongest response I heard was that I was brutally foisting a scientific epistemology upon an unwilling theology.* Theologians had always defined God as simple. Who was I, a scientist, to dictate to theologians that their God had to be complex? Scientific arguments, such as those I was accustomed to deploying in my own field, were inappropriate since theologians had always maintained that God lay outside science.

I did not gain the impression that the theologians who mounted this evasive defence were being wilfully dishonest. I think they were sincere. Nevertheless, I was irresistibly reminded of Peter Medawar's comment on Father Teilhard de Chardin's *The Phenomenon of Man*, in the course of what is possibly the greatest negative book review of all time: 'its author can be excused of dishonesty only on the grounds that

---

* This accusation is reminiscent of 'NOMA', whose overblown claims I dealt with in Chapter 2.

before deceiving others he has taken great pains to deceive himself'.[72] The theologians of my Cambridge encounter were *defining* themselves into an epistemological Safe Zone where rational argument could not reach them because they had *declared by fiat* that it could not. Who was I to say that rational argument was the only admissible kind of argument? There are other ways of knowing besides the scientific, and it is one of these other ways of knowing that must be deployed to know God.

The most important of these other ways of knowing turned out to be personal, subjective experience of God. Several discussants at Cambridge claimed that God spoke to them, inside their heads, just as vividly and as personally as another human might. I have dealt with illusion and hallucination in Chapter 3 ('The argument from personal experience'), but at the Cambridge conference I added two points. First, that if God really did communicate with humans that fact would emphatically not lie outside science. God comes bursting through from whatever other-worldly domain is his natural abode, crashing through into our world where his messages can be intercepted by human brains – and that phenomenon has nothing to do with science? Second, a God who is capable of sending intelligible signals to millions of people simultaneously, and of receiving messages from all of them simultaneously, cannot be, whatever else he might be, simple. Such bandwidth! God may not have a brain made of neurones, or a CPU made of silicon, but if he has the powers attributed to him he must have something far more elaborately and non-randomly constructed than the largest brain or the largest computer we know.

Time and again, my theologian friends returned to the point that there had to be a reason why there is something rather than nothing. There must have been a first cause of everything, and we might as well give it the name God. Yes, I said, but it must have been simple and therefore, whatever else we call it, God is not an appropriate name (unless we very explicitly divest it of all the baggage that the word 'God' carries in the minds of most religious believers). The first cause that we seek must have been the simple basis for a self-bootstrapping crane which eventually raised the world as we know it into its present complex existence. To suggest that the original prime mover was complicated enough to indulge in intelligent design, to say nothing of mindreading millions of humans simultaneously, is tantamount to dealing yourself a perfect hand at bridge. Look around at the world of life, at the Amazon rainforest with its rich interlacement of lianas, bromeliads, roots and flying buttresses; its army ants and its jaguars, its tapirs and peccaries, treefrogs and parrots. What you are looking at is the statistical equivalent of a perfect hand of cards (think of all the other ways you could permute the parts, none of which would work) – except that we know how it came about: by the gradualistic crane of natural selection. It is not just scientists who revolt at mute acceptance of such improbability arising spontaneously; common sense baulks too. To suggest that the first cause, the great unknown which is responsible for something existing rather than nothing, is a being capable of designing the universe and of talking to a million people simultaneously, is a total abdication of the responsibility to find an explanation. It is a dreadful exhibition of self-indulgent, thought-denying skyhookery.

I am not advocating some sort of narrowly scientistic way of thinking. But the very least that any honest quest for truth must have in setting out to explain such monstrosities of improbability as a rainforest, a coral reef, or a universe is a crane and not a skyhook. The crane doesn't have to be natural selection. Admittedly, nobody has ever thought of a better one. But there could be others yet to be discovered. Maybe the 'inflation' that physicists postulate as occupying some fraction of the first yoctosecond of the universe's existence will turn out, when it is better understood, to be a cosmological crane to stand alongside Darwin's biological one. Or maybe the elusive crane that cosmologists seek will be a version of Darwin's idea itself: either Smolin's model or something similar. Or maybe it will be the multiverse plus anthropic principle espoused by Martin Rees and others. It may even be a superhuman designer – but, if so, it will most certainly *not* be a designer who just popped into existence, or who always existed. If (which I don't believe for a moment) our universe was designed, and *a fortiori* if the designer reads our thoughts and hands out omniscient advice, forgiveness and redemption, the designer himself must be the end product of some kind of cumulative escalator or crane, perhaps a version of Darwinism in another universe.

The last-ditch defence by my critics in Cambridge was attack. My whole world-view was condemned as 'nineteenth-century'. This is such a bad argument that I almost omitted to mention it. But regrettably I encounter it rather frequently. Needless to say, to call an argument nineteenth-century is not the same as explaining what is wrong with it. Some nineteenth-century ideas were very good ideas, not least Darwin's own dangerous idea. In any case, this particular piece of namecalling seemed a bit rich coming, as it did, from an individual (a distinguished Cambridge geologist, surely well advanced along the Faustian road to a future Templeton Prize) who justified his own Christian belief by invoking what he called the historicity of the New Testament. It was precisely in the nineteenth century that theologians, especially in Germany, called into grave doubt that alleged historicity, using the evidence-based methods of history to do so. This was, indeed, swiftly pointed out by the theologians at the Cambridge conference.

In any case, I know the 'nineteenth-century' taunt of old. It goes with the 'village atheist' gibe. It goes with 'Contrary to what you seem to think Ha Ha Ha we don't believe in an old man with a long white beard any more Ha Ha Ha.' All three jokes are code for something else, just as, when I lived in America in the late 1960s, 'law and order' was politicians' code for anti-black prejudice.* What, then, is the coded meaning of

* In Britain 'inner cities' had the equivalent coded meaning, prompting Auberon Waugh's wickedly hilarious reference to 'inner cities of both sexes'.

'You are so nineteenth-century' in the context of an argument about religion? It is code for: 'You are so crude and unsubtle, how could you be so insensitive and ill-mannered as to ask me a direct, point-blank question like "Do you believe in miracles?" or "Do you believe Jesus was born of a virgin?" Don't you know that in polite society we don't ask such questions? That sort of question went out in the nineteenth century.' But think about why it is impolite to ask such direct, factual questions of religious people today. It is because it is embarrassing! But it is the answer that is embarrassing, if it is yes.

The nineteenth-century connection is now clear. The nineteenth century is the last time when it was possible for an educated person to admit to believing in miracles like the virgin birth without embarrassment. When pressed, many educated Christians today are too loyal to deny the virgin birth and the resurrection. But it embarrasses them because their rational minds know it is absurd, so they would much rather not be asked. Hence, if somebody like me insists on asking the question, it is I who am accused of being 'nineteenth-century'. It is really quite funny, when you think about it.

I left the conference stimulated and invigorated, and reinforced in my conviction that the argument from improbability – the 'Ultimate 747' gambit – is a very serious argument against the existence of God, and one to which I have yet to hear a theologian give a convincing answer despite numerous opportunities and invitations to do so. Dan Dennett rightly describes it as 'an unrebuttable refutation, as devastating today as when Philo used it to trounce Cleanthes in Hume's Dialogues two centuries earlier. A skyhook would at best simply postpone the solution to the problem, but Hume couldn't think of any cranes, so he caved in.'[73] Darwin, of course, supplied the vital crane. How Hume would have loved it.

This chapter has contained the central argument of my book, and so, at the risk of sounding repetitive, I shall summarize it as a series of six numbered points.

1. One of the greatest challenges to the human intellect, over the centuries, has been to explain how the complex, improbable appearance of design in the universe arises.
2. The natural temptation is to attribute the appearance of design to actual design itself. In the case of a man-made artefact such as a watch, the designer really was an intelligent engineer. It is tempting to apply the same logic to an eye or a wing, a spider or a person.
3. The temptation is a false one, because the designer hypothesis immediately raises the larger problem of who designed the designer. The whole problem we started out with was the problem of explaining statistical improbability. It is obviously no solution to postulate something even more improbable. We need a 'crane', not a 'skyhook', for only a crane can do the business of working up gradually and plausibly from simplicity to otherwise improbable complexity.
4. The most ingenious and powerful crane so far discovered is Darwinian evolution by natural selection. Darwin and his successors have shown how living creatures, with their spectacular statistical improbability and appearance of design, have evolved by slow, gradual degrees from simple beginnings. We can now safely say that the illusion of design in living creatures is just that – an illusion.
5. We don't yet have an equivalent crane for physics. Some kind of multiverse theory could in principle do for physics the same explanatory work as Darwinism does for biology. This kind of explanation is superficially less satisfying than the biological version of Darwinism, because it makes heavier demands on luck. But the anthropic principle entitles us to postulate far more luck than our limited human intuition is comfortable with.
6. We should not give up hope of a better crane arising in

> physics, something as powerful as Darwinism is for biology. But even in the absence of a strongly satisfying crane to match the biological one, the relatively weak cranes we have at present are, when abetted by the anthropic principle, self-evidently better than the self-defeating skyhook hypothesis of an intelligent designer.

If the argument of this chapter is accepted, the factual premise of religion – the God Hypothesis – is untenable. God almost certainly does not exist. This is the main conclusion of the book so far. Various questions now follow. Even if we accept that God doesn't exist, doesn't religion still have a lot going for it? Isn't it consoling? Doesn't it motivate people to do good? If it weren't for religion, how would we know what is good? Why, in any case, be so hostile? Why, if it is false, does every culture in the world have religion? True or false, religion is ubiquitous, so where does it come from? It is to this last question that we turn next.

CHAPTER 5

The roots of religion

To an evolutionary psychologist, the universal extravagance of religious rituals, with their costs in time, resources, pain and privation, should suggest as vividly as a mandrill's bottom that religion may be adaptive.

MAREK KOHN

THE DARWINIAN IMPERATIVE

Everybody has their own pet theory of where religion comes from and why all human cultures have it. It gives consolation and comfort. It fosters togetherness in groups. It satisfies our yearning to understand why we exist. I shall come to explanations of this kind in a moment, but I want to begin with a prior question, one that takes precedence for reasons we shall see: a Darwinian question about natural selection.

Knowing that we are products of Darwinian evolution, we should ask what pressure or pressures exerted by natural selection originally favoured the impulse to religion. The question gains urgency from standard Darwinian considerations of economy. Religion is so wasteful, so extravagant; and Darwinian selection habitually targets and eliminates waste. Nature is a miserly accountant, grudging the pennies, watching the clock, punishing the smallest extravagance. Unrelentingly and unceasingly, as Darwin explained, 'natural selection is daily and hourly scrutinising, throughout the world, every variation, even the slightest; rejecting that which is bad, preserving and adding up all that is good; silently and insensibly working, whenever and wherever opportunity offers, at the improvement of each organic being'. If a wild animal habitually performs some useless activity, natural selection will favour rival individuals who devote the time and energy, instead, to surviving and reproducing. Nature cannot afford frivolous *jeux d'esprit*. Ruthless utilitarianism trumps, even if it doesn't always seem that way.

On the face of it, the tail of a peacock is a *jeu d'esprit par excellence*. It surely does no favours to the survival of its possessor. But it does benefit the genes that distinguish him from his less spectacular rivals. The tail is an advertisement, which buys its place in the economy of nature by attracting females. The same is true of the labour and time that a male bower bird devotes to his bower: a sort of external tail built of grass, twigs, colourful berries, flowers and, when available, beads, baubles and bottle caps. Or, to choose an example that doesn't involve advertising, there is 'anting': the odd habit of birds, such as jays, of 'bathing' in an ants' nest or otherwise applying ants to the feathers. Nobody is sure what the benefit of anting is – perhaps some kind of hygiene, cleaning out parasites from the feathers; there are various other hypotheses, none of them strongly supported by evidence. But uncertainty as to details doesn't – nor should it – stop Darwinians from presuming, with great confidence, that anting must be 'for' something. In this case common sense might agree, but Darwinian logic has a particular reason for thinking that, if the birds didn't do it, their statistical prospects of genetic success would be damaged, even if we don't yet know the precise route of the damage. The conclusion follows from the twin premises that natural selection punishes wastage of time and energy, and that birds are consistently observed to devote time and energy to anting. If there is a one-sentence manifesto of this 'adaptationist' principle, it was expressed – admittedly in somewhat extreme and exaggerated terms – by the distinguished Harvard geneticist Richard Lewontin: 'That is the one point which I think all evolutionists are agreed upon, that it is virtually impossible to do a better job than an organism is doing in its own environment.'[74] If anting wasn't positively useful for survival and reproduction, natural selection would long ago have favoured individuals who refrained from it. A Darwinian might be tempted to say the same of religion; hence the need for this discussion.

To an evolutionist, religious rituals 'stand out like peacocks in a sunlit glade' (Dan Dennett's phrase). Religious behaviour is a writ-large human equivalent of anting or bower-building. It is time-consuming, energy-consuming, often as extravagantly ornate as the plumage of a bird of paradise. Religion can endanger the life of the pious individual, as well as the lives of others. Thousands of people have been tortured for their loyalty to a religion, persecuted by zealots for what is in many cases a scarcely distinguishable alternative faith. Religion devours resources, sometimes on a massive scale. A medieval cathedral could consume a hundred man-centuries in its construction, yet was never used as a dwelling, or for any recognizably useful purpose. Was it some kind of architectural peacock's tail? If so, at whom was the advertisement aimed? Sacred music and devotional paintings largely monopolized medieval and Renaissance talent. Devout people have died for their gods and killed for them; whipped blood from their backs, sworn themselves to a lifetime of celibacy or to lonely silence, all in the service of religion. What is it all for? What is the benefit of religion?

By 'benefit', the Darwinian normally means some enhancement to the survival of the individual's genes. What is missing from this is the important point that Darwinian benefit is not restricted to the genes of the individual organism. There are three possible alternative targets of benefit. One arises from the theory of group selection, and I'll come to that. The second follows from the theory that I advocated in *The Extended Phenotype*: the individual you are watching may be working under the manipulative influence of genes in another individual, perhaps a parasite. Dan Dennett reminds us that the common cold is universal to all human peoples in much the same way as religion is, yet we would not want to suggest that colds benefit us. Plenty of examples are known of animals manipulated into behaving in such a way as to benefit the transmission of a parasite to its next host. I encapsulated the point in my 'central theorem of the extended phenotype': 'An animal's behaviour tends to maximize the survival of the genes "for" that behaviour, whether or not those genes happen to be in the body of the particular animal performing it.'

Third, the 'central theorem' may substitute for 'genes' the more general term 'replicators'. The fact that religion is ubiquitous probably means that it has worked to the benefit of something, but it may not be us or our genes. It may be to the benefit of only the religious ideas themselves, to the extent that they behave in a somewhat gene-like way, as replicators. I shall deal with this below, under the heading 'Tread softly, because you tread on my memes'. Meanwhile, I press on with more traditional interpretations of Darwinism, in which 'benefit' is assumed to mean benefit to individual survival and reproduction.

Hunter-gatherer peoples such as Australian aboriginal tribes presumably live in something like the way our distant ancestors did. The New Zealand/Australian philosopher of science Kim Sterelny points up a dramatic contrast in their lives. On the one hand aboriginals are superb survivors under conditions that test their practical skills to the uttermost. But, Sterelny goes on, intelligent as our species might be, we are *perversely* intelligent. The very same peoples who are so savvy about the natural world and how to survive in it simultaneously clutter their minds with beliefs that are palpably false and for which the word 'useless' is a generous understatement. Sterelny himself is familiar with aboriginal peoples of Papua New Guinea. They survive under arduous conditions where food is hard to come by, by dint of 'a legendarily accurate understanding of their biological environment. But they combine this understanding with deep and destructive obsessions about female menstrual pollution and about witchcraft. Many of the local cultures are tormented by fears of witchcraft and magic, and by the violence that accompanies those fears.' Sterelny challenges us to explain 'how we can be simultaneously so smart and so dumb'.[75]

Though the details differ across the world, no known culture lacks some version of the time-consuming, wealth-consuming, hostility-provoking rituals, the anti-factual, counter-productive fantasies of religion. Some educated individuals may have abandoned religion, but all were brought up in a religious culture from which they usually had to make a conscious decision to depart. The old Northern Ireland joke, 'Yes, but are you a Protestant atheist or a Catholic atheist?', is spiked with bitter truth. Religious behaviour can be called a human universal in the same way as heterosexual behaviour can. Both generalizations allow individual exceptions, but all those exceptions understand only too well the rule from which they have departed. Universal features of a species demand a Darwinian explanation.

Obviously, there is no difficulty in explaining the Darwinian advantage of sexual behaviour. It is about making babies, even on those occasions where contraception or homosexuality seems to belie it. But what about religious behaviour? Why do humans fast, kneel, genuflect, self-flagellate, nod maniacally towards a wall, crusade, or otherwise indulge in costly practices that can consume life and, in extreme cases, terminate it?

DIRECT ADVANTAGES OF RELIGION

There is a little evidence that religious belief protects people from stress-related diseases. The evidence is not strong, but it would not be surprising if it were true, for the same kind of reason as faith-healing might turn out to work in a few cases. I wish it were not necessary to add that such beneficial effects in no way boost the truth value of religion's claims. In George Bernard Shaw's words, 'The fact that a believer is happier than a skeptic is no more to the point than the fact that a drunken man is happier than a sober one.'

Part of what a doctor can give a patient is consolation and reassurance. This is not to be dismissed out of hand. My doctor doesn't literally practise faith-healing by laying on of hands. But many's the time I've been instantly 'cured' of some minor ailment by a reassuring voice from an intelligent face surmounting a stethoscope. The placebo effect is well documented and not even very mysterious. Dummy pills, with no pharmacological activity at all, demonstrably improve health. That is why double-blind drug trials must use placebos as controls. It's why homoeopathic remedies appear to work, even though they are so dilute that they have the same amount of active ingredient as the placebo control – zero molecules. Incidentally, an unfortunate by-product of the encroachment by lawyers on doctors' territory is that doctors are now afraid to prescribe placebos in normal practice. Or bureaucracy may oblige them to identify the placebo in written notes to which the patient has access, which of course defeats the object. Homoeopaths may be achieving relative success because they, unlike orthodox practitioners, are still allowed to administer placebos – under another name. They also have more time to devote to talking and simply being kind to the patient. In the early part of its long history, moreover, homoeopathy's reputation was inadvertently enhanced by the fact that its remedies did nothing at all – by contrast with orthodox medical practices, such as bloodletting, which did active harm.

Is religion a placebo that prolongs life by reducing stress? Possibly, although the theory must run a gauntlet of sceptics who point out the many circumstances in which religion causes rather than relieves stress. It is hard to believe, for example, that health is improved by the semi-permanent state of morbid guilt suffered by a Roman Catholic possessed of normal human frailty and less than normal intelligence. Perhaps it is unfair to single out the Catholics. The American comedian Cathy Ladman observes that 'All religions are the same: religion is basically guilt, with different holidays.' In any case, I find the placebo theory unworthy of the massively pervasive worldwide phenomenon of religion. I don't think the reason we have religion is that it reduced the stress levels of our ancestors. That's not a big enough theory for the job, although it may have played a subsidiary role. Religion is a large phenomenon and it needs a large theory to explain it.

Other theories miss the point of Darwinian explanations altogether. I'm talking about suggestions like 'religion satisfies our curiosity about the universe and our place in it', or 'religion is consoling'. There may be some psychological truth here, as we shall see in Chapter 10, but neither is in itself a Darwinian explanation. As Steven Pinker pointedly said of the consolation theory, in *How the Mind Works*: 'it only raises the question of *why* a mind would evolve to find comfort in beliefs it can plainly see are false. A freezing person finds no comfort in believing he is warm; a person face-to-face with a lion is not put at ease by the conviction that it is a rabbit.' At the very least, the consolation theory needs to be translated into Darwinian terms, and that is harder than you might think. Psychological explanations to the effect that people find some belief agreeable or disagreeable are proximate, not ultimate, explanations.

Darwinians make much of this distinction between proximate and ultimate. The proximate explanation for the explosion in the cylinder of an internal combustion engine invokes the sparking plug. The ultimate explanation concerns the purpose for which the explosion was designed: to impel a piston from the cylinder, thereby turning a crankshaft. The proximate cause of religion might be hyperactivity in a particular node of the brain. I shall not pursue the neurological idea of a 'god centre' in the brain because I am not concerned here with proximate questions. That is not to belittle them. I recommend Michael Shermer's *How We Believe: The Search for God in an Age of Science* for a succinct discussion, which includes the suggestion by Michael Persinger and others that visionary religious experiences are related to temporal lobe epilepsy.

But my preoccupation in this chapter is with Darwinian

ultimate explanations. If neuroscientists find a 'god centre' in the brain, Darwinian scientists like me will still want to understand the natural selection pressure that favoured it. Why did those of our ancestors who had a genetic tendency to grow a god centre survive to have more grandchildren than rivals who didn't? The Darwinian ultimate question is not a better question, not a more profound question, not a more scientific question than the neurological proximate question. But it is the one I am talking about here.

Nor are Darwinians satisfied by political explanations, such as 'Religion is a tool used by the ruling class to subjugate the underclass.' It is surely true that black slaves in America were consoled by promises of another life, which blunted their dissatisfaction with this one and thereby benefited their owners. The question of whether religions are deliberately designed by cynical priests or rulers is an interesting one, to which historians should attend. But it is not, in itself, a Darwinian question. The Darwinian still wants to know why people are *vulnerable* to the charms of religion and therefore open to exploitation by priests, politicians and kings.

A cynical manipulator might use sexual lust as a tool of political power, but we still need the Darwinian explanation of why it works. In the case of sexual lust, the answer is easy: our brains are set up to enjoy sex because sex, in the natural state, makes babies. Or a political manipulator might use torture to achieve his ends. Once again, the Darwinian must supply the explanation for why torture is effective; why we will do almost anything to avoid intense pain. Again it seems obvious to the point of banality, but the Darwinian still needs to spell it out: natural selection has set up the perception of pain as a token of life-threatening bodily damage, and programmed us to avoid it. Those rare individuals who cannot feel pain, or don't care about it, usually die young of injuries which the rest of us would have taken steps to avoid. Whether it is cynically exploited, or whether it just manifests itself spontaneously, what ultimately explains the lust for gods?

GROUP SELECTION

Some alleged ultimate explanations turn out to be – or avowedly are – 'group-selection' theories. Group selection is the controversial idea that Darwinian selection chooses among species or other *groups* of individuals. The Cambridge archaeologist Colin Renfrew suggests that Christianity survived by a form of group selection because it fostered the idea of in-group loyalty and in-group brotherly love, and this helped religious groups to survive at the expense of less religious groups. The American group-selection apostle D. S. Wilson independently developed a similar suggestion at more length, in *Darwin's Cathedral*.

Here's an invented example, to show what a group-selection theory of religion might look like. A tribe with a stirringly belligerent 'god of battles' wins wars against rival tribes whose gods urge peace and harmony, or tribes with no gods at all. Warriors who unshakeably believe that a martyr's death will send them straight to paradise fight bravely, and willingly give up their lives. So tribes with this kind of religion are more likely to survive in inter-tribal warfare, steal the conquered tribe's livestock and seize their women as concubines. Such successful tribes prolifically spawn daughter tribes that go off and propagate more daughter tribes, all worshipping the same tribal god. The idea of a group spawning daughter groups, like a beehive throwing off swarms, is not implausible, by the way. The anthropologist Napoleon Chagnon mapped just such fissioning of villages in his celebrated study of the 'Fierce People', the Yanomamö of the South American jungle.[76]

Chagnon is not a supporter of group selection, and nor am I. There are formidable objections to it. A partisan in the controversy, I must beware of riding off on my pet steed Tangent, far from the main track of this book. Some biologists betray a confusion between true group selection, as in my hypothetical example of the god of battles, and something else which they *call* group selection but which turns out on closer inspection to be either kin selection or reciprocal altruism (see Chapter 6).

Those of us who belittle group selection admit that in principle it can happen. The question is whether it amounts to a significant force in evolution. When it is pitted against selection at lower levels – as when group selection is advanced as an explanation for individual self-sacrifice – lower-level selection is likely to be stronger. In our hypothetical tribe, imagine a single self-interested warrior in an army dominated by aspiring martyrs eager to die for the tribe and earn a heavenly reward. He will be only slightly less likely to end up on the winning side as a result of hanging back in the battle to save his own skin. The martyrdom of his comrades will benefit him more than it benefits each one of them on average, because they will be dead. He is more likely to reproduce than they are, and his genes for refusing to be martyred are more likely to be reproduced into the next generation. Hence tendencies towards martyrdom will decline in future generations.

This is a simplified toy example, but it illustrates a perennial problem with group selection. Group-selection theories of individual self-sacrifice are always vulnerable to subversion from within. Individual deaths and reproductions occur on a faster timescale and with greater frequency than group extinctions and fissionings. Mathematical models can be crafted to come up with special conditions under which group selection might be evolutionarily powerful. These special conditions are usually unrealistic in nature, but it can be argued that religions in human tribal groupings foster just such otherwise unrealistic special conditions. This is an interesting line of theory, but I shall not pursue it here except to concede that Darwin himself, though he was normally a staunch advocate of selection at the level of the individual organism, came as close as he ever came to group selectionism in his discussion of human tribes:

> When two tribes of primeval man, living in the same country, came into competition, if the one tribe included (other circumstances being equal) a greater

> number of courageous, sympathetic, and faithful members, who were always ready to warn each other of danger, to aid and defend each other, this tribe would without doubt succeed best and conquer the other . . . Selfish and contentious people will not cohere, and without coherence nothing can be effected. A tribe possessing the above qualities in a high degree would spread and be victorious over other tribes; but in the course of time it would, judging from all past history, be in turn overcome by some other and still more highly-endowed tribe.[77]

To satisfy any biological specialists who might be reading this, I should add that Darwin's idea was not strictly group selection, in the true sense of successful groups spawning daughter groups whose frequency might be counted in a metapopulation of groups. Rather, Darwin visualized tribes with altruistically cooperative members spreading and becoming more numerous in terms of numbers of individuals. Darwin's model is more like the spread of the grey squirrel in Britain at the expense of the red: ecological replacement, not true group selection.

RELIGION AS A BY-PRODUCT OF SOMETHING ELSE

In any case, I want now to set aside group selection and turn to my own view of the Darwinian survival value of religion. I am one of an increasing number of biologists who see religion as a *by-product* of something else. More generally, I believe that we who speculate about Darwinian survival value need to 'think by-product'. When we ask about the survival value of anything, we may be asking the wrong question. We need to rewrite the question in a more helpful way. Perhaps the feature we are interested in (religion in this case) doesn't have a direct survival value of its own, but is a by-product of something else that does. I find it helpful to introduce the by-product idea with an analogy from my own field of animal behaviour.

Moths fly into the candle flame, and it doesn't look like an accident. They go out of their way to make a burnt offering of themselves. We could label it 'self-immolation behaviour' and, under that provocative name, wonder how on earth natural selection could favour it. My point is that we must rewrite the question before we can even attempt an intelligent answer. It isn't suicide. Apparent suicide emerges as an inadvertent side-effect or by-product of something else. A by-product of . . . what? Well, here's one possibility, which will serve to make the point.

Artificial light is a recent arrival on the night scene. Until recently, the only night lights on view were the moon and the stars. They are at optical infinity, so rays coming from them are parallel. This fits them for use as compasses. Insects are known to use celestial objects such as the sun and the moon to steer accurately in a straight line, and they can use the same compass, with reversed sign, for returning home after a foray. The insect nervous system is adept at setting up a temporary rule of thumb of this kind: 'Steer a course such that the light rays hit your eye at an angle of 30 degrees.' Since insects have compound eyes (with straight tubes or light guides radiating out from the centre of the eye like the spines of a hedgehog), this might amount in practice to something as simple as keeping the light in one particular tube or ommatidium.

But the light compass relies critically on the celestial object being at optical infinity. If it isn't, the rays are not parallel but diverge like the spokes of a wheel. A nervous system applying a 30-degree (or any acute angle) rule of thumb to a nearby candle, as though it were the moon at optical infinity, will steer the moth, via a spiral trajectory, into the flame. Draw it out for yourself, using some particular acute angle such as 30 degrees, and you'll produce an elegant logarithmic spiral into the candle.

Though fatal in this particular circumstance, the moth's rule of thumb is still, on average, a good one because, for a moth, sightings of candles are rare compared with sightings of the moon. We don't notice the hundreds of moths that are silently and effectively steering by the moon or a bright star, or even the glow from a distant city. We see only moths wheeling into our candle, and we ask the wrong question: Why are all these moths committing suicide? Instead, we should ask why they have nervous systems that steer by maintaining a fixed angle to light rays, a tactic that we notice only where it goes wrong. When the question is rephrased, the mystery evaporates. It never was right to call it suicide. It is a misfiring by-product of a normally useful compass.

Now, apply the by-product lesson to religious behaviour in humans. We observe large numbers of people – in many areas it amounts to 100 per cent – who hold beliefs that flatly contradict demonstrable scientific facts as well as rival religions followed by others. People not only hold these beliefs with passionate certitude, but devote time and resources to costly activities that flow from holding them. They die for them, or kill for them. We marvel at this, just as we marvelled at the 'self-immolation behaviour' of the moths. Baffled, we ask why. But my point is that we may be asking the wrong question. The religious behaviour may be a misfiring, an unfortunate by-product of an underlying psychological propensity which in other circumstances is, or once was, useful. On this view, the propensity that was naturally selected in our ancestors was not religion *per se*; it had some other benefit, and it only incidentally manifests itself as religious behaviour. We shall understand religious behaviour only after we have renamed it.

If, then, religion is a by-product of something else, what is that something else? What is the counterpart to the moth habit of navigating by celestial light compasses? What is the primitively advantageous trait that sometimes misfires to generate religion? I shall offer one suggestion by way of illustration, but I must stress that it is only an example of the *kind* of thing I mean, and I shall come on to parallel suggestions made by others. I am much more wedded to the general principle that the question should be properly put, and if necessary rewritten, than I am to any particular answer.

My specific hypothesis is about children. More than any other species, we survive by the accumulated experience of previous generations, and that experience needs to be passed on to children for their protection and well-being. Theoretically, children might learn from personal experience not to go too near a cliff edge, not to eat untried red berries, not to swim in crocodile-infested waters. But, to say the least, there will be a selective advantage to child brains that possess the rule of thumb: believe, without question, whatever your grown-ups tell you. Obey your parents; obey the tribal elders, especially when they adopt a solemn, minatory tone. Trust your elders without question. This is a generally valuable rule for a child. But, as with the moths, it can go wrong.

I have never forgotten a horrifying sermon, preached in my school chapel when I was little. Horrifying in retrospect, that is: at the time, my child brain accepted it in the spirit intended by the preacher. He told us a story of a squad of soldiers, drilling beside a railway line. At a critical moment the drill sergeant's attention was distracted, and he failed to give the order to halt. The soldiers were so well schooled to obey orders without question that they carried on marching, right into the path of an oncoming train. Now, of course, I don't believe the story and I hope the preacher didn't either. But I believed it when I was nine, because I heard it from an adult in authority over me. And whether he believed it or not, the preacher wished us children to admire and model ourselves on the soldiers' slavish and unquestioning obedience to an order, however preposterous, from an authority figure. Speaking for myself, I think we *did* admire it. As an adult I find it almost impossible to credit that my childhood self wondered whether I would have had the courage to do my duty by marching under the train. But that, for what it is worth, is how I remember my feelings. The sermon obviously made a deep impression on me, for I have remembered it and passed it on to you.

To be fair, I don't think the preacher thought he was serving up a religious message. It was probably more military than religious, in the spirit of Tennyson's 'Charge of the Light Brigade', which he may well have quoted:

> 'Forward the Light Brigade!'
> Was there a man dismayed?
> Not though the soldiers knew
> Some one had blundered:
> Theirs not to make reply,
> Theirs not to reason why,
> Theirs but to do and die:
> Into the valley of Death
> Rode the six hundred.

(One of the earliest and scratchiest recordings of the human voice ever made is of Lord Tennyson himself reading this poem, and the impression of hollow declaiming down a long, dark tunnel from the depths of the past seems eerily appropriate.) From the high command's point of view it would be madness to allow each individual soldier discretion over whether or not to obey orders. Nations whose infantrymen act on their own initiative rather than following orders will tend to lose wars. From the nation's point of view, this remains a good rule of thumb even if it sometimes leads to individual disasters. Soldiers are drilled to become as much like automata, or computers, as possible.

Computers do what they are told. They slavishly obey any instructions given in their own programming language. This is how they do useful things like word processing and spreadsheet calculations. But, as an inevitable by-product, they are equally robotic in obeying bad instructions. They have no way of telling whether an instruction will have a good effect or a bad. They simply obey, as soldiers are supposed to. It is their unquestioning obedience that makes computers useful, and exactly the same thing makes them inescapably vulnerable to infection by software viruses and worms. A maliciously designed program that says, 'Copy me and send me to every address that you find on this hard disk' will simply be obeyed, and then obeyed again by the other computers down the line to which it is sent, in exponential expansion. It is difficult, perhaps impossible, to design a computer which is usefully obedient and at the same time immune to infection.

If I have done my softening-up work well, you will already have completed my argument about child brains and religion. Natural selection builds child brains with a tendency to believe whatever their parents and tribal elders tell them. Such trusting obedience is valuable for survival: the analogue of steering by the moon for a moth. But the flip side of trusting obedience is slavish gullibility. The inevitable by-product is vulnerability to infection by mind viruses. For excellent reasons related to Darwinian survival, child brains need to trust parents, and elders whom parents tell them to trust. An automatic consequence is that the truster has no way of distinguishing good advice from bad. The child cannot know that 'Don't paddle in the crocodile-infested Limpopo' is good advice but 'You must sacrifice a goat at the time of the full moon, otherwise the rains will fail' is at best a waste of time and goats. Both admonitions sound equally trustworthy. Both come from a respected source and are delivered with a solemn earnestness that commands respect and demands obedience. The same goes for propositions about the world, about the cosmos, about morality and about human nature. And, very likely, when the child grows up and has children of her own, she will naturally pass the whole lot on to her own children – nonsense as well as sense – using the same infectious gravitas of manner.

On this model we should expect that, in different geographical regions, different arbitrary beliefs, none of which have any factual basis, will be handed down, to be believed with the same conviction as useful pieces of traditional wisdom such as the belief that manure is good for the crops. We should also expect that superstitions and other non-factual beliefs will locally evolve – change over generations – either by random drift or by some sort of analogue of Darwinian selection, eventually showing a pattern of significant divergence from common ancestry. Languages drift apart from a common progenitor given sufficient time in geographical separation (I shall return to this point in a moment). The same seems to be true of baseless and arbitrary beliefs and injunctions, handed down the generations – beliefs that were perhaps given a fair wind by the useful programmability of the child brain.

Religious leaders are well aware of the vulnerability of the child brain, and the importance of getting the indoctrination in early. The Jesuit boast, 'Give me the child for his first seven years, and I'll give you the man,' is no less accurate (or sinister) for being hackneyed. In more recent times, James Dobson, founder of today's infamous 'Focus on the Family' movement,* is equally acquainted with the principle: 'Those who control what young people are taught, and what they experience – what they see, hear, think, and believe – will determine the future course for the nation.'[78]

But remember, my specific suggestion about the useful gullibility of the child mind is only an example of the *kind* of thing that might be the analogue of moths navigating by the moon or the stars. The ethologist Robert Hinde, in *Why Gods Persist*, and the anthropologists Pascal Boyer, in *Religion Explained*, and Scott Atran, in *In Gods We Trust*, have independently promoted the general idea of religion as a by-product of normal psychological dispositions – many by-products, I should say, for the anthropologists especially are concerned to emphasize the diversity of the world's religions as well as what they have in common. The findings of anthropologists seem

---

* I was amused when I saw 'Focus on your own damn family' on a car bumper sticker in Colorado, but it now seems to me less funny. Maybe some children need to be protected from indoctrination by their own parents (see Chapter 9).

weird to us only because they are unfamiliar. All religious beliefs seem weird to those not brought up in them. Boyer did research on the Fang people of Cameroon, who believe . . .

> . . . that witches have an extra internal animal-like organ that flies away at night and ruins other people's crops or poisons their blood. It is also said that these witches sometimes assemble for huge banquets, where they devour their victims and plan future attacks. Many will tell you that a friend of a friend actually saw witches flying over the village at night, sitting on a banana leaf and throwing magical darts at various unsuspecting victims.

Boyer continues with a personal anecdote:

> I was mentioning these and other exotica over dinner in a Cambridge college when one of our guests, a prominent Cambridge theologian, turned to me and said: 'That is what makes anthropology so fascinating and so difficult too. You have to explain *how people can believe such nonsense*.' Which left me dumbfounded. The conversation had moved on before I could find a pertinent response – to do with kettles and pots.

Assuming that the Cambridge theologian was a mainstream Christian, he probably believed some combination of the following:

- In the time of the ancestors, a man was born to a virgin mother with no biological father being involved.
- The same fatherless man called out to a friend called Lazarus, who had been dead long enough to stink, and Lazarus promptly came back to life.

- The fatherless man himself came alive after being dead and buried three days.
- Forty days later, the fatherless man went up to the top of a hill and then disappeared bodily into the sky.
- If you murmur thoughts privately in your head, the fatherless man, and his 'father' (who is also himself) will hear your thoughts and may act upon them. He is simultaneously able to hear the thoughts of everybody else in the world.
- If you do something bad, or something good, the same fatherless man sees all, even if nobody else does. You may be rewarded or punished accordingly, including after your death.
- The fatherless man's virgin mother never died but was 'assumed' bodily into heaven.
- Bread and wine, if blessed by a priest (who must have testicles), 'become' the body and blood of the fatherless man.

What would an objective anthropologist, coming fresh to this set of beliefs while on fieldwork in Cambridge, make of them?

PSYCHOLOGICALLY PRIMED FOR RELIGION

The idea of psychological by-products grows naturally out of the important and developing field of evolutionary psychology.[79] Evolutionary psychologists suggest that, just as the eye is an evolved organ for seeing, and the wing an evolved organ for flying, so the brain is a collection of organs (or 'modules') for dealing with a set of specialist data-processing needs. There is a module for dealing with kinship, a module for dealing with reciprocal exchanges, a module for dealing with empathy, and so on. Religion can be seen as a by-product of the misfiring of several of these modules, for example the modules for forming theories of other minds, for forming coalitions, and for discriminating in favour of in-group members and against strangers. Any of these could serve as the human equivalent of the moths' celestial navigation, vulnerable to misfiring in the same kind of way as I suggested for childhood gullibility. The psychologist Paul Bloom, another advocate of the 'religion is a by-product' view, points out that children have a natural tendency towards a *dualistic* theory of mind. Religion, for him, is a by-product of such instinctive dualism. We humans, he suggests, and especially children, are natural born dualists.

A dualist acknowledges a fundamental distinction between matter and mind. A monist, by contrast, believes that mind is a manifestation of matter – material in a brain or perhaps a computer – and cannot exist apart from matter. A dualist believes the mind is some kind of disembodied spirit that *inhabits* the body and therefore conceivably could leave the body and exist somewhere else. Dualists readily interpret mental illness as 'possession by devils', those devils being spirits whose residence in the body is temporary, such that they might be 'cast out'. Dualists personify inanimate physical objects at the slightest opportunity, seeing spirits and demons even in waterfalls and clouds.

F. Anstey's 1882 novel *Vice Versa* makes sense to a dualist, but strictly should be incomprehensible to a dyed-in-the-wool monist like me. Mr Bultitude and his son mysteriously find that they have swapped bodies. The father, much to the son's glee, is obliged to go to school in the son's body; while the son, in the father's body, almost ruins the father's business through his immature decisions. A similar plotline was used by P. G. Wodehouse in *Laughing Gas*, where the Earl of Havershot and a child movie star go under the anaesthetic at the same moment in neighbouring dentist's chairs, and wake up in each other's bodies. Once again, the plot makes sense only to a dualist. There has to be something corresponding to Lord Havershot which is no part of his body, otherwise how could he wake up in the body of a child actor?

Like most scientists, I am not a dualist, but I am nevertheless easily capable of enjoying *Vice Versa* and *Laughing Gas*. Paul Bloom would say this is because, even though I have learned to be an intellectual monist, I am a human animal and therefore evolved as an instinctive dualist. The idea that there is a me perched somewhere behind my eyes and capable, at least in fiction, of migrating into somebody else's head, is deeply ingrained in me and in every other human being, whatever our intellectual pretensions to monism. Bloom supports his contention with experimental evidence that children are even more likely to be dualists than adults are, especially extremely young children. This suggests that a tendency to dualism is built into the brain and, according to Bloom, provides a natural predisposition to embrace religious ideas.

Bloom also suggests that we are innately predisposed to be creationists. Natural selection 'makes no intuitive sense'. Children are especially likely to assign purpose to everything, as the psychologist Deborah Keleman tells us in her article 'Are children "intuitive theists"?'[80] Clouds are 'for raining'. Pointy rocks are 'so that animals could scratch on them when they get itchy'. The assignment of purpose to everything is called teleology. Children are native teleologists, and many never grow out of it.

Native dualism and native teleology predispose us, given the right conditions, to religion, just as my moths' light-compass reaction predisposed them to inadvertent 'suicide'. Our innate dualism prepares us to believe in a 'soul' which inhabits the body rather than being integrally part of the body. Such a disembodied spirit can easily be imagined to move on somewhere else after the death of the body. We can also easily imagine the existence of a deity as pure spirit, not an emergent property of complex matter but existing independently of matter. Even more obviously, childish teleology sets us up for religion. If everything has a purpose, whose purpose is it? God's, of course.

But what is the counterpart of the *usefulness* of the moths' light compass? Why might natural selection have favoured dualism and teleology in the brains of our ancestors and their children? So far, my account of the 'innate dualists' theory has simply posited that humans are natural born dualists and teleologists. But what would the Darwinian advantage be? Predicting the behaviour of entities in our world is important for our survival, and we would expect natural selection to have shaped our brains to do it efficiently and fast. Might dualism and teleology serve us in this capacity? We may understand this hypothesis better in the light of what Daniel Dennett has called the intentional stance.

Dennett has offered a helpful three-way classification of the 'stances' that we adopt in trying to understand and hence predict the behaviour of entities such as animals, machines or each other.[81] They are the physical stance, the design stance and the intentional stance. The *physical stance* always works in principle, because everything ultimately obeys the laws of physics. But working things out using the physical stance can be very slow. By the time we have sat down to calculate all the interactions of a complicated object's moving parts, our prediction of its behaviour will probably be too late. For an object that really is designed, like a washing machine or a crossbow, the *design stance* is an economical short cut. We can guess how the object will behave by going over the head of physics and appealing directly to design. As Dennett says,

> Almost anyone can predict when an alarm clock will sound on the basis of the most casual inspection of its exterior. One does not know or care to know whether it is spring wound, battery driven, sunlight powered, made of brass wheels and jewel bearings or silicon chips – one just assumes that it is designed so that the alarm will sound when it is set to sound.

Living things are not designed, but Darwinian natural selection licenses a version of the design stance for them. We get a short cut to understanding the heart if we assume that it is 'designed' to pump blood. Karl von Frisch was led to investigate colour vision in bees (in the face of orthodox opinion that they were colour-blind) because he assumed that the bright colours of flowers were 'designed' to attract them. The quotation marks are designed to scare off mendacious creationists who might otherwise claim the great Austrian zoologist as one of their own. Needless to say, he was perfectly capable of translating the design stance into proper Darwinian terms.

The *intentional stance* is another short cut, and it goes one better than the design stance. An entity is assumed not merely to be designed for a purpose but to be, or contain, an *agent* with intentions that guide its actions. When you see a tiger, you had better not delay your prediction of its probable behaviour. Never mind the physics of its molecules, and never mind the design of its limbs, claws and teeth. That cat intends to eat you, and it will deploy its limbs, claws and teeth in flexible and resourceful ways to carry out its intention. The quickest way to second-guess its behaviour is to forget physics and physiology and cut to the intentional chase. Note that, just as the design stance works even for things that were not actually designed as well as things that were, so the intentional stance works for things that don't have deliberate conscious intentions as well as things that do.

It seems to me entirely plausible that the intentional stance has survival value as a brain mechanism that speeds up decision-making in dangerous circumstances, and in crucial social situations. It is less immediately clear that dualism is a necessary concomitant of the intentional stance. I shan't pursue the matter here, but I think a case could be developed that some kind of theory of other minds, which could fairly be described as dualistic, is likely to underlie the intentional stance – especially in complicated social situations, and even more especially where *higher-order* intentionality comes into play.

Dennett speaks of *third-order intentionality* (the man believed that the woman knew he wanted her), *fourth-order* (the woman realized that the man believed that the woman knew he wanted her) and even *fifth-order* intentionality (the shaman guessed that the woman realized that the man believed that the woman knew he wanted her). Very high orders of intentionality are probably confined to fiction, as satirized in Michael Frayn's hilarious novel *The Tin Men*: 'Watching Nunopoulos, Rick knew that he was almost certain that Anna felt a passionate contempt for Fiddlingchild's failure to understand her feelings about Fiddlingchild, and she knew too that Nina knew she knew about Nunopoulos's knowledge . . .' But the fact that we can laugh at such contortions of other-mind inference in fiction is probably telling us something important about the way our minds have been naturally selected to work in the real world.

In its lower orders at least, the intentional stance, like the design stance, saves time that might be vital to survival. Consequently, natural selection shaped brains to deploy the intentional stance as a short cut. We are biologically programmed to impute intentions to entities whose behaviour matters to us. Once again, Paul Bloom quotes experimental evidence that children are especially likely to adopt the intentional stance. When small babies see an object apparently following another object (for example, on a computer screen), they assume that they are witnessing an active chase by an intentional agent, and they demonstrate the fact by registering surprise when the putative agent fails to pursue the chase.

The design stance and the intentional stance are useful brain mechanisms, important for speeding up the second-guessing of entities that really matter for survival, such as predators or potential mates. But, like other brain mechanisms, these stances can misfire. Children, and primitive peoples, impute intentions to the weather, to waves and currents, to falling rocks. All of us are prone to do the same thing with machines, especially when they let us down. Many will remember with affection the day

Basil Fawlty's car broke down during his vital mission to save Gourmet Night from disaster. He gave it fair warning, counted to three, then got out of the car, seized a tree branch and thrashed it to within an inch of its life. Most of us have been there, at least momentarily, with a computer if not with a car. Justin Barrett coined the acronym HADD, for hyperactive agent detection device. We hyperactively detect agents where there are none, and this makes us suspect malice or benignity where, in fact, nature is only indifferent. I catch myself momentarily harbouring savage resentment against some blameless inanimate such as my bicycle chain. There was a poignant recent report of a man who tripped over his untied shoelace in the Fitzwilliam Museum in Cambridge, fell down the stairs, and smashed three priceless Qing Dynasty vases: 'He landed in the middle of the vases and they splintered into a million pieces. He was still sitting there stunned when staff appeared. Everyone stood around in silence, as if in shock. The man kept pointing to his shoelace, saying, "There it is; that's the culprit." '[82]

Other by-product explanations of religion have been proposed by Hinde, Shermer, Boyer, Atran, Bloom, Dennett, Keleman and others. One especially intriguing possibility mentioned by Dennett is that the irrationality of religion is a by-product of a particular built-in irrationality mechanism in the brain: our tendency, which presumably has genetic advantages, to fall in love.

The anthropologist Helen Fisher, in *Why We Love*, has beautifully expressed the insanity of romantic love, and how over-the-top it is compared with what might seem strictly necessary. Look at it this way. From the point of view of a man, say, it is unlikely that any one woman of his acquaintance is a hundred times more lovable than her nearest competitor, yet that is how he is likely to describe her when 'in love'. Rather than the fanatically monogamous devotion to which we are susceptible, some sort of 'polyamory' is on the face of it more rational. (Polyamory is the belief that one can simultaneously love several members of the opposite sex, just as one can love more than one wine, composer, book or sport.) We happily accept that we can love more than one child, parent, sibling, teacher, friend or pet. When you think of it like that, isn't the total exclusiveness that we expect of spousal love positively weird? Yet it *is* what we expect, and it is what we set out to achieve. There must be a reason.

Helen Fisher and others have shown that being in love is accompanied by unique brain states, including the presence of neurally active chemicals (in effect, natural drugs) that are highly specific and characteristic of the state. Evolutionary psychologists agree with her that the irrational *coup de foudre* could be a mechanism to ensure loyalty to one co-parent, lasting for long enough to rear a child together. From a Darwinian point of view it is, no doubt, important to choose a good partner, for all sorts of reasons. But, once having made a choice – even a poor one – and conceived a child, it is more important to stick with that one choice through thick and thin, at least until the child is weaned.

Could irrational religion be a by-product of the irrationality mechanisms that were originally built into the brain by selection for falling in love? Certainly, religious faith has something of the same character as falling in love (and both have many of the attributes of being high on an addictive drug*). The neuropsychiatrist John Smythies cautions that there are significant differences between the brain areas activated by the two kinds of mania. Nevertheless, he notes some similarities too:

> One facet of the many faces of religion is intense love focused on one supernatural person, i.e. God, plus reverence for icons of that person. Human life is driven largely by our selfish genes and by the processes of reinforcement. Much positive reinforcement derives from religion: warm and comforting

---

* See my exposé of the dangerous narcotic Gerin Oil: R. Dawkins, 'Gerin Oil', *Free Inquiry* 24: 1, 2003, 9–11.

> feelings of being loved and protected in a dangerous world, loss of fear of death, help from the hills in response to prayer in difficult times, etc. Likewise, romantic love for another real person (usually of the other sex) exhibits the same intense concentration on the other and related positive reinforcements. These feelings can be triggered by icons of the other, such as letters, photographs, and even, as in Victorian times, locks of hair. The state of being in love has many physiological accompaniments, such as sighing like a furnace.[83]

I made the comparison between falling in love and religion in 1993, when I noted that the symptoms of an individual infected by religion 'may be startlingly reminiscent of those more ordinarily associated with sexual love. This is an extremely potent force in the brain, and it is not surprising that some viruses have evolved to exploit it' ('viruses' here is a metaphor for religions: my article was called 'Viruses of the mind'). St Teresa of Avila's famously orgasmic vision is too notorious to need quoting again. More seriously, and on a less crudely sensual plane, the philosopher Anthony Kenny provides moving testimony to the pure delight that awaits those who manage to believe in the mystery of the transubstantiation. After describing his ordination as a Roman Catholic priest, empowered by laying on of hands to celebrate mass, he goes on that he vividly recalls

> the exaltation of the first months during which I had the power to say Mass. Normally a slow and sluggish riser, I would leap early out of bed, fully awake and full of excitement at the thought of the momentous act I was privileged to perform . . .
>
> It was touching the body of Christ, the closeness of the priest to Jesus, which most enthralled me. I would gaze on the Host after the words of con-

> secration, soft-eyed like a lover looking into the eyes of his beloved . . . Those early days as a priest remain in my memory as days of fulfilment and tremulous happiness; something precious, and yet too fragile to last, like a romantic love-affair brought up short by the reality of an ill-assorted marriage.

The equivalent of the moth's light-compass reaction is the apparently irrational but useful habit of falling in love with one, and only one, member of the opposite sex. The misfiring by-product – equivalent to flying into the candle flame – is falling in love with Yahweh (or with the Virgin Mary, or with a wafer, or with Allah) and performing irrational acts motivated by such love.

The biologist Lewis Wolpert, in *Six Impossible Things Before Breakfast*, makes a suggestion that can be seen as a generalization of the idea of constructive irrationality. His point is that irrationally strong conviction is a guard against fickleness of mind: 'if beliefs that saved lives were not held strongly, it would have been disadvantageous in early human evolution. It would be a severe disadvantage, for example, when hunting or making tools, to keep changing one's mind.' The implication of Wolpert's argument is that, at least under some circumstances, it is better to persist in an irrational belief than to vacillate, even if new evidence or ratiocination favours a change. It is easy to see the 'falling in love' argument as a special case, and it is correspondingly easy to see Wolpert's 'irrational persistence' as yet another useful psychological predisposition that could explain important aspects of irrational religious behaviour: yet another by-product.

In his book *Social Evolution*, Robert Trivers enlarged on his 1976 evolutionary theory of self-deception. Self-deception is

> hiding the truth from the conscious mind the better to hide it from others. In our own species we

> recognize that shifty eyes, sweaty palms and croaky voices may indicate the stress that accompanies conscious knowledge of attempted deception. By becoming unconscious of its deception, the deceiver hides these signs from the observer. He or she can lie without the nervousness that accompanies deception.

The anthropologist Lionel Tiger says something similar in *Optimism: The Biology of Hope.* The connection to the sort of constructive irrationality we have just been discussing is seen in Trivers's paragraph about 'perceptual defense':

> There is a tendency for humans consciously to see what they wish to see. They literally have difficulty seeing things with negative connotations while seeing with increasing ease items that are positive. For example, words that evoke anxiety, either because of an individual's personal history or because of experimental manipulation, require greater illumination before first being perceived.

The relevance of this to the wishful thinking of religion should need no spelling out.

The general theory of religion as an accidental by-product – a misfiring of something useful – is the one I wish to advocate. The details are various, complicated and disputable. For the sake of illustration, I shall continue to use my 'gullible child' theory as representative of 'by-product' theories in general. This theory – that the child brain is, for good reasons, vulnerable to infection by mental 'viruses' – will strike some readers as incomplete. Vulnerable the mind may be, but why should it be infected by *this* virus rather than that? Are some viruses especially proficient at infecting vulnerable minds? Why does 'infection' manifest itself as religion rather than as . . . well, what? Part of what I want to say is that it doesn't matter what particular style of nonsense infects the child brain. Once infected, the child will grow up and infect the next generation with the same nonsense, whatever it happens to be.

An anthropological survey such as Frazer's *Golden Bough* impresses us with the diversity of irrational human beliefs. Once entrenched in a culture they persist, evolve and diverge, in a manner reminiscent of biological evolution. Yet Frazer discerns certain general principles, for example 'homoeopathic magic', whereby spells and incantations borrow some symbolic aspect of the real-world object they are intended to influence. An instance with tragic consequences is the belief that powdered rhinoceros horn has aphrodisiac properties. Fatuous as it is, the legend stems from the horn's supposed resemblance to a virile penis. The fact that 'homoeopathic magic' is so widespread suggests that the nonsense that infects vulnerable brains is not entirely random, arbitrary nonsense.

It is tempting to pursue the biological analogy to the point of wondering whether something corresponding to natural selection is at work. Are some ideas more spreadable than others, because of intrinsic appeal or merit, or compatibility with existing psychological dispositions, and could this account for the nature and properties of actual religions as we see them, in something like the way we use natural selection to account for living organisms? It is important to understand that 'merit' here means only ability to survive and spread. It doesn't mean deserving of a positive value judgement – something of which we might be humanly proud.

Even on an evolutionary model, there doesn't have to be any natural selection. Biologists acknowledge that a gene may spread through a population not because it is a good gene but simply because it is a lucky one. We call this genetic drift. How important it is *vis-à-vis* natural selection has been controversial. But it is now widely accepted in the form of the so-called neutral theory of molecular genetics. If a gene mutates to a different version of itself which has an identical effect, the difference is neutral, and selection cannot favour one or the other. Nevertheless, by what statisticians call sampling error over generations, the new mutant form can eventually replace the original form in the gene pool. This is a true evolutionary change at the molecular level (even if no change is observed in the world of whole organisms). It is a neutral evolutionary change that owes nothing to selective advantage.

The cultural equivalent of genetic drift is a persuasive option, one that we cannot neglect when thinking about the evolution of religion. Language evolves in a quasi-biological way and the direction its evolution takes looks undirected, pretty much like random drift. It is handed down by a cultural analogue of genetics, changing slowly over the centuries, until eventually various strands have diverged to the point of mutual unintelligibility. It is possible that some of the evolution of language is guided by a kind of natural selection, but that argument doesn't seem very persuasive. I'll explain below that some such idea has been proposed for major trends in language, such as the Great Vowel Shift which took place in English from the fifteenth to the eighteenth century. But such a functional hypothesis is not necessary to explain most of what we observe. It seems probable that language normally evolves by the cultural equivalent of random genetic drift. In different parts of Europe, Latin drifted to become Spanish, Portuguese, Italian, French, Romansche and the various dialects of these languages. It is, to say the least, not obvious that these evolutionary shifts reflect local advantages or 'selection pressures'.

I surmise that religions, like languages, evolve with sufficient randomness, from beginnings that are sufficiently arbitrary, to generate the bewildering – and sometimes dangerous – richness of diversity that we observe. At the same time, it is possible that a form of natural selection, coupled with the fundamental uniformity of human psychology, sees to it that the diverse religions share significant features in common. Many religions, for example, teach the objectively implausible but subjectively appealing doctrine that our personalities survive our bodily death. The idea of immortality itself survives and spreads because it caters to wishful thinking. And wishful thinking counts, because human psychology has a near-universal tendency to let belief be coloured by desire ('Thy wish was father, Harry, to that thought', as Henry IV Part II said to his son*).

There seems to be no doubt that many of the attributes of religion are well fitted to helping the religion's own survival, and the survival of the attributes concerned, in the stew of human culture. The question now arises of whether the good fit is achieved by 'intelligent design' or by natural selection. The answer is probably both. On the side of design, religious leaders are fully capable of verbalizing the tricks that aid the survival of religion. Martin Luther was well aware that reason was religion's arch-enemy, and he frequently warned of its dangers: 'Reason is the greatest enemy that faith has; it never comes to the aid of spiritual things, but more frequently than not struggles against the divine Word, treating with contempt all that emanates from God.'[84] Again: 'Whoever wants to be a Christian should tear the eyes out of his reason.' And again: 'Reason should be destroyed in all Christians.' Luther would have had no difficulty in intelligently designing unintelligent aspects of a religion to help it survive. But that doesn't necessarily mean that he, or anyone else, did design it. It could also have evolved by a (non-genetic) form of natural selection, with Luther not its designer but a shrewd observer of its efficacy.

Even though conventional Darwinian selection of genes might have favoured psychological predispositions that produce religion as a by-product, it is unlikely to have shaped the details. I have already hinted that, if we are going to apply some form of selection theory to those details, we should look not to genes but to their cultural equivalents. Are religions such stuff as memes are made on?

---

* Not my joke: *1066 and All That.*

TREAD SOFTLY, BECAUSE YOU TREAD ON MY MEMES

> *Truth, in matters of religion, is simply the opinion that has survived.*
>
> OSCAR WILDE

This chapter began with the observation that, because Darwinian natural selection abhors waste, any ubiquitous feature of a species – such as religion – must have conferred some advantage or it wouldn't have survived. But I hinted that the advantage doesn't have to redound to the survival or reproductive success of the individual. As we saw, advantage to the genes of the cold virus sufficiently explains the ubiquity of that miserable complaint among our species.* And it doesn't even have to be genes that benefit. Any *replicator* will do. Genes are only the most obvious examples of replicators. Other candidates are computer viruses, and memes – units of cultural inheritance and the topic of this section. If we are to understand memes, we have first to look a little more carefully at exactly how natural selection works.

In its most general form, natural selection must choose between alternative replicators. A replicator is a piece of coded information that makes exact copies of itself, along with occasional inexact copies or 'mutations'. The point about this is the Darwinian one. Those varieties of replicator that happen to be good at getting copied become more numerous at the expense of alternative replicators that are bad at getting copied. That, at its most rudimentary, is natural selection. The archetypal replicator is a gene, a stretch of DNA that is duplicated, nearly always with extreme accuracy, through an

---

* Especially my nation, according to national stereotyping legend: '*Voici l'anglais avec son sang froid habituel*' (Here is the Englishman with his habitual bloody cold). This comes from *Fractured French* by F. S. Pearson, along with other gems such as '*coup de grâce*' (lawnmower).

indefinite number of generations. The central question for meme theory is whether there are units of cultural imitation which behave as true replicators, like genes. I am not saying that memes necessarily *are* close analogues of genes, only that the more like genes they are, the better will meme theory work; and the purpose of this section is to *ask* whether meme theory might work for the special case of religion.

In the world of genes, the occasional flaws in replication (mutations) see to it that the gene pool contains alternative variants of any given gene – 'alleles' – which may therefore be seen as competing with each other. Competing for what? For the particular chromosomal slot or 'locus' that belongs to that set of alleles. And how do they compete? Not by direct molecule-to-molecule combat but by proxy. The proxies are their 'phenotypic traits' – things like leg length or fur colour: manifestations of genes fleshed out as anatomy, physiology, biochemistry or behaviour. A gene's fate is normally bound up with the bodies in which it successively sits. To the extent that it influences those bodies, it affects its own chances of surviving in the gene pool. As the generations go by, genes increase or decrease in frequency in the gene pool by virtue of their phenotypic proxies.

Might the same be true of memes? One respect in which they are not like genes is that there is nothing obviously corresponding to chromosomes or loci or alleles or sexual recombination. The meme pool is less structured and less organized than the gene pool. Nevertheless, it is not obviously silly to speak of a meme pool, in which particular memes might have a 'frequency' which can change as a consequence of competitive interactions with alternative memes.

Some people have objected to memetic explanations, on various grounds that usually stem from the fact that memes are not entirely like genes. The exact physical nature of a gene is now known (it is a sequence of DNA) whereas that of memes is not, and different memeticists confuse one another by switching from one physical medium to another. Do memes exist only in brains? Or is every paper copy and electronic copy of, say, a particular limerick also entitled to be called a meme? Then again, genes replicate with very high fidelity, whereas, if memes replicate at all, don't they do so with low accuracy?

These alleged problems of memes are exaggerated. The most important objection is the allegation that memes are copied with insufficiently high fidelity to function as Darwinian replicators. The suspicion is that if the 'mutation rate' in every generation is high, the meme will mutate itself out of existence before Darwinian selection can have an impact on its frequency in the meme pool. But the problem is illusory. Think of a master carpenter, or a prehistoric flint-knapper, demonstrating a particular skill to a young apprentice. If the apprentice faithfully reproduced every hand movement of the master, you would indeed expect to see the meme mutate out of all recognition in a few 'generations' of master/apprentice transmission. But of course the apprentice does not faithfully reproduce every hand movement. It would be ridiculous to do so. Instead, he notes the goal that the master is trying to achieve, and imitates that. Drive in the nail until the head is flush, using as many hammer blows as it takes, which may not be the same number as the master used. It is such rules that can pass unmutated down an indefinite number of imitation 'generations'; no matter that the details of their execution may vary from individual to individual, and from case to case. Stitches in knitting, knots in ropes or fishing nets, origami folding patterns, useful tricks in carpentry or pottery: all can be reduced to discrete elements that really do have the opportunity to pass down an indefinite number of imitation generations without alteration. The details may wander idiosyncratically, but the essence passes down unmutated, and that is all that is needed for the analogy of memes with genes to work.

In my foreword to Susan Blackmore's *The Meme Machine* I developed the example of an origami procedure for making a model Chinese junk. It is quite a complicated recipe, involving thirty-two folding (or similar) operations. The end result (the Chinese junk itself) is a pleasing object, as are at least three intermediate stages in the 'embryology', namely the 'catamaran', the 'box with two lids' and the 'picture frame'. The whole performance does indeed remind me of the foldings and invaginations that the membranes of an embryo undergo as it morphs itself from blastula to gastrula to neurula. I learned to make the Chinese junk as a boy from my father who, at about the same age, had acquired the skill at his boarding school. A craze for making Chinese junks, initiated by the school matron, had spread through the school in his time like a measles epidemic, then died away, also like a measles epidemic. Twenty-six years later, when that matron was long gone, I went to the same school. I reintroduced the craze and it again spread, like another measles epidemic, and then again died away. The fact that such a teachable skill can spread like an epidemic tells us something important about the high fidelity of memetic transmission. We may be sure that the junks made by my father's generation of schoolboys in the 1920s were in no general respect different from those made by my generation in the 1950s.

We could investigate the phenomenon more systematically by the following experiment: a variant of the childhood game of Chinese Whispers (American children call it Telephone). Take two hundred people who have never made a Chinese junk before, and line them up in twenty teams of ten people each. Gather the heads of the twenty teams around a table and teach them, by demonstration, how to make a Chinese junk. Now send each one off to find the second person in his own team, and teach that person alone, again by demonstration, to make a Chinese junk. Each second 'generation' person then teaches the third person in her own team, and so on until the tenth member of every team has been reached. Keep all the junks made along the way, and label them by their team and 'generation' number for subsequent inspection.

I haven't done the experiment yet (I'd like to), but I have a strong prediction of what the result will be. My prediction is that not all of the twenty teams will succeed in passing the skill intact down the line to their tenth members, but that a significant number of them will. In some of the teams there will be mistakes: perhaps a weak link in the chain will forget some vital step in the procedure, and everyone downstream of the mistake will then obviously fail. Perhaps team 4 gets as far as the 'catamaran' but falters thereafter. Perhaps the eighth member of team 13 produces a 'mutant' somewhere between the 'box with two lids' and the 'picture frame' and the ninth and tenth members of his team then copy the mutated version.

Now, of those teams in which the skill is transferred successfully to the tenth generation, I make a further prediction. If you rank the junks in order of 'generation' you will not see a systematic deterioration of quality with generation number. If, on the other hand, you were to run an experiment identical in all respects except that the skill transferred was not origami but copying a *drawing* of a junk, there would definitely be a systematic deterioration in the accuracy with which the generation 1 pattern 'survived' to generation 10.

In the drawing version of the experiment, all the generation 10 drawings would bear some slight resemblance to the generation 1 drawing. And within each team, the resemblance would more or less steadily deteriorate as you proceed down the generations. In the origami version of the experiment, by contrast, the mistakes would be all-or-none: they'd be 'digital' mutations. Either a team would make no mistakes and the generation 10 junk would be no worse, and no better, on average than that produced by generation 5 or generation 1; or there would be a 'mutation' in some particular generation and all downstream efforts would be complete failures, often faithfully reproducing the mutation.

What is the crucial difference between the two skills? It is that the origami skill consists of a series of discrete actions, none of which is difficult to perform in itself. Mostly the operations are things like 'Fold both sides into the middle.' A particular team member may execute the step ineptly, but it will be clear to the next team member down the line what he is *trying* to do. The origami steps are 'self-normalizing'. It is this that makes them 'digital'. It is like my master carpenter, whose intention to flatten the nail head in the wood is obvious to his apprentice, regardless of the details of the hammer blows. Either you get a given step of the origami recipe right or you don't. The drawing skill, by contrast, is an analogue skill. Everybody can have a go, but some people copy a drawing more accurately than others, and nobody copies it perfectly. The accuracy of the copy depends, too, on the amount of time and care devoted to it, and these are continuously variable quantities. Some team members, moreover, will embellish and 'improve', rather than strictly copy, the preceding model.

Words – at least when they are understood – are self-normalizing in the same kind of way as origami operations. In the original game of Chinese Whispers (Telephone) the first child is told a story, or a sentence, and is asked to pass it on to the next child, and so on. If the sentence is less than about seven words, in the native language of all the children, there is a good chance that it will survive, unmutated, down ten generations. If it is in an unknown foreign language, so that the children are forced to imitate phonetically rather than word by word, the message does not survive. The pattern of decay down the generations is then the same as for a drawing, and it will become garbled. When the message makes sense in the children's own language, and doesn't contain any unfamiliar words like 'phenotype' or 'allele', it survives. Instead of mimicking the sounds phonetically, each child recognizes each word as a member of a finite vocabulary and selects the same word, although very probably pronounced in a different accent, when passing it on to the next child. Written language is also self-normalizing because the squiggles on paper, no matter how much they may differ in detail, are all drawn from a finite alphabet of (say) twenty-six letters.

The fact that memes can sometimes display very high

fidelity, due to self-normalizing processes of this kind, is enough to answer some of the commonest objections that are raised to the meme/gene analogy. In any case, the main purpose of meme theory, at this early stage of its development, is not to supply a comprehensive theory of culture, on a par with Watson–Crick genetics. My original purpose in advocating memes, indeed, was to counter the impression that the gene was the only Darwinian game in town – an impression that *The Selfish Gene* was otherwise at risk of conveying. Peter Richerson and Robert Boyd emphasize the point in the title of their valuable and thoughtful book *Not by Genes Alone*, although they give reasons for not adopting the word 'meme' itself, preferring 'cultural variants'. Stephen Shennan's *Genes, Memes and Human History* was partly inspired by an earlier excellent book by Boyd and Richerson, *Culture and the Evolutionary Process.* Other book-length treatments of memes include Robert Aunger's *The Electric Meme*, Kate Distin's *The Selfish Meme*, and *Virus of the Mind: The New Science of the Meme* by Richard Brodie.

But it is Susan Blackmore, in *The Meme Machine*, who has pushed memetic theory further than anyone. She repeatedly visualizes a world full of brains (or other receptacles or conduits, such as computers or radio frequency bands) and memes jostling to occupy them. As with genes in a gene pool, the memes that prevail will be the ones that are good at getting themselves copied. This may be because they have direct appeal, as, presumably, the immortality meme has for some people. Or it may be because they flourish in the presence of other memes that have already become numerous in the meme pool. This gives rise to meme complexes or 'memeplexes'. As usual with memes, we gain understanding by going back to the genetic origin of the analogy.

For didactic purposes, I treated genes as though they were isolated units, acting independently. But of course they are not independent of one another, and this fact shows itself in two ways. First, genes are linearly strung along chromosomes, and so tend to travel through generations in the company of particular other genes that occupy neighbouring chromosomal loci. We doctors call that kind of linkage *linkage*, and I shall say no more about it because memes don't have chromosomes, alleles or sexual recombination. The other respect in which genes are not independent is very different from genetic linkage, and here there is a good memetic analogy. It concerns embryology which – the fact is often misunderstood – is completely distinct from genetics. Bodies are not jigsawed together as mosaics of phenotypic pieces, each one contributed by a different gene. There is no one-to-one mapping between genes and units of anatomy or behaviour. Genes 'collaborate' with hundreds of other genes in programming the developmental *processes* that culminate in a body, in the same kind of way as the words of a recipe collaborate in a cookery process that culminates in a dish. It is not the case that each word of the recipe corresponds to a different morsel of the dish.

Genes, then, cooperate in cartels to build bodies, and that is one of the important principles of embryology. It is tempting to say that natural selection favours cartels of genes in a kind of group selection between alternative cartels. That is confusion. What really happens is that the other genes of the gene pool constitute a major part of the *environment* in which each gene is selected versus its alleles. Because each is selected to be successful in the presence of the others – which are also being selected in a similar way – cartels of cooperating genes *emerge*. We have here something more like a free market than a planned economy. There is a butcher and a baker, but perhaps a gap in the market for a candlestick maker. The invisible hand of natural selection fills the gap. That is different from having a central planner who favours the troika of butcher + baker + candlestick maker. The idea of cooperating cartels assembled by the invisible hand will turn out to be central to our understanding of religious memes and how they work.

Different kinds of gene cartel emerge in different gene pools. Carnivore gene pools have genes that program prey-detecting sense organs, prey-catching claws, carnassial teeth, meat-digesting enzymes and many other genes, all fine-tuned to cooperate with each other. At the same time, in herbivore gene pools, different sets of mutually compatible genes are favoured for their cooperation with each other. We are familiar with the idea that a gene is favoured for the compatibility of its phenotype with the external environment of the species: desert, woodland or whatever it is. The point I am now making is that it is also favoured for its compatibility with the other genes of its particular gene pool. A carnivore gene would not survive in a herbivore gene pool, and vice versa. In the long gene's-eye-view, the gene pool of the species – the set of genes that are shuffled and reshuffled by sexual reproduction – constitutes the genetic environment in which each gene is selected for its capacity to cooperate. Although meme pools are less regimented and structured than gene pools, we can still speak of a meme pool as an important part of the 'environment' of each meme in the memeplex.

A memeplex is a set of memes which, while not necessarily being good survivors on their own, are good survivors in the presence of other members of the memeplex. In the previous section I doubted that the details of language evolution are favoured by any kind of natural selection. I guessed that language evolution is instead governed by random drift. It is just conceivable that certain vowels or consonants carry better than others through mountainous terrain, and therefore might become characteristic of, say, Swiss, Tibetan and Andean dialects, while other sounds are suitable for whispering in dense forests and are therefore characteristic of Pygmy and Amazonian languages. But the one example I cited of language being naturally selected – the theory that the Great Vowel Shift might have a functional explanation – is not of this type. Rather, it has to do with memes fitting in with mutually compatible memeplexes. One vowel shifted first, for reasons unknown – perhaps fashionable imitation of an admired or powerful individual, as is alleged to be the origin of the Spanish lisp. Never mind how the Great Vowel Shift started: according to this theory, once the first vowel had changed, other vowels had to shift in its train, to reduce ambiguity, and so on in cascade. In this second stage of the process, memes were selected against the background of already existing meme pools, building up a new memeplex of mutually compatible memes.

We are finally equipped to turn to the memetic theory of religion. Some religious ideas, like some genes, might survive because of absolute merit. These memes would survive in any meme pool, regardless of the other memes that surround them. (I must repeat the vitally important point that 'merit' in this sense means only 'ability to survive in the pool'. It carries no value judgement apart from that.) Some religious ideas survive because they are compatible with other memes that are already numerous in the meme pool – as part of a memeplex. The following is a partial list of religious memes that might plausibly have survival value in the meme pool, either because of absolute 'merit' or because of compatibility with an existing memeplex:

- You will survive your own death.
- If you die a martyr, you will go to an especially wonderful part of paradise where you will enjoy seventy-two virgins (spare a thought for the unfortunate virgins).
- Heretics, blasphemers and apostates should be killed (or otherwise punished, for example by ostracism from their families).
- Belief in God is a supreme virtue. If you find your belief wavering, work hard at restoring it, and beg God to help your unbelief. (In my discussion of Pascal's Wager I mentioned the odd assumption that the one thing God really wants of us is belief. At the time I treated it as an oddity. Now we have an explanation for it.)

- Faith (belief without evidence) is a virtue. The more your beliefs defy the evidence, the more virtuous you are. Virtuoso believers who can manage to believe something really weird, unsupported and insupportable, in the teeth of evidence and reason, are especially highly rewarded.
- Everybody, even those who do not hold religious beliefs, must respect them with a higher level of automatic and unquestioned respect than that accorded to other kinds of belief (we met this in Chapter 1).
- There are some weird things (such as the Trinity, transubstantiation, incarnation) that we are not *meant* to understand. Don't even *try* to understand one of these, for the attempt might destroy it. Learn how to gain fulfilment in calling it a *mystery*. Remember Martin Luther's virulent condemnations of reason, quoted on page 221, and think how protective of meme survival they would be.
- Beautiful music, art and scriptures are themselves self-replicating tokens of religious ideas.*

Some of the above list probably have absolute survival value and would flourish in any memeplex. But, as with genes, some memes survive only against the right background of other memes, leading to the build-up of alternative memeplexes. Two different religions might be seen as two alternative memeplexes. Perhaps Islam is analogous to a carnivorous gene complex, Buddhism to a herbivorous one. The ideas of one religion are not 'better' than those of the other in any absolute

* Different schools and genres of art can be analysed as alternative memeplexes, as artists copy ideas and motifs from earlier artists, and new motifs survive only if they mesh with others. Indeed, the whole academic discipline of History of Art, with its sophisticated tracing of iconographies and symbolisms, could be seen as an elaborate study in memeplexity. Details will have been favoured or disfavoured by the presence of existing members of the meme pool, and these will often include religious memes.

sense, any more than carnivorous genes are 'better' than herbivorous ones. Religious memes of this kind don't necessarily have any absolute aptitude for survival; nevertheless, they are good in the sense that they flourish in the presence of other memes of their own religion, but not in the presence of memes of the other religion. On this model, Roman Catholicism and Islam, say, were not necessarily designed by individual people, but evolved separately as alternative collections of memes that flourish in the presence of other members of the same memeplex.

Organized religions are organized by people: by priests and bishops, rabbis, imams and ayatollahs. But, to reiterate the point I made with respect to Martin Luther, that doesn't mean they were conceived and designed by people. Even where religions have been exploited and manipulated to the benefit of powerful individuals, the strong possibility remains that the detailed form of each religion has been largely shaped by unconscious evolution. Not by genetic natural selection, which is too slow to account for the rapid evolution and divergence of religions. The role of genetic natural selection in the story is to provide the brain, with its predilections and biases – the hardware platform and low-level system software which form the background to memetic selection. Given this background, memetic natural selection of some kind seems to me to offer a plausible account of the detailed evolution of particular religions. In the early stages of a religion's evolution, before it becomes organized, simple memes survive by virtue of their universal appeal to human psychology. This is where the meme theory of religion and the psychological by-product theory of religion overlap. The later stages, where a religion becomes organized, elaborate and arbitrarily different from other religions, are quite well handled by the theory of memeplexes – cartels of mutually compatible memes. This doesn't rule out the additional role of deliberate manipulation by priests and others. Religions probably are, at least in part, intelligently designed, as are schools and fashions in art.

One religion that was intelligently designed, almost in its entirety, is Scientology, but I suspect that it is exceptional. Another candidate for a purely designed religion is Mormonism. Joseph Smith, its enterprisingly mendacious inventor, went to the lengths of composing a complete new holy book, the Book of Mormon, inventing from scratch a whole new bogus American history, written in bogus seventeenth-century English. Mormonism, however, has evolved since it was fabricated in the nineteenth century and has now become one of the respectable mainstream religions of America – indeed, it claims to be the fastest-growing one, and there is talk of fielding a presidential candidate.

Most religions evolve. Whatever theory of religious evolution we adopt, it has to be capable of explaining the astonishing speed with which the process of religious evolution, given the right conditions, can take off. A case study follows.

CARGO CULTS

In *The Life of Brian*, one of the many things the Monty Python team got right was the extreme rapidity with which a new religious cult can get started. It can spring up almost overnight and then become incorporated into a culture, where it plays a disquietingly dominant role. The 'cargo cults' of Pacific Melanesia and New Guinea provide the most famous real life example. The entire history of some of these cults, from initiation to expiry, is wrapped up within living memory. Unlike the cult of Jesus, the origins of which are not reliably attested, we can see the whole course of events laid out before our eyes (and even here, as we shall see, some details are now lost). It is fascinating to guess that the cult of Christianity almost certainly began in very much the same way, and spread initially at the same high speed.

My main authority for the cargo cults is David Attenborough's *Quest in Paradise*, which he very kindly presented to me. The pattern is the same for all of them, from the earliest cults in the nineteenth century to the more famous ones that grew up in the aftermath of the Second World War. It seems that in every case the islanders were bowled over by the wondrous possessions of the white immigrants to their islands, including administrators, soldiers and missionaries. They were perhaps the victims of (Arthur C.) Clarke's Third Law, which I quoted in Chapter 2: 'Any sufficiently advanced technology is indistinguishable from magic.'

The islanders noticed that the white people who enjoyed these wonders never made them themselves. When articles needed repairing they were sent away, and new ones kept arriving as 'cargo' in ships or, later, planes. No white man was ever seen to make or repair anything, nor indeed did they do anything that could be recognized as useful work of any kind (sitting behind a desk shuffling papers was obviously some kind of religious devotion). Evidently, then, the 'cargo' must be of supernatural origin. As if in corroboration of this, the white men did do certain things that could only have been ritual ceremonies:

> They build tall masts with wires attached to them; they sit listening to small boxes that glow with light and emit curious noises and strangled voices; they persuade the local people to dress up in identical clothes, and march them up and down – and it would hardly be possible to devise a more useless occupation than that. And then the native realizes that he has stumbled on the answer to the mystery. It is these incomprehensible actions that are the rituals employed by the white man to persuade the gods to send the cargo. If the native wants the cargo, then he too must do these things.

It is striking that similar cargo cults sprang up independently on islands that were widely separated both geographically and culturally. David Attenborough tells us that

> Anthropologists have noted two separate outbreaks in New Caledonia, four in the Solomons, four in Fiji, seven in the New Hebrides, and over fifty in New Guinea, most of them being quite independent and unconnected with one another. The majority of these religions claim that one particular messiah will bring the cargo when the day of the apocalypse arrives.

The independent flowering of so many independent but similar cults suggests some unifying features of human psychology in general.

One famous cult on the island of Tanna in the New Hebrides (known as Vanuatu since 1980) is still extant. It is centred on a messianic figure called John Frum. References to John Frum in official government records go back only as far as 1940 but, even for so recent a myth, it is not known for certain whether he ever existed as a real man. One legend described him as a little man with a high-pitched voice and bleached hair, wearing a coat with shining buttons. He made strange prophecies, and he went out of his way to turn the people against the missionaries. Eventually he returned to the ancestors, after promising a triumphal second coming, bearing bountiful cargo. His apocalyptic vision included a 'great cataclysm; the mountains would fall flat and the valleys would be filled;* old people would regain their youth and sickness would vanish; the white people would be expelled from the island never to return; and cargo would arrive in great quantity so that everybody would have as much as he wanted'.

Most worryingly for the government, John Frum also prophesied that, on his second coming, he would bring a new coinage, stamped with the image of a coconut. The people must

---

* Compare Isaiah 40: 4: 'Every valley shall be exalted, and every mountain and hill shall be made low.' This similarity doesn't necessarily indicate any fundamental feature of the human psyche, or Jungian 'collective unconscious'. These islands had long been infested with missionaries.

therefore get rid of all their money of the white man's currency. In 1941 this led to a wild spending spree; the people stopped working and the island's economy was seriously damaged. The colonial administrators arrested the ringleaders but nothing that they could do would kill the cult, and the mission churches and schools became deserted.

A little later, a new doctrine grew up that John Frum was King of America. Providentially, American troops arrived in the New Hebrides around this time and, wonder of wonders, they included black men who were not poor like the islanders but

> as richly endowed with cargo as the white soldiers. Wild excitement overwhelmed Tanna. The day of the apocalypse was imminent. It seemed that everyone was preparing for the arrival of John Frum. One of the leaders said that John Frum would be coming from America by aeroplane and hundreds of men began to clear the bush in the centre of the island so that the plane might have an airstrip on which to land.

The airstrip had a bamboo control tower with 'air traffic controllers' wearing dummy headphones made of wood. There were dummy planes on the 'runway' to act as decoys, designed to lure down John Frum's plane.

In the 1950s, the young David Attenborough sailed to Tanna with a cameraman, Geoffrey Mulligan, to investigate the cult of John Frum. They found plenty of evidence of the religion and were eventually introduced to its high priest, a man called Nambas. Nambas referred to his messiah familiarly as John, and claimed to speak regularly to him, by 'radio'. This ('radio belong John') consisted of an old woman with an electric wire around her waist who would fall into a trance and talk gibberish, which Nambas interpreted as the words of John Frum. Nambas claimed to have known in advance that Attenborough was coming to see him, because John Frum had told him on the 'radio'. Attenborough asked to see the 'radio' but was (understandably) refused. He changed the subject and asked whether Nambas had seen John Frum:

> Nambas nodded vigorously. 'Me see him plenty time.'
>
> 'What does he look like?'
>
> Nambas jabbed his finger at me. ''E look like you. 'E got white face. 'E tall man. 'E live 'long South America.'

This detail contradicts the legend referred to above that John Frum was a short man. Such is the way with evolving legends.

It is believed that the day of John Frum's return will be 15 February, but the year is unknown. Every year on 15 February his followers assemble for a religious ceremony to welcome him. So far he has not returned, but they are not downhearted. David Attenborough said to one cult devotee, called Sam:

> 'But, Sam, it is nineteen years since John say that the cargo will come. He promise and he promise, but still the cargo does not come. Isn't nineteen years a long time to wait?'
>
> Sam lifted his eyes from the ground and looked at me. 'If you can wait two thousand years for Jesus Christ to come an' 'e no come, then I can wait more than nineteen years for John.'

Robert Buckman's book *Can We Be Good Without God?* quotes the same admirable retort by a John Frum disciple, this time to a Canadian journalist some forty years after David Attenborough's encounter.

The Queen and Prince Philip visited the area in 1974, and the Prince subsequently became deified in a rerun of a John-Frum-type cult (once again, note how rapidly the details in religious evolution can change). The Prince is a handsome man who would have cut an imposing figure in his white naval uniform and plumed helmet, and it is perhaps not surprising that he, rather than the Queen, was elevated in this way, quite apart from the fact that the culture of the islanders made it difficult for them to accept a female deity.

I don't want to make too much of the cargo cults of the South Pacific. But they do provide a fascinating contemporary model for the way religions spring up from almost nothing. In particular, they suggest four lessons about the origin of religions generally, and I'll set them out briefly here. First is the amazing speed with which a cult can spring up. Second is the speed with which the origination process covers its tracks. John Frum, if he existed at all, did so within living memory. Yet, even for so recent a possibility, it is not certain whether he lived at all. The third lesson springs from the independent emergence of similar cults on different islands. The systematic study of these similarities can tell us something about human psychology and its susceptibility to religion. Fourth, the cargo cults are similar, not just to each other but to older religions. Christianity and other ancient religions that have spread worldwide presumably began as local cults like that of John Frum. Indeed, scholars such as Geza Vermes, Professor of Jewish Studies at Oxford University, have suggested that Jesus was one of many such charismatic figures who emerged in Palestine around his time, surrounded by similar legends. Most of those cults died away. The one that survived, on this view, is the one that we encounter today. And, as the centuries go by, it has been honed by further evolution (memetic selection, if you like that way of putting it; not if you don't) into the sophisticated system – or rather diverging sets of descendant systems – that dominate large parts of the world today. The deaths of charismatic modern figures such as Haile Selassie, Elvis Presley and Princess Diana offer other opportunities to study the rapid rise of cults and their subsequent memetic evolution.

That is all I want to say about the roots of religion itself, apart from a brief reprise in Chapter 10 when I discuss the 'imaginary friend' phenomenon of childhood under the heading of the psychological 'needs' that religion fulfils.

Morality is often thought to have its roots in religion, and in the next chapter I want to question this view. I shall argue that the origin of morality can itself be the subject of a Darwinian question. Just as we asked: What is the Darwinian survival value of religion?, so we can ask the same question of morality. Morality, indeed, probably predated religion. Just as with religion we drew back from the question and rephrased it, so with morality we shall find that it is best seen as a *by-product* of something else.

CHAPTER 6

The roots of morality: why are we good?

Strange is our situation here on Earth. Each of us comes for a short visit, not knowing why, yet sometimes seeming to divine a purpose. From the standpoint of daily life, however, there is one thing we do know: that man is here for the sake of other men – above all for those upon whose smiles and well-being our own happiness depends.

ALBERT EINSTEIN

Many religious people find it hard to imagine how, without religion, one can be good, or would even want to be good. I shall discuss such questions in this chapter. But the doubts go further, and drive some religious people to paroxysms of hatred against those who don't share their faith. This is important, because moral considerations lie hidden behind religious attitudes to other topics that have no real link with morality. A great deal of the opposition to the teaching of evolution has no connection with evolution itself, or with anything scientific, but is spurred on by moral outrage. This ranges from the naïve 'If you teach children that they evolved from monkeys, then they will act like monkeys' to the more sophisticated underlying motivation for the whole 'wedge' strategy of 'intelligent design', as it is mercilessly laid bare by Barbara Forrest and Paul Gross in *Creationism's Trojan Horse: The Wedge of Intelligent Design*.

I receive a large number of letters from readers of my

books,* most of them enthusiastically friendly, some of them helpfully critical, a few nasty or even vicious. And the nastiest of all, I am sorry to report, are almost invariably motivated by religion. Such unchristian abuse is commonly experienced by those who are perceived as enemies of Christianity. Here, for example, is a letter, posted on the Internet and addressed to Brian Flemming, author and director of *The God Who Wasn't There*,[85] a sincere and moving film advocating atheism. Titled 'Burn while we laugh' and dated 21 December 2005, the letter to Flemming reads as follows:

> You've definitely got some nerve. I'd love to take a knife, gut you fools, and scream with joy as your insides spill out in front of you. You are attempting to ignite a holy war in which some day I, and others like me, may have the pleasure of taking action like the above mentioned.

The writer at this point seems to come to a belated recognition that his language is not very Christian, for he goes on, more charitably:

> However, GOD teaches us not to seek vengeance, but to pray for those like you all.

His charity is short-lived, however:

> I'll get comfort in knowing that the punishment GOD will bring to you will be 1000 times worse than anything I can inflict. The best part is that you WILL suffer for eternity for these sins that you're completely ignorant about. The Wrath of GOD will show no mercy. For your sake, I hope the truth is revealed to you before the knife connects with your flesh. Merry CHRISTMAS!!!

* More than I can hope adequately to reply to, for which I apologize.

> PS You people really don't have a clue as to what is in store for you . . . I thank GOD I'm not you.

I find it genuinely puzzling that a mere difference of theological opinion can generate such venom. Here's a sample (original spelling preserved) from the postbag of the Editor of the magazine *Freethought Today*, published by the Freedom from Religion Foundation (FFRF), which campaigns peacefully against the undermining of the constitutional separation of church and state:

> Hello, cheese-eating scumbags. Their are way more of us Christians than you losers. Their is NO separation of church and state and you heathens will lose . . .

What is it with cheese? American friends have suggested to me a connection with the notoriously liberal state of Wisconsin – home of the FFRF and centre of the dairy industry – but surely there must be more to it than that? And how about those French 'cheese-eating surrender-monkeys'? What is the semiotic iconography of cheese? To continue:

> Satan worshiping scum . . . Please die and go to hell . . . I hope you get a painful disease like rectal cancer and die a slow painful death, so you can meet your God, SATAN . . . Hey dude this freedom from religion thing sux . . . So you fags and dykes take it easy and watch where you go cuz whenever you least expect it god will get you . . . If you don't like this country and what it was founded on & for, get the fuck out of it and go straight to hell . . . PS Fuck you, you comunist whore . . . Get your black asses out of the U.S.A. . . . You are without excuse. Creation is more than enough evidence of the LORD JESUS CHRIST'S omnipotent power.

Why not Allah's omnipotent power? Or Lord Brahma's? Or even Yahweh's?

> We will not go quietly away. If in the future that requires violence just remember you brought it on. My rifle is loaded.

Why, I can't help wondering, is God thought to need such ferocious defence? One might have supposed him amply capable of looking after himself. Bear in mind, through all this, that the Editor being abused and threatened so viciously is a gentle and charming young woman.

Perhaps because I don't live in America, most of my hate mail is not quite in the same league, but nor does it display to advantage the charity for which the founder of Christianity was notable. The following, dated May 2005, from a British medical doctor, while it is certainly hateful, strikes me as more tormented than nasty, and reveals how the whole issue of morality is a deep wellspring of hostility towards atheism. After some preliminary paragraphs excoriating evolution (and sarcastically asking whether a 'Negro' is 'still in the process of evolving'), insulting Darwin personally, misquoting Huxley as an anti-evolutionist, and encouraging me to read a book (I have read it) which argues that the world is only eight thousand years old (can he *really* be a doctor?) he concludes:

> Your own books, your prestige in Oxford, everything you love in life, and have ever achieved, are an exercise in total futility . . . Camus' question-challenge becomes inescapable: Why don't we all commit suicide? Indeed, your world view has that sort of effect on students and many others . . . that we all evolved by blind chance, from nothing, and return to nothing. Even if religion were not true, it is better, much, much better, to believe a noble myth, like Plato's, if it leads to peace of mind while we live. But

> *your* world view leads to anxiety, drug addiction, violence, nihilism, hedonism, Frankenstein science, and hell on earth, and World War III . . . I wonder how happy *you* are in your personal relationships? Divorced? Widowed? Gay? Those like you are never happy, or they would not try so hard to prove there *is* no happiness nor meaning in anything.

The sentiment of this letter, if not its tone, is typical of many. Darwinism, this person believes, is inherently nihilistic, teaching that we evolved by blind chance (for the umpteenth time, natural selection is the very *opposite* of a chance process) and are annihilated when we die. As a direct consequence of such alleged negativity, all manner of evils follow. Presumably he didn't *really* mean to suggest that widowhood could follow directly from my Darwinism, but his letter, by this point, had reached that level of frenzied malevolence which I repeatedly recognize among my Christian correspondents. I have devoted a whole book (*Unweaving the Rainbow*) to ultimate meaning, to the poetry of science, and to rebutting, specifically and at length, the charge of nihilistic negativity, so I shall restrain myself here. This chapter is about evil, and its opposite, good; about morality: where it comes from, why we should embrace it, and whether we need religion to do so.

DOES OUR MORAL SENSE HAVE A DARWINIAN ORIGIN?

Several books, including Robert Hinde's *Why Good is Good*, Michael Shermer's *The Science of Good and Evil*, Robert Buckman's *Can We Be Good Without God?*, and Marc Hauser's *Moral Minds*, have argued that our sense of right and wrong can be derived from our Darwinian past. This section is my own version of the argument.

On the face of it, the Darwinian idea that evolution is driven by natural selection seems ill-suited to explain such goodness as we possess, or our feelings of morality, decency, empathy and pity. Natural selection can easily explain hunger, fear and sexual lust, all of which straightforwardly contribute to our survival or the preservation of our genes. But what about the wrenching compassion we feel when we see an orphaned child weeping, an old widow in despair from loneliness, or an animal whimpering in pain? What gives us the powerful urge to send an anonymous gift of money or clothes to tsunami victims on the other side of the world whom we shall never meet, and who are highly unlikely to return the favour? Where does the Good Samaritan in us come from? Isn't goodness incompatible with the theory of the 'selfish gene'? No. This is a common misunderstanding of the theory – a distressing (and, with hindsight, foreseeable) misunderstanding.* It is necessary to put the stress on the right word. The selfish *gene* is the correct emphasis, for it makes the contrast with the selfish organism, say, or the selfish species. Let me explain.

The logic of Darwinism concludes that the unit in the hierarchy of life which survives and passes through the filter of natural selection will tend to be selfish. The units that survive in the world will be the ones that succeeded in surviving at the expense of their rivals at their own level in the hierarchy. That, precisely, is what selfish means in this context. The question is, what is the level of the action? The whole idea of the selfish gene, with the stress properly applied to the last word, is that the unit of natural selection (i.e. the unit of self-interest) is not the selfish organism, nor the selfish group or selfish species or

---

* I was mortified to read in the *Guardian* ('Animal Instincts', 27 May 2006) that *The Selfish Gene* is the favourite book of Jeff Skilling, CEO of the infamous Enron Corporation, and that he derived inspiration of a Social Darwinist character from it. The *Guardian* journalist Richard Conniff gives a good explanation of the misunderstanding: http://money.guardian.co.uk/workweekly/story/0,,1783900,00.html. I have tried to forestall similar misunderstandings in my new preface to the thirtieth-anniversary edition of *The Selfish Gene*, just brought out by Oxford University Press.

selfish ecosystem, but the selfish *gene*. It is the gene that, in the form of information, either survives for many generations or does not. Unlike the gene (and arguably the meme), the organism, the group and the species are not the right kind of entity to serve as a unit in this sense, because they do not make exact copies of themselves, and do not compete in a pool of such self-replicating entities. That is precisely what genes do, and that is the – essentially logical – justification for singling the gene out as the unit of 'selfishness' in the special Darwinian sense of selfish.

The most obvious way in which genes ensure their own 'selfish' survival relative to other genes is by programming individual organisms to be selfish. There are indeed many circumstances in which survival of the individual organism will favour the survival of the genes that ride inside it. But different circumstances favour different tactics. There are circumstances – not particularly rare – in which genes ensure their own selfish survival by influencing organisms to behave altruistically. Those circumstances are now fairly well understood and they fall into two main categories. A gene that programs individual organisms to favour their genetic kin is statistically likely to benefit copies of itself. Such a gene's frequency can increase in the gene pool to the point where kin altruism becomes the norm. Being good to one's own children is the obvious example, but it is not the only one. Bees, wasps, ants, termites and, to a lesser extent, certain vertebrates such as naked mole rats, meerkats and acorn woodpeckers, have evolved societies in which elder siblings care for younger siblings (with whom they are likely to share the genes for doing the caring). In general, as my late colleague W. D. Hamilton showed, animals tend to care for, defend, share resources with, warn of danger, or otherwise show altruism towards close kin because of the statistical likelihood that kin will share copies of the same genes.

The other main type of altruism for which we have a well-worked-out Darwinian rationale is reciprocal altruism ('You scratch my back and I'll scratch yours'). This theory, first introduced to evolutionary biology by Robert Trivers and often expressed in the mathematical language of game theory, does not depend upon shared genes. Indeed, it works just as well, probably even better, between members of widely different species, when it is often called symbiosis. The principle is the basis of all trade and barter in humans too. The hunter needs a spear and the smith wants meat. The asymmetry brokers a deal. The bee needs nectar and the flower needs pollinating. Flowers can't fly so they pay bees, in the currency of nectar, for the hire of their wings. Birds called honeyguides can find bees' nests but can't break into them. Honey badgers (ratels) can break into bees' nests, but lack wings with which to search for them. Honeyguides lead ratels (and sometimes men) to honey by a special enticing flight, used for no other purpose. Both sides benefit from the transaction. A crock of gold may lie under a large stone, too heavy for its discoverer to move. He enlists the help of others even though he then has to share the gold, because without their help he would get none. The living kingdoms are rich in such mutualistic relationships: buffaloes and oxpeckers, red tubular flowers and hummingbirds, groupers and cleaner wrasses, cows and their gut micro-organisms. Reciprocal altruism works because of asymmetries in needs and in capacities to meet them. That is why it works especially well between different species: the asymmetries are greater.

In humans, IOUs and money are devices that permit delays in the transactions. The parties to the trade don't hand over the goods simultaneously but can hold a debt over to the future, or even trade the debt on to others. As far as I know, no non-human animals in the wild have any direct equivalent of money. But memory of individual identity plays the same role more informally. Vampire bats learn which other individuals of their social group can be relied upon to pay their debts (in regurgitated blood) and which individuals cheat. Natural selection favours genes that predispose individuals, in relationships of asymmetric need and opportunity, to give when they can, and to solicit giving when they can't. It also favours tendencies to remember obligations, bear grudges, police exchange relationships and punish cheats who take, but don't give when their turn comes.

For there will always be cheats, and stable solutions to the game-theoretic conundrums of reciprocal altruism always involve an element of punishment of cheats. Mathematical theory allows two broad classes of stable solution to 'games' of this kind. 'Always be nasty' is stable in that, if everybody else is doing it, a single nice individual cannot do better. But there is another strategy which is also stable. ('Stable' means that, once it exceeds a critical frequency in the population, no alternative does better.) This is the strategy, 'Start out being nice, and give others the benefit of the doubt. Then repay good deeds with good, but avenge bad deeds.' In game theory language, this strategy (or family of related strategies) goes under various names, including Tit-for-Tat, Retaliator and Reciprocator. It is evolutionarily stable under some conditions in the sense that, given a population dominated by reciprocators, no single nasty individual, and no single unconditionally nice individual, will do better. There are other, more complicated variants of Tit-for-Tat which can in some circumstances do better.

I have mentioned kinship and reciprocation as the twin pillars of altruism in a Darwinian world, but there are secondary structures which rest atop those main pillars. Especially in human society, with language and gossip, reputation is important. One individual may have a reputation for kindness and generosity. Another individual may have a reputation for unreliability, for cheating and reneging on deals. Another may have a reputation for generosity when trust has been built up, but for ruthless punishment of cheating. The unadorned theory of reciprocal altruism expects animals of any species to base their behaviour upon unconscious responsiveness to such traits in their fellows. In human societies we add the power of language to spread reputations, usually in the form of gossip. You don't need to have suffered personally from X's failure to buy his round at the pub. You hear 'on the grapevine' that X is a tightwad, or – to add an ironic complication to the example – that Y is a terrible gossip. Reputation is important, and biologists can acknowledge a Darwinian survival value in not just being a good reciprocator but fostering a *reputation* as a good reciprocator too. Matt Ridley's *The Origins of Virtue*, as well as being a lucid account of the whole field of Darwinian morality, is especially good on reputation.*

The Norwegian-American economist Thorstein Veblen and, in a rather different way, the Israeli zoologist Amotz Zahavi have added a further fascinating idea. Altruistic giving may be an advertisement of dominance or superiority. Anthropologists know it as the Potlatch Effect, named after the custom whereby rival chieftains of Pacific north-west tribes vie with each other in duels of ruinously generous feasts. In extreme cases, bouts of retaliatory entertaining continue until one side is reduced to penury, leaving the winner not much better off. Veblen's concept of 'conspicuous consumption' strikes a chord with many observers of the modern scene. Zahavi's contribution, unregarded by biologists for many years until vindicated by brilliant mathematical models from the evolutionary theorist Alan Grafen, has been to provide an evolutionary version of the potlatch idea. Zahavi studies Arabian babblers, little brown birds who live in social groups and breed cooperatively. Like many small birds, babblers give warning cries, and they also donate food to each other. A standard Darwinian investigation of such altruistic acts would look, first, for reciprocation and kinship relationships among the birds. When a babbler feeds a companion, is it in the expectation of being fed at a later date?

---

* Reputation is not confined to humans. It has recently been shown to apply to one of the classic cases of reciprocal altruism in animals, the symbiotic relationship between small cleaner fish and their large fish clients. In an ingenious experiment, individual cleaner wrasse, *Labroides dimidiatus*, that had been observed by a would-be client to be diligent cleaners were more likely to be chosen by the client than rival *Labroides* that had been observed neglecting to clean. See R. Bshary and A. S. Grutter, 'Image scoring and cooperation in a cleaner fish mutualism', *Nature* 441, 22 June 2006, 975–8.

Or is the recipient of the favour a close genetic relative? Zahavi's interpretation is radically unexpected. Dominant babblers assert their dominance by feeding subordinates. To use the sort of anthropomorphic language Zahavi delights in, the dominant bird is saying the equivalent of, 'Look how superior I am to you, I can afford to give you food.' Or 'Look how superior I am, I can afford to make myself vulnerable to hawks by sitting on a high branch, acting as a sentinel to warn the rest of the flock feeding on the ground.' The observations of Zahavi and his colleagues suggest that babblers actively compete for the dangerous role of sentinel. And when a subordinate babbler attempts to offer food to a dominant individual, the apparent generosity is violently rebuffed. The essence of Zahavi's idea is that advertisements of superiority are authenticated by their cost. Only a genuinely superior individual can afford to advertise the fact by means of a costly gift. Individuals buy success, for example in attracting mates, through costly demonstrations of superiority, including ostentatious generosity and public-spirited risk-taking.

We now have four good Darwinian reasons for individuals to be altruistic, generous or 'moral' towards each other. First, there is the special case of genetic kinship. Second, there is reciprocation: the repayment of favours given, and the giving of favours in 'anticipation' of payback. Following on from this there is, third, the Darwinian benefit of acquiring a reputation for generosity and kindness. And fourth, if Zahavi is right, there is the particular additional benefit of conspicuous generosity as a way of buying unfakeably authentic advertising.

Through most of our prehistory, humans lived under conditions that would have strongly favoured the evolution of all four kinds of altruism. We lived in villages, or earlier in discrete roving bands like baboons, partially isolated from neighbouring bands or villages. Most of your fellow band members would have been kin, more closely related to you than members of other bands – plenty of opportunities for kin altruism to evolve. And, whether kin or not, you would tend to meet the same individuals again and again throughout your life – ideal conditions for the evolution of reciprocal altruism. Those are also the ideal conditions for building a reputation for altruism, and the very same ideal conditions for advertising conspicuous generosity. By any or all of the four routes, genetic tendencies towards altruism would have been favoured in early humans. It is easy to see why our prehistoric ancestors would have been good to their own in-group but bad – to the point of xenophobia – towards other groups. But why – now that most of us live in big cities where we are no longer surrounded by kin, and where every day we meet individuals whom we are never going to meet again – why are we still so good to each other, even sometimes to others who might be thought to belong to an out-group?

It is important not to mis-state the reach of natural selection. Selection does not favour the evolution of a cognitive awareness of what is good for your genes. That awareness had to wait for the twentieth century to reach a cognitive level, and even now full understanding is confined to a minority of scientific specialists. What natural selection favours is rules of thumb, which work in practice to promote the genes that built them. Rules of thumb, by their nature, sometimes misfire. In a bird's brain, the rule 'Look after small squawking things in your nest, and drop food into their red gapes' typically has the effect of preserving the genes that built the rule, because the squawking, gaping objects in an adult bird's nest are normally its own offspring. The rule misfires if another baby bird somehow gets into the nest, a circumstance that is positively engineered by cuckoos. Could it be that our Good Samaritan urges are misfirings, analogous to the misfiring of a reed warbler's parental instincts when it works itself to the bone for a young cuckoo? An even closer analogy is the human urge to adopt a child. I must rush to add that 'misfiring' is intended only in a strictly Darwinian sense. It carries no suggestion of the pejorative.

The 'mistake' or 'by-product' idea, which I am espousing, works like this. Natural selection, in ancestral times when we lived in small and stable bands like baboons, programmed into our brains altruistic urges, alongside sexual urges, hunger urges, xenophobic urges and so on. An intelligent couple can read their Darwin and know that the ultimate reason for their sexual urges is procreation. They know that the woman cannot conceive because she is on the pill. Yet they find that their sexual desire is in no way diminished by the knowledge. Sexual desire is sexual desire and its force, in an individual's psychology, is independent of the ultimate Darwinian pressure that drove it. It is a strong urge which exists independently of its ultimate rationale.

I am suggesting that the same is true of the urge to kindness – to altruism, to generosity, to empathy, to pity. In ancestral times, we had the opportunity to be altruistic only towards close kin and potential reciprocators. Nowadays that restriction is no longer there, but the rule of thumb persists. Why would it not? It is just like sexual desire. We can no more help ourselves feeling pity when we see a weeping unfortunate (who is unrelated and unable to reciprocate) than we can help ourselves feeling lust for a member of the opposite sex (who may be infertile or otherwise unable to reproduce). Both are misfirings, Darwinian mistakes: blessed, precious mistakes.

Do not, for one moment, think of such Darwinizing as demeaning or reductive of the noble emotions of compassion and generosity. Nor of sexual desire. Sexual desire, when channelled through the conduits of linguistic culture, emerges as great poetry and drama: John Donne's love poems, say, or *Romeo and Juliet*. And of course the same thing happens with the misfired redirection of kin- and reciprocation-based compassion. Mercy to a debtor is, when seen out of context, as un-Darwinian as adopting someone else's child:

> The quality of mercy is not strained.
> It droppeth as the gentle rain from heaven
> Upon the place beneath.

Sexual lust is the driving force behind a large proportion of human ambition and struggle, and much of it constitutes a misfiring. There is no reason why the same should not be true of the lust to be generous and compassionate, if this is the misfired consequence of ancestral village life. The best way for natural selection to build in both kinds of lust in ancestral times was to install rules of thumb in the brain. Those rules still influence us today, even where circumstances make them inappropriate to their original functions.

Such rules of thumb influence us still, not in a Calvinistically deterministic way but filtered through the civilizing influences of literature and custom, law and tradition – and, of course, religion. Just as the primitive brain rule of sexual lust passes through the filter of civilization to emerge in the love scenes of *Romeo and Juliet*, so primitive brain rules of us-versus-them vendetta emerge in the form of the running battles between Capulets and Montagues; while primitive brain rules of altruism and empathy end up in the misfiring that cheers us in the chastened reconciliation of Shakespeare's final scene.

A CASE STUDY IN THE ROOTS OF MORALITY

If our moral sense, like our sexual desire, is indeed rooted deep in our Darwinian past, predating religion, we should expect that research on the human mind would reveal some moral universals, crossing geographical and cultural barriers, and also, crucially, religious barriers. The Harvard biologist Marc Hauser, in his book *Moral Minds: How Nature Designed our Universal Sense of Right and Wrong*, has enlarged upon a fruitful line of thought experiments originally suggested by moral philosophers. Hauser's study will serve the additional purpose of introducing the way moral philosophers think. A hypothetical moral dilemma is posed, and the difficulty we experience in answering it tells us something about our sense of right and wrong. Where Hauser goes beyond the philosophers is that he actually does statistical surveys and psychological experiments, using questionnaires on the Internet, for example, to investigate the moral sense of real people. From the present point of view, the interesting thing is that most people come to the same decisions when faced with these dilemmas, and their agreement over the decisions themselves is stronger than their ability to articulate their reasons. This is what we should expect if we have a moral sense which is built into our brains, like our sexual instinct or our fear of heights or, as Hauser himself prefers to say, like our capacity for language (the details vary from culture to culture, but the underlying deep structure of grammar is universal). As we shall see, the way people respond to these moral tests, and their inability to articulate their reasons, seems largely independent of their religious beliefs or lack of them. The message of Hauser's book, to anticipate it in his own words, is this: 'Driving our moral judgments is a universal moral grammar, a faculty of the mind that evolved over millions of years to include a set of principles for building a range of possible moral systems. As with language, the principles that make up our moral grammar fly beneath the radar of our awareness.'

Typical of Hauser's moral dilemmas are variations on the theme of a runaway truck or 'trolley' on a railway line which threatens to kill a number of people. The simplest story imagines a person, Denise, standing by a set of points and in a position to divert the trolley onto a siding, thereby saving the lives of five people trapped on the main line ahead. Unfortunately there is a man trapped on the siding. But since he is only one, outnumbered by the five people trapped on the main track, most people agree that it is morally permissible, if not obligatory, for Denise to throw the switch and save the five by killing the one. We ignore hypothetical possibilities such as that the one man on the siding might be Beethoven, or a close friend.

Elaborations of the thought experiment present a series of increasingly teasing moral conundrums. What if the trolley can be stopped by dropping a large weight in its path from a bridge overhead? That's easy: obviously we must drop the weight. But what if the only large weight available is a very fat man sitting on the bridge, admiring the sunset? Almost everybody agrees that it is immoral to push the fat man off the bridge, even though, from one point of view, the dilemma might seem parallel to Denise's, where throwing the switch kills one to save five. Most of us have a strong intuition that there is a crucial difference between the two cases, though we may not be able to articulate what it is.

Pushing the fat man off the bridge is reminiscent of another dilemma considered by Hauser. Five patients in a hospital are dying, each with a different organ failing. Each would be saved if a donor could be found for their particular faulty organ, but none is available. Then the surgeon notices that there is a healthy man in the waiting-room, all five of whose organs are in good working order and suitable for transplanting. In this case, almost nobody can be found who is prepared to say that the moral act is to kill the one to save the five.

As with the fat man on the bridge, the intuition that most of us share is that an innocent bystander should not suddenly be dragged into a bad situation and used for the sake of others without his consent. Immanuel Kant famously articulated the principle that a rational being should never be used as merely an unconsenting means to an end, even the end of benefiting others. This seems to provide the crucial difference between the case of the fat man on the bridge (or the man in the hospital waiting-room) and the man on Denise's siding. The fat man on the bridge is being positively used as the means to stop the runaway trolley. This clearly violates the Kantian principle. The person on the siding is not being used to save the lives of the five people on the line. It is the siding that is being used, and he just has the bad luck to be standing on it. But, when you put the distinction like that, why does it satisfy us? For Kant, it was a moral absolute. For Hauser it is built into us by our evolution.

The hypothetical situations involving the runaway trolley become increasingly ingenious, and the moral dilemmas correspondingly tortuous. Hauser contrasts the dilemmas faced by hypothetical individuals called Ned and Oscar. Ned is standing by the railway track. Unlike Denise, who could divert the trolley onto a siding, Ned's switch diverts it onto a side loop which joins the main track again just before the five people. Simply switching the points doesn't help: the trolley will plough into the five anyway when the diversion rejoins the main track. However, as it happens, there is an extremely fat man on the diversionary track who is heavy enough to stop the trolley. Should Ned change the points and divert the train? Most people's intuition is that he should not. But what is the difference between Ned's dilemma, and Denise's? Presumably people are intuitively applying Kant's principle. Denise diverts the trolley from ploughing into the five people, and the unfortunate casualty on the siding is 'collateral damage', to use the charmingly Rumsfeldian phrase. He is not being used by Denise to save the others. Ned is actually *using* the fat man to stop the trolley, and most people (perhaps unthinkingly), along with Kant (thinking it out in great detail), see this as a crucial difference.

The difference is brought out again by the dilemma of Oscar. Oscar's situation is identical to Ned's, except that there is a large iron weight on the diversionary loop of track, heavy enough to stop the trolley. Clearly Oscar should have no problem deciding to pull the points and divert the trolley. Except that there happens to be a hiker walking in front of the iron weight. He will certainly be killed if Oscar pulls the switch, just as surely as Ned's fat man. The difference is that Oscar's hiker is not being used to stop the trolley: he is collateral damage, as in Denise's dilemma. Like Hauser, and like most of Hauser's experimental subjects, I feel that Oscar is permitted to throw the switch but Ned is not. But I also find it quite hard to justify my intuition. Hauser's point is that such moral intuitions are often not well thought out but that we feel them strongly anyway, because of our evolutionary heritage.

In an intriguing venture into anthropology, Hauser and his colleagues adapted their moral experiments to the Kuna, a small Central American tribe with little contact with Westerners and no formal religion. The researchers changed the 'trolley on a line' thought experiment to locally suitable equivalents, such as crocodiles swimming towards canoes. With corresponding minor differences, the Kuna show the same moral judgements as the rest of us.

Of particular interest for this book, Hauser also wondered whether religious people differ from atheists in their moral intuitions. Surely, if we get our morality from religion, they should differ. But it seems that they don't. Hauser, working with the moral philosopher Peter Singer,[86] focused on three hypothetical dilemmas and compared the verdicts of atheists with those of religious people. In each case, the subjects were asked to choose whether a hypothetical action is morally 'obligatory', 'permissible' or 'forbidden'. The three dilemmas were:

1 Denise's dilemma. Ninety per cent of people said it was permissible to divert the trolley, killing the one to save the five.

2 You see a child drowning in a pond and there is no other help in sight. You can save the child, but your trousers will be ruined in the process. Ninety-seven per cent agreed that you should save the child (amazingly, 3 per cent apparently would prefer to save their trousers).

3 The organ transplant dilemma described above. Ninety-seven per cent of subjects agreed that it is morally forbidden to seize the healthy person in the waiting-room and kill him for his organs, thereby saving five other people.

The main conclusion of Hauser and Singer's study was that there is no statistically significant difference between atheists and religious believers in making these judgements. This seems compatible with the view, which I and many others hold, that we do not need God in order to be good – or evil.

IF THERE IS NO GOD, WHY BE GOOD?

Posed like that, the question sounds positively ignoble. When a religious person puts it to me in this way (and many of them do), my immediate temptation is to issue the following challenge: 'Do you really mean to tell me the only reason you try to be good is to gain God's approval and reward, or to avoid his disapproval and punishment? That's not morality, that's just sucking up, apple-polishing, looking over your shoulder at the great surveillance camera in the sky, or the still small wiretap inside your head, monitoring your every move, even your every base thought.' As Einstein said, 'If people are good only because they fear punishment, and hope for reward, then we are a sorry lot indeed.' Michael Shermer, in *The Science of Good and Evil*, calls it a debate stopper. If you agree that, in the absence of God, you would 'commit robbery, rape, and murder', you reveal yourself as an immoral person, 'and we would be well advised to steer a wide course around you'. If, on the other hand, you admit that you would continue to be a good person even when not under divine surveillance, you have fatally undermined your claim that God is necessary for us to be good. I suspect that quite a lot of religious people do think religion is what motivates them to be good, especially if they belong to one of those faiths that systematically exploits personal guilt.

It seems to me to require quite a low self-regard to think that, should belief in God suddenly vanish from the world, we would all become callous and selfish hedonists, with no kindness, no charity, no generosity, nothing that would deserve the name of goodness. It is widely believed that Dostoevsky was of that opinion, presumably because of some remarks he put into the mouth of Ivan Karamazov:

> [Ivan] solemnly observed that there was absolutely no law of nature to make man love humanity, and that if love did exist and had existed at all in the world up to now, then it was not by virtue of the

> natural law, but entirely because man believed in his own immortality. He added as an aside that it was precisely that which constituted the natural law, namely, that once man's faith in his own immortality was destroyed, not only would his capacity for love be exhausted, but so would the vital forces that sustained life on this earth. And furthermore, nothing would be immoral then, everything would be permitted, even anthropophagy. And finally, as though all this were not enough, he declared that for every individual, such as you and me, for example, who does not believe either in God or in his own immortality, the natural law is bound immediately to become the complete opposite of the religion-based law that preceded it, and that egoism, even extending to the perpetration of crime, would not only be permissible but would be recognized as the essential, the most rational, and even the noblest *raison d'être* of the human condition.[87]

Perhaps naïvely, I have inclined towards a less cynical view of human nature than Ivan Karamazov. Do we really need policing – whether by God or by each other – in order to stop us from behaving in a selfish and criminal manner? I dearly want to believe that I do not need such surveillance – and nor, dear reader, do you. On the other hand, just to weaken our confidence, listen to Steven Pinker's disillusioning experience of a police strike in Montreal, which he describes in *The Blank Slate*:

> As a young teenager in proudly peaceable Canada during the romantic 1960s, I was a true believer in Bakunin's anarchism. I laughed off my parents' argument that if the government ever laid down its arms all hell would break loose. Our competing predictions were put to the test at 8:00 A.M. on

> October 17, 1969, when the Montreal police went on strike. By 11:20 A.M. the first bank was robbed. By noon most downtown stores had closed because of looting. Within a few more hours, taxi drivers burned down the garage of a limousine service that competed with them for airport customers, a rooftop sniper killed a provincial police officer, rioters broke into several hotels and restaurants, and a doctor slew a burglar in his suburban home. By the end of the day, six banks had been robbed, a hundred shops had been looted, twelve fires had been set, forty carloads of storefront glass had been broken, and three million dollars in property damage had been inflicted, before city authorities had to call in the army and, of course, the Mounties to restore order. This decisive empirical test left my politics in tatters . . .

Perhaps I, too, am a Pollyanna to believe that people would remain good when unobserved and unpoliced by God. On the other hand, the majority of the population of Montreal presumably believed in God. Why didn't the fear of God restrain them when earthly policemen were temporarily removed from the scene? Wasn't the Montreal strike a pretty good natural experiment to test the hypothesis that belief in God makes us good? Or did the cynic H. L. Mencken get it right when he tartly observed: 'People say we need religion when what they really mean is we need police.'

Obviously, not everybody in Montreal behaved badly as soon as the police were off the scene. It would be interesting to know whether there was any statistical tendency, however slight, for religious believers to loot and destroy less than unbelievers. My uninformed prediction would have been opposite. It is often cynically said that there are no atheists in foxholes. I'm inclined to suspect (with some evidence, although it may be simplistic to draw conclusions from it) that there are very few atheists in prisons. I am not necessarily claiming that atheism increases morality, although humanism – the ethical system that often goes with atheism – probably does. Another good possibility is that atheism is correlated with some third factor, such as higher education, intelligence or reflectiveness, which might counteract criminal impulses. Such research evidence as there is certainly doesn't support the common view that religiosity is positively correlated with morality. Correlational evidence is never conclusive, but the following data, described by Sam Harris in his *Letter to a Christian Nation*, are nevertheless striking.

> While political party affiliation in the United States is not a perfect indicator of religiosity, it is no secret that the 'red [Republican] states' are primarily red due to the overwhelming political influence of conservative Christians. If there were a strong correlation between Christian conservatism and societal health, we might expect to see some sign of it in red-state America. We don't. Of the twenty-five cities with the lowest rates of violent crime, 62 percent are in 'blue' [Democrat] states, and 38 percent are in 'red' [Republican] states. Of the twenty-five most dangerous cities, 76 percent are in red states, and 24 percent are in blue states. In fact, three of the five most dangerous cities in the U.S. are in the pious state of Texas. The twelve states with the highest rates of burglary are red. Twenty-four of the twenty-nine states with the highest rates of theft are red. Of the twenty-two states with the highest rates of murder, seventeen are red.*

Systematic research if anything tends to support such correlational data. Gregory S. Paul, in the *Journal of Religion*

---

* Note that these colour conventions in America are exactly the opposite of those in Britain, where blue is the colour of the Conservative Party, and red, as in the rest of the world, is the colour traditionally associated with the political left.

and Society (2005), systematically compared seventeen economically developed nations, and reached the devastating conclusion that 'higher rates of belief in and worship of a creator correlate with higher rates of homicide, juvenile and early mortality, STD infection rates, teen pregnancy and abortion in the prosperous democracies'. Dan Dennett, in *Breaking the Spell*, sardonically comments on such studies generally:

> Needless to say, these results strike so hard at the standard claims of greater moral virtue among the religious that there has been a considerable surge of further research initiated by religious organizations attempting to refute them . . . one thing we can be sure of is that *if* there is a significant positive relationship between moral behaviour and religious affiliation, practice, or belief, it will soon be discovered, since so many religious organizations are eager to confirm their traditional beliefs about this scientifically. (They are quite impressed with the truth-finding power of science when it supports what they already believe.) Every month that passes without such a demonstration underlines the suspicion that it just isn't so.

Most thoughtful people would agree that morality in the absence of policing is somehow more truly moral than the kind of false morality that vanishes as soon as the police go on strike or the spy camera is switched off, whether the spy camera is a real one monitored in the police station or an imaginary one in heaven. But it is perhaps unfair to interpret the question 'If there is no God, why bother to be good?' in such a cynical way.* A religious thinker could offer a more genuinely moral interpretation, along the lines of the following statement from an

---

* H. L. Mencken, again with characteristic cynicism, defined conscience as the inner voice that warns us that someone may be looking.

imaginary apologist. 'If you don't believe in God, you don't believe there are any absolute standards of morality. With the best will in the world you may intend to be a good person, but how do you decide what is good and what is bad? Only religion can ultimately provide your standards of good and evil. Without religion you have to make it up as you go along. That would be morality without a rule book: morality flying by the seat of its pants. If morality is merely a matter of choice, Hitler could claim to be moral by his own eugenically inspired standards, and all the atheist can do is make a personal choice to live by different lights. The Christian, the Jew or the Muslim, by contrast, can claim that evil has an absolute meaning, true for all time and in all places, according to which Hitler was absolutely evil.'

Even if it were true that we need God to be moral, it would of course not make God's existence more likely, merely more desirable (many people cannot tell the difference). But that is not the issue here. My imaginary religious apologist has no need to admit that sucking up to God is the religious motive for doing good. Rather, his claim is that, wherever the *motive* to be good comes from, without God there would be no standard for *deciding* what is good. We could each make up our own definition of good, and behave accordingly. Moral principles that are based only upon religion (as opposed to, say, the 'golden rule', which is often associated with religions but can be derived from elsewhere) may be called absolutist. Good is good and bad is bad, and we don't mess around deciding particular cases by whether, for example, somebody suffers. My religious apologist would claim that only religion can provide a basis for deciding what is good.

Some philosophers, notably Kant, have tried to derive absolute morals from non-religious sources. Though a religious man himself, as was almost inevitable in his time,* Kant tried

---

* This is the standard interpretation of Kant's views. However, the noted philosopher A. C. Grayling has plausibly argued (*New Humanist*, July–Aug. 2006) that, although Kant publicly went along with the religious conventions of his time, he was really an atheist.

to base a morality on duty for duty's sake, rather than for God's. His famous categorical imperative enjoins us to 'act only on that maxim whereby thou canst at the same time will that it should become a universal law'. This works tidily for the example of telling lies. Imagine a world in which people told lies as a matter of principle, where lying was regarded as a good and moral thing to do. In such a world, lying itself would cease to have any meaning. Lying needs a presumption of truth for its very definition. If a moral principle is something we should wish everybody to follow, lying cannot be a moral principle because the principle itself would break down in meaninglessness. Lying, as a rule for life, is inherently unstable. More generally, selfishness, or free-riding parasitism on the goodwill of others, may work for me as a lone selfish individual and give me personal satisfaction. But I cannot wish that everybody would adopt selfish parasitism as a moral principle, if only because then I would have nobody to parasitize.

The Kantian imperative seems to work for truth-telling and some other cases. It is not so easy to see how to broaden it to morality generally. Kant notwithstanding, it is tempting to agree with my hypothetical apologist that absolutist morals are usually driven by religion. Is it always wrong to put a terminally ill patient out of her misery at her own request? Is it always wrong to make love to a member of your own sex? Is it always wrong to kill an embryo? There are those who believe so, and their grounds are absolute. They brook no argument or debate. Anybody who disagrees deserves to be shot: metaphorically of course, not literally – except in the case of some doctors in American abortion clinics (see next chapter). Fortunately, however, morals do not have to be absolute.

Moral philosophers are the professionals when it comes to thinking about right and wrong. As Robert Hinde succinctly put it, they agree that 'moral precepts, while not necessarily constructed by reason, should be defensible by reason'.[88] They classify themselves in many ways, but in modern terminology the major divide is between 'deontologists' (such as Kant) and 'consequentialists' (including 'utilitarians' such as Jeremy Bentham, 1748–1832). Deontology is a fancy name for the belief that morality consists in the obeying of rules. It is literally the science of duty, from the Greek for 'that which is binding'. Deontology is not quite the same thing as moral absolutism, but for most purposes in a book about religion there is no need to dwell on the distinction. Absolutists believe there are absolutes of right and wrong, imperatives whose rightness makes no reference to their consequences. Consequentialists more pragmatically hold that the morality of an action should be judged by its consequences. One version of consequentialism is utilitarianism, the philosophy associated with Bentham, his friend James Mill (1773–1836) and Mill's son John Stuart Mill (1806–73). Utilitarianism is often summed up in Bentham's unfortunately imprecise catchphrase: 'the greatest happiness of the greatest number is the foundation of morals and legislation'.

Not all absolutism is derived from religion. Nevertheless, it is pretty hard to defend absolutist morals on grounds other than religious ones. The only competitor I can think of is patriotism, especially in times of war. As the distinguished Spanish film director Luis Buñuel said, 'God and Country are an unbeatable team; they break all records for oppression and bloodshed.' Recruiting officers rely heavily on their victims' sense of patriotic duty. In the First World War, women handed out white feathers to young men not in uniform.

> Oh, we don't want to lose you, but we think you ought to go,
> For your King and your country both need you so.

People despised conscientious objectors, even those of the enemy country, because patriotism was held to be an absolute virtue. It is hard to get much more absolute than the 'My country right or wrong' of the professional soldier, for the slogan commits you to kill whomever the politicians of some future date might choose to call enemies. Consequentialist reasoning may influence the political decision to go to war but, once war is declared, absolutist patriotism takes over with a force and a power not otherwise seen outside religion. A soldier who allows his own thoughts of consequentialist morality to persuade him not to go over the top would likely find himself court-martialled and even executed.

The springboard for this discussion of moral philosophy was a hypothetical religious claim that, without a God, morals are relative and arbitrary. Kant and other sophisticated moral philosophers apart, and with due recognition given to patriotic fervour, the preferred source of absolute morality is usually a holy book of some kind, interpreted as having an authority far beyond its history's capacity to justify. Indeed, adherents of scriptural authority show distressingly little curiosity about the (normally highly dubious) historical origins of their holy books. The next chapter will demonstrate that, in any case, people who claim to derive their morals from scripture do not really do so in practice. And a very good thing too, as they themselves, on reflection, should agree.

CHAPTER 7

The 'Good' Book and the changing moral *Zeitgeist*

Politics has slain its thousands, but religion has slain its tens of thousands.

SEAN O'CASEY

There are two ways in which scripture might be a source of morals or rules for living. One is by direct instruction, for example through the Ten Commandments, which are the subject of such bitter contention in the culture wars of America's boondocks. The other is by example: God, or some other biblical character, might serve as – to use the contemporary jargon – a role model. Both scriptural routes, if followed through religiously (the adverb is used in its metaphoric sense but with an eye to its origin), encourage a system of morals which any civilized modern person, whether religious or not, would find – I can put it no more gently – obnoxious.

To be fair, much of the Bible is not systematically evil but just plain weird, as you would expect of a chaotically cobbled-together anthology of disjointed documents, composed, revised, translated, distorted and 'improved' by hundreds of anonymous authors, editors and copyists, unknown to us and mostly unknown to each other, spanning nine centuries.[89] This may explain some of the sheer strangeness of the Bible. But unfortunately it is this same weird volume that religious zealots hold up to us as the inerrant source of our morals and rules for living. Those who wish to base their morality literally on the Bible have either not read it or not understood it, as Bishop John Shelby Spong, in *The Sins of Scripture*, rightly observed. Bishop Spong, by the way, is a nice example of a liberal bishop whose beliefs are so advanced as to be almost unrecognizable to the majority of those who call themselves Christians. A British counterpart is Richard Holloway, recently retired as Bishop of Edinburgh. Bishop Holloway even describes himself as a 'recovering Christian'. I had a public discussion with him in Edinburgh, which was one of the most stimulating and interesting encounters I have had.[90]

The Old Testament

Begin in Genesis with the well-loved story of Noah, derived from the Babylonian myth of Uta-Napisthim and known from the older mythologies of several cultures. The legend of the animals going into the ark two by two is charming, but the moral of the story of Noah is appalling. God took a dim view of humans, so he (with the exception of one family) drowned the lot of them including children and also, for good measure, the rest of the (presumably blameless) animals as well.

Of course, irritated theologians will protest that we don't take the book of Genesis literally any more. But that is my whole point! We pick and choose which bits of scripture to believe, which bits to write off as symbols or allegories. Such picking and choosing is a matter of personal decision, just as much, or as little, as the atheist's decision to follow this moral precept or that was a personal decision, without an absolute foundation. If one of these is 'morality flying by the seat of its pants', so is the other.

In any case, despite the good intentions of the sophisticated theologian, a frighteningly large number of people still do take their scriptures, including the story of Noah, literally. According to Gallup, they include approximately 50 per cent of the US electorate. Also, no doubt, many of those Asian holy men who blamed the 2004 tsunami not on a plate tectonic shift but on human sins,[91] ranging from drinking and dancing in bars to breaking some footling sabbath rule. Steeped in the story of Noah, and ignorant of all except biblical learning, who can blame them? Their whole education has led them to view natural disasters as bound up with human affairs, paybacks for human misdemeanours rather than anything so impersonal as plate tectonics. By the way, what presumptuous egocentricity to believe that earth-shaking events, on the scale at which a god (or a tectonic plate) might operate, must always have a human connection. Why should a divine being, with creation and eternity on his mind, care a fig for petty human malefactions? We humans give ourselves such airs, even aggrandizing our poky little 'sins' to the level of cosmic significance!

When I interviewed for television the Reverend Michael Bray, a prominent American anti-abortion activist, I asked him why evangelical Christians were so obsessed with private sexual inclinations such as homosexuality, which didn't interfere with anybody else's life. His reply invoked something like self-defence. Innocent citizens are at risk of becoming collateral damage when God chooses to strike a town with a natural disaster because it houses sinners. In 2005, the fine city of New Orleans was catastrophically flooded in the aftermath of a hurricane, Katrina. The Reverend Pat Robertson, one of America's best-known televangelists and a former presidential candidate, was reported as blaming the hurricane on a lesbian comedian who happened to live in New Orleans.* You'd think

---

* It is unclear whether the story, which originated at http://datelinehollywood.com/archives/2005/09/05/robertson-blames-hurricane-on-choice-of-ellen-deneres-to-host-emmys/ [link no longer active, 2016] is true. Whether true or not, it is widely believed, no doubt because it is entirely typical of utterances by evangelical clergy, including Robertson, on disasters such as Katrina. See, for example, http://thinkprogress.org/lgbt/2012/10/29/1104901/anti-gay-preacher-blames-hurricane-sandy-on-homosexuality-and-marriage-equality/. The website that says the Katrina story is untrue (www.snopes.com/katrina/satire/robertson.asp) also quotes Robertson as saying, of an earlier Gay Pride march in Orlando, Florida, 'I would warn Orlando that you're right in the way of some serious hurricanes, and I don't think I'd be waving those flags in God's face if I were you.'

an omnipotent God would adopt a slightly more targeted approach to zapping sinners: a judicious heart attack, perhaps, rather than the wholesale destruction of an entire city just because it happened to be the domicile of one lesbian comedian.

In November 2005, the citizens of Dover, Pennsylvania voted off their local school board the entire slate of fundamentalists who had brought the town notoriety, not to say ridicule, by attempting to enforce the teaching of 'intelligent design'. When Pat Robertson heard that the fundamentalists had been democratically defeated at the ballot, he offered a stern warning to Dover:

> I'd like to say to the good citizens of Dover, if there is a disaster in your area, don't turn to God. You just rejected him from your city, and don't wonder why he hasn't helped you when problems begin, if they begin, and I'm not saying they will. But if they do, just remember you just voted God out of your city. And if that's the case, then don't ask for his help, because he might not be there.[92]

Pat Robertson would be harmless comedy, were he less typical of those who today hold power and influence in the United States.

In the destruction of Sodom and Gomorrah, the Noah equivalent, chosen to be spared with his family because he was uniquely righteous, was Abraham's nephew Lot. Two male angels were sent to Sodom to warn Lot to leave the city before the brimstone arrived. Lot hospitably welcomed the angels into his house, whereupon all the men of Sodom gathered around and demanded that Lot should hand the angels over so that they could (what else?) sodomize them: 'Where are the men which came in to thee this night? Bring them out unto us, that we may know them' (Genesis 19: 5). Yes, 'know' has the Authorized Version's usual euphemistic meaning, which is very funny in the context. Lot's gallantry in refusing the demand suggests that God might have been onto something when he singled him out as the only good man in Sodom. But Lot's halo is tarnished by the terms of his refusal: 'I pray you, brethren, do not so wickedly. Behold now, I have two daughters which have not known man; let me, I pray you, bring them out unto you, and do ye to them as is good in your eyes: only unto these men do nothing; for therefore came they under the shadow of my roof' (Genesis 19: 7–8).

Whatever else this strange story might mean, it surely tells us something about the respect accorded to women in this intensely religious culture. As it happened, Lot's bargaining away of his daughters' virginity proved unnecessary, for the angels succeeded in repelling the marauders by miraculously striking them blind. They then warned Lot to decamp immediately with his family and his animals, because the city was about to be destroyed. The whole household escaped, with the exception of Lot's unfortunate wife, whom the Lord turned into a pillar of salt because she committed the offence – comparatively mild, one might have thought – of looking over her shoulder at the fireworks display.

Lot's two daughters make a brief reappearance in the story. After their mother was turned into a pillar of salt, they lived with their father in a cave up a mountain. Starved of male company, they decided to make their father drunk and copulate with him. Lot was beyond noticing when his elder daughter arrived in his bed or when she left, but he was not too drunk to impregnate her. The next night the two daughters agreed it was the younger one's turn. Again Lot was too drunk to notice, and he impregnated her too (Genesis 19: 31–6). If this dysfunctional family was the best Sodom had to offer by way of morals, some might begin to feel a certain sympathy with God and his judicial brimstone.

The story of Lot and the Sodomites is eerily echoed in chapter 19 of the book of Judges, where an unnamed Levite (priest) was travelling with his concubine in Gibeah. They spent the night in the house of a hospitable old man. While they were eating their supper, the men of the city came and beat on the door, demanding that the old man should hand over his male guest 'so that we may know him'. In almost exactly the same words as Lot, the old man said: 'Nay, my brethren, nay, I pray you, do not so wickedly; seeing that this man is come into mine house do not this folly. Behold, here is my daughter a maiden, and his concubine; them I will bring out now, and humble ye them, and do with them what seemeth good unto you; but unto this man do not so vile a thing' (Judges 19: 23–4). Again, the misogynistic ethos comes through, loud and clear. I find the phrase 'humble ye them' particularly chilling. Enjoy yourselves by humiliating and raping my daughter and this priest's concubine, but show a proper respect for my guest who is, after all, male. In spite of the similarity between the two stories, the *dénouement* was less happy for the Levite's concubine than for Lot's daughters.

The Levite handed her over to the mob, who gang-raped her all night: 'They knew her and abused her all the night until the morning: and when the day began to spring, they let her go. Then came the woman in the dawning of the day, and fell down at the door of the man's house where her lord was, till it was light' (Judges 19: 25–6). In the morning, the Levite found his concubine lying prostrate on the doorstep and said – with what we today might see as callous abruptness – 'Up, and let us be going.' But she didn't move. She was dead. So he 'took a knife, and laid hold on his concubine, and divided her, together with her bones, into twelve pieces, and sent her into all the coasts of Israel'. Yes, you read correctly. Look it up in Judges 19: 29. Let's charitably put it down again to the ubiquitous weirdness of the Bible. Actually, it is not quite as loopy as it sounds. There was a motive – to provoke revenge – and it succeeded, for the incident provoked a war of vengeance against the tribe of Benjamin in which, so Judges chapter 20 lovingly records, more than 60,000 men were killed. The story of the Levite's concubine is so similar to that of Lot, one can't help wondering whether a fragment of manuscript became accidentally misplaced in some long-forgotten scriptorium: an illustration of the erratic provenance of sacred texts.

Lot's uncle Abraham was the founding father of all three 'great' monotheistic religions. His patriarchal status renders him only somewhat less likely than God to be taken as a role model. But what modern moralist would wish to follow him? Relatively early in his long life, Abraham went to Egypt to tough out a famine with his wife Sarah. He realized that such a beautiful woman would be desirable to the Egyptians and that therefore his own life, as her husband, might be endangered. So he decided to pass her off as his sister. In this capacity she was taken into Pharaoh's harem, and Abraham consequently became rich in Pharaoh's favour. God disapproved of this cosy arrangement, and sent plagues on Pharaoh and his house (why not on Abraham?). An understandably aggrieved Pharaoh demanded to know why Abraham had not told him Sarah was his wife. He then handed her back to Abraham and kicked them both out of Egypt (Genesis 12: 18–19). Weirdly, it seems that the couple later tried to pull the same stunt again, this time with Abimelech the King of Gerar. He too was induced by Abraham to marry Sarah, again having been led to believe she was Abraham's sister, not his wife (Genesis 20: 2–5). He too expressed his indignation, in almost identical terms to Pharaoh's, and one can't help sympathizing with both of them. Is the similarity another indicator of textual unreliability?

Such unpleasant episodes in Abraham's story are mere peccadilloes compared with the infamous tale of the sacrificing of his son Isaac (Muslim scripture tells the same story about Abraham's other son, Ishmael). God ordered Abraham to make a burnt offering of his longed-for son. Abraham built an altar, put firewood upon it, and trussed Isaac up on top of the wood. His murdering knife was already in his hand when an angel dramatically intervened with the news of a last-minute change of plan: God was only joking after all, 'tempting' Abraham, and testing his faith. A modern moralist cannot help but wonder how a child could ever recover from such psychological trauma. By the standards of modern morality, this disgraceful story is an example simultaneously of child abuse, bullying in two asymmetrical power relationships, and the first recorded use of the Nuremberg defence: 'I was only obeying orders.' Yet the legend is one of the great foundational myths of all three monotheistic religions.

Once again, modern theologians will protest that the story of Abraham sacrificing Isaac should not be taken as literal fact. And, once again, the appropriate response is twofold. First, many many people, even to this day, do take the whole of their scripture to be literal fact, and they have a great deal of political power over the rest of us, especially in the United States and in the Islamic world. Second, if not as literal fact, how should we take the story? As an allegory? Then an allegory for what? Surely nothing praiseworthy. As a moral lesson? But what kind of morals could one derive from this appalling story? Remember, all I am trying to establish for the moment is that we do not, as a matter of fact, derive our morals from scripture. Or, if we do, we pick and choose among the scriptures for the nice bits and reject the nasty. But then we must have some independent criterion for deciding which are the moral bits: a criterion which, wherever it comes from, cannot come from scripture itself and is presumably available to all of us whether we are religious or not.

Apologists even seek to salvage some decency for the God character in this deplorable tale. Wasn't it good of God to spare Isaac's life at the last minute? In the unlikely event that any of my readers are persuaded by this obscene piece of special pleading, I refer them to another story of human sacrifice, which ended more unhappily. In Judges, chapter 11, the military leader Jephthah made a bargain with God that, if God would guarantee Jephthah's victory over the Ammonites, Jephthah would, without fail, sacrifice as a burnt offering 'whatsoever cometh forth of the doors of my house to meet me, when I return'. Jephthah did indeed defeat the Ammonites ('with a very great slaughter', as is par for the course in the book of Judges) and he returned home victorious. Not surprisingly, his daughter, his only child, came out of the house to greet him (with timbrels and dances) and – alas – she was the first living thing to do so. Understandably Jephthah rent his clothes, but there was nothing he could do about it. God was obviously looking forward to the promised burnt offering, and in the circumstances the daughter very decently agreed to be sacrificed. She asked only that she should be allowed to go into the mountains for two months to bewail her virginity. At the end of this time she meekly returned, and Jephthah cooked her. God did not see fit to intervene on this occasion.

God's monumental rage whenever his chosen people flirted with a rival god resembles nothing so much as sexual jealousy of the worst kind, and again it should strike a modern moralist as far from good role-model material. The temptation to sexual infidelity is readily understandable even to those who do not succumb, and it is a staple of fiction and drama, from Shakespeare to bedroom farce. But the apparently irresistible temptation to whore with foreign gods is something we moderns find harder to empathize with. To my naïve eyes, 'Thou shalt have no other gods but me' would seem an easy enough commandment to keep: a doddle, one might think, compared with 'Thou shalt not covet thy neighbour's wife'. Or her ass. (Or her ox.) Yet throughout the Old Testament, with the same predictable regularity as in bedroom farce, God had only to turn his back for a moment and the Children of Israel would be off and at it with Baal, or some trollop of a graven image.* Or, on one calamitous occasion, a golden calf . . .

Moses, even more than Abraham, is a likely role model for followers of all three monotheistic religions. Abraham may be the original patriarch, but if anybody should be called the

* This richly comic idea was suggested to me by Jonathan Miller who, surprisingly, never included it in a *Beyond the Fringe* sketch. I also thank him for recommending the scholarly book upon which it is based: Halbertal and Margalit (1992).

doctrinal founder of Judaism and its derivative religions, it is Moses. On the occasion of the golden calf episode, Moses was safely out of the way up Mount Sinai, communing with God and getting tablets of stone graven by him. The people down below (who were on pain of death to refrain from so much as *touching* the mountain) didn't waste any time:

> When the people saw that Moses delayed to come down out of the mount, the people gathered themselves together unto Aaron, and said unto him, Up, make us gods, which shall go before us; for as for this Moses, the man that brought us up out of the land of Egypt, we wot not what is become of him. (Exodus 32: 1)

Aaron got everybody to pool their gold, melted it down and made a golden calf, for which newly invented deity he then built an altar so they could all start sacrificing to it.

Well, they should have known better than to fool around behind God's back like that. He might be up a mountain but he was, after all, omniscient and he lost no time in dispatching Moses as his enforcer. Moses raced hotfoot down the mountain, carrying the stone tablets on which God had written the Ten Commandments. When he arrived and saw the golden calf he was so furious that he dropped the tablets and broke them (God later gave him a replacement set, so that was all right). Moses seized the golden calf, burned it, ground it to powder, mixed it with water and made the people swallow it. Then he told everybody in the priestly tribe of Levi to pick up a sword and kill as many people as possible. This amounted to about three thousand which, one might have hoped, would have been enough to assuage God's jealous sulk. But no, God wasn't finished yet. In the last verse of this terrible chapter his parting shot was to send a plague upon what was left of the people 'because they made the calf, which Aaron made'.

The book of Numbers tells how God incited Moses to attack

the Midianites. His army made short work of slaying all the men, and they burned all the Midianite cities, but they spared the women and children. This merciful restraint by his soldiers infuriated Moses, and he gave orders that all the boy children should be killed, and all the women who were not virgins. 'But all the women children, that have not known a man by lying with him, keep alive for yourselves' (Numbers 31: 18). No, Moses was not a great role model for modern moralists.

In so far as modern religious writers attach any kind of symbolic or allegorical meaning to the massacre of the Midianites, the symbolism is aimed in precisely the wrong direction. The unfortunate Midianites, so far as one can tell from the biblical account, were the victims of genocide in their own country. Yet their name lives on in Christian lore only in that favourite hymn (which I can still sing from memory after fifty years, to two different tunes, both in grim minor keys):

> Christian, dost thou see them
> On the holy ground?
> How the troops of Midian
> Prowl and prowl around?
> Christian, up and smite them,
> Counting gain but loss;
> Smite them by the merit
> Of the holy cross.

Alas, poor slandered, slaughtered Midianites, to be remembered only as poetic symbols of universal evil in a Victorian hymn.

The rival god Baal seems to have been a perennially seductive tempter to wayward worship. In Numbers, chapter 25, many of the Israelites were lured by Moabite women to sacrifice to Baal. God reacted with characteristic fury. He ordered Moses to 'Take all the heads of the people and hang them up before the Lord against the sun, that the fierce anger of the Lord may be turned away from Israel.' One cannot help, yet again, marvelling at the extraordinarily draconian view taken of the sin of flirting with rival gods. To our modern sense of values and justice it seems a trifling sin compared to, say, offering your daughter for a gang rape. It is yet another example of the disconnect between scriptural and modern (one is tempted to say civilized) morals. Of course, it is easily enough understood in terms of the theory of memes, and the qualities that a deity needs in order to survive in the meme pool.

The tragi-farce of God's maniacal jealousy against alternative gods recurs continually throughout the Old Testament. It motivates the first of the Ten Commandments (the ones on the tablets that Moses broke: Exodus 20, Deuteronomy 5), and it is even more prominent in the (otherwise rather different) substitute commandments that God provided to replace the broken tablets (Exodus 34). Having promised to drive out of their homelands the unfortunate Amorites, Canaanites, Hittites, Perizzites, Hivites and Jebusites, God gets down to what really matters: rival *gods*!

> . . . ye shall destroy their altars, break their images, and cut down their groves. For thou shalt worship no other god: for the Lord, whose name is Jealous, is a jealous God. Lest thou make a covenant with the inhabitants of the land, and they go a whoring after their gods, and do sacrifice unto their gods, and one call thee, and thou eat of his sacrifice; And thou take of their daughters unto thy sons, and their daughters go a whoring after their gods, and make thy sons go a whoring after their gods. Thou shalt make thee no molten gods (Exodus 34: 13–17)

I know, yes, of course, of course, times have changed, and no religious leader today (apart from the likes of the Taliban or the American Christian equivalent) thinks like Moses. But that is my whole point. All I am establishing is that modern morality, wherever else it comes from, does not come from the Bible.

Apologists cannot get away with claiming that religion provides them with some sort of inside track to defining what is good and what is bad – a privileged source unavailable to atheists. They cannot get away with it, not even if they employ that favourite trick of interpreting selected scriptures as 'symbolic' rather than literal. By what criterion do you *decide* which passages are symbolic, which literal?

The ethnic cleansing begun in the time of Moses is brought to bloody fruition in the book of Joshua, a text remarkable for the bloodthirsty massacres it records and the xenophobic relish with which it does so. As the charming old song exultantly has it, 'Joshua fit the battle of Jericho, and the walls came a-tumbling down . . . There's none like good old Joshuay, at the battle of Jericho.' Good old Joshua didn't rest until 'they utterly destroyed all that was in the city, both man and woman, young and old, and ox, and sheep, and ass, with the edge of the sword' (Joshua 6: 21).

Yet again, theologians will protest, it didn't happen. Well, no – the story has it that the walls came tumbling down at the mere sound of men shouting and blowing horns, so indeed it didn't happen – but that is not the point. The point is that, whether true or not, the Bible is held up to us as the source of our morality. And the Bible story of Joshua's destruction of Jericho, and the invasion of the Promised Land in general, is morally indistinguishable from Hitler's invasion of Poland, or Saddam Hussein's massacres of the Kurds and the Marsh Arabs. The Bible may be an arresting and poetic work of fiction, but it is not the sort of book you should give your children to form their morals. As it happens, the story of Joshua in Jericho is the subject of an interesting experiment in child morality, to be discussed later in this chapter.

Do not think, by the way, that the God character in the story nursed any doubts or scruples about the massacres and genocides that accompanied the seizing of the Promised Land. On the contrary, his orders, for example in Deuteronomy 20, were ruthlessly explicit. He made a clear distinction between the people who lived in the land that was needed, and those who lived a long way away. The latter should be invited to surrender peacefully. If they refused, all the men were to be killed and the women carried off for breeding. In contrast to this relatively humane treatment, see what was in store for those tribes unfortunate enough to be already in residence in the promised *Lebensraum*: 'But of the cities of these people, which the Lord thy God doth give thee for an inheritance, thou shalt save alive nothing that breatheth: But thou shalt utterly destroy them; namely, the Hittites, and the Amorites, the Canaanites, and the Perizzites, the Hivites and the Jebusites; as the Lord thy God hath commanded thee.'

Do those people who hold up the Bible as an inspiration to moral rectitude have the slightest notion of what is actually written in it? The following offences merit the death penalty, according to Leviticus 20: cursing your parents; committing adultery; making love to your stepmother or your daughter-in-law; homosexuality; marrying a woman and her daughter; bestiality (and, to add injury to insult, the unfortunate beast is to be killed too). You also get executed, of course, for working on the sabbath: the point is made again and again throughout the Old Testament. In Numbers 15, the children of Israel found a man in the wilderness gathering sticks on the forbidden day. They arrested him and then asked God what to do with him. As it turned out, God was in no mood for half-measures that day. 'And the Lord said unto Moses, The man shall surely be put to death: all the congregation shall stone him with stones without the camp. And all the congregation brought him without the camp, and stoned him with stones, and he died.' Did this harmless gatherer of firewood have a wife and children to grieve for him? Did he whimper with fear as the first stones flew, and scream with pain as the fusillade crashed into his head? What shocks me today about such stories is not that they really happened. They probably didn't. What makes my jaw drop is that people today should base their lives on such an appalling role model as Yahweh – and, even worse, that they should bossily try to force the same evil monster (whether fact or fiction) on the rest of us.

The political power of America's Ten Commandment tablet-toters is especially regrettable in that great republic whose constitution, after all, was drawn up by men of the Enlightenment in explicitly secular terms. If we took the Ten Commandments seriously, we would rank the worship of the wrong gods, and the making of graven images, as first and second among sins. Rather than condemn the unspeakable vandalism of the Taliban, who dynamited the 150-foot-high Bamiyan Buddhas in the mountains of Afghanistan, we would praise them for their righteous piety. What we think of as their vandalism was certainly motivated by sincere religious zeal. This is vividly attested by a truly bizarre story, which was the lead in the (London) *Independent* of 6 August 2005. Under the front-page headline, 'The destruction of Mecca', the *Independent* reported:

> Historic Mecca, the cradle of Islam, is being buried in an unprecedented onslaught by religious zealots. Almost all of the rich and multi-layered history of the holy city is gone . . . Now the actual birthplace of the Prophet Muhammad is facing the bulldozers, with the connivance of Saudi religious authorities whose hardline interpretation of Islam is compelling them to wipe out their own heritage . . . The motive behind the destruction is the Wahhabists' fanatical fear that places of historical and religious interest could give rise to idolatry or polytheism, the worship of multiple and potentially equal gods. The practice of idolatry in Saudi Arabia remains, in principle, punishable by beheading.*

* 'We all fund this torrent of Saudi bigotry' by Johann Hari is an exposé of the insidious influence of Saudi Wahhabism in Britain today. Originally published in the *Independent* on 8 Feb. 2007, it is reproduced on various websites including http://richarddawkins.net.

I do not believe there is an atheist in the world who would bulldoze Mecca – or Chartres, York Minster or Notre Dame, the Shwe Dagon, the temples of Kyoto or, of course, the Buddhas of Bamiyan. As the Nobel Prize-winning American physicist Steven Weinberg said, 'Religion is an insult to human dignity. With or without it, you'd have good people doing good things and evil people doing evil things. But for good people to do evil things, it takes religion.' Blaise Pascal (he of the wager) said something similar: 'Men never do evil so completely and cheerfully as when they do it from religious conviction.'

My main purpose here has not been to show that we *shouldn't* get our morals from scripture (although that is my opinion). My purpose has been to demonstrate that we (and that includes most religious people) as a matter of fact *don't* get our morals from scripture. If we did, we would strictly observe the sabbath and think it just and proper to execute anybody who chose not to. We would stone to death any new bride who couldn't prove she was a virgin, if her husband pronounced himself unsatisfied with her. We would execute disobedient children. We would . . . but wait. Perhaps I have been unfair. Nice Christians will have been protesting throughout this section: everyone knows the Old Testament is pretty unpleasant. The New Testament of Jesus undoes the damage and makes it all right. Doesn't it?

Is the New Testament any better?

Well, there's no denying that, from a moral point of view, Jesus is a huge improvement over the cruel ogre of the Old Testament. Indeed Jesus, if he existed (or whoever wrote his script if he didn't) was surely one of the great ethical innovators of history. The Sermon on the Mount is way ahead of its time. His 'turn the other cheek' anticipated Gandhi and Martin Luther King by two thousand years. It was not for nothing that I wrote an article called 'Atheists for Jesus' (and was later delighted to be presented with a T-shirt bearing the legend).[93]

But the moral superiority of Jesus precisely bears out my point. Jesus was not content to derive his ethics from the scriptures of his upbringing. He explicitly departed from them, for example when he deflated the dire warnings about breaking the sabbath. 'The sabbath was made for man, not man for the sabbath' has been generalized into a wise proverb. Since a principal thesis of this chapter is that we do not, and should not, derive our morals from scripture, Jesus has to be honoured as a model for that very thesis.

Jesus' family values, it has to be admitted, were not such as one might wish to focus on. He was short, to the point of brusqueness, with his own mother, and he encouraged his disciples to abandon their families to follow him. 'If any man come to me and hate not his father, and mother, and wife, and children, and brethren, and sisters, yea and his own life also, he cannot be my disciple.' The American comedian Julia Sweeney expressed her bewilderment in her one-woman stage show, *Letting Go of God*:[94] 'Isn't that what cults do? Get you to reject your family in order to inculcate you?'[95]

Notwithstanding his somewhat dodgy family values, Jesus' ethical teachings were – at least by comparison with the ethical disaster area that is the Old Testament – admirable; but there are other teachings in the New Testament that no good person should support. I refer especially to the central doctrine of Christianity: that of 'atonement' for 'original sin'. This teaching, which lies at the heart of New Testament theology, is almost as morally obnoxious as the story of Abraham setting out to barbecue Isaac, which it resembles – and that is no accident, as Geza Vermes makes clear in *The Changing Faces of Jesus*. Original sin itself comes straight from the Old Testament myth of Adam and Eve. Their sin – eating the fruit of a forbidden tree – seems mild enough to merit a mere reprimand. But the symbolic nature of the fruit (knowledge of good and evil, which in practice turned out to be knowledge that they were naked) was enough to turn their scrumping escapade into the mother and father of all sins.* They and all their descendants were banished forever from the Garden of Eden, deprived of the gift of eternal life, and condemned to generations of painful labour, in the field and in childbirth respectively.

So far, so vindictive: par for the Old Testament course. New Testament theology adds a new injustice, topped off by a new sadomasochism whose viciousness even the Old Testament barely exceeds. It is, when you think about it, remarkable that a religion should adopt an instrument of torture and execution as its sacred symbol, often worn around the neck. Lenny Bruce rightly quipped that 'If Jesus had been killed twenty years ago, Catholic school children would be wearing little electric chairs around their necks instead of crosses.' But the theology and punishment-theory behind it is even worse. The sin of Adam and Eve is thought to have passed down the male line – transmitted in the semen according to Augustine. What kind of ethical philosophy is it that condemns every child, even before it is born, to inherit the sin of a remote ancestor? Augustine, by the way, who rightly regarded himself as something of a personal authority on sin, was responsible for coining the phrase 'original sin'. Before him it was known as 'ancestral sin'. Augustine's pronouncements and debates epitomize, for me, the unhealthy preoccupation of early Christian theologians with sin. They could have devoted their pages and their sermons to extolling the sky splashed with stars, or mountains and green forests, seas and dawn choruses. These are occasionally mentioned, but the Christian focus is overwhelmingly on sin sin sin sin sin sin sin. What a nasty little preoccupation to have dominating your life. Sam Harris is magnificently scathing

---

* I am aware that 'scrumping' will not be familiar to American readers. But I enjoy reading unfamiliar American words and looking them up to broaden my vocabulary. I have deliberately used a few other region-specific words for this reason. Scrumping itself is a *mot juste* of unusual economy. It doesn't just mean stealing: it specifically means stealing *apples* and *only* apples. It is hard for a *mot* to get more *juste* than that. Admittedly the Genesis story doesn't specify that the fruit was an apple, but tradition has long held it so.

in his *Letter to a Christian Nation*: 'Your principal concern appears to be that the Creator of the universe will take offense at something people do while naked. This prudery of yours contributes daily to the surplus of human misery.'

But now, the sado-masochism. God incarnated himself as a man, Jesus, in order that he should be tortured and executed in *atonement* for the hereditary sin of Adam. Ever since Paul expounded this repellent doctrine, Jesus has been worshipped as the *redeemer* of all our sins. Not just the past sin of Adam: *future* sins as well, whether future people decided to commit them or not!

As another aside, it has occurred to various people, including Robert Graves in his epic novel *King Jesus*, that poor Judas Iscariot has received a bad deal from history, given that his 'betrayal' was a necessary part of the cosmic plan. The same could be said of Jesus' alleged murderers. If Jesus wanted to be betrayed and then murdered, in order that he could redeem us all, isn't it rather unfair of those who consider themselves redeemed to take it out on Judas and on Jews down the ages? I have already mentioned the long list of non-canonical gospels. A manuscript purporting to be the lost Gospel of Judas has recently been translated and has received publicity in consequence.[96] The circumstances of its discovery are disputed, but it seems to have turned up in Egypt some time in the 1970s or 60s. It is in Coptic script on sixty-two pages of papyrus, carbon-dated to around AD 300 but probably based on an earlier Greek manuscript. Whoever the author was, the gospel is seen from the point of view of Judas Iscariot and makes the case that Judas betrayed Jesus only because Jesus asked him to play that role. It was all part of the plan to get Jesus crucified so that he could redeem humankind. Obnoxious as that doctrine is, it seems to compound the unpleasantness that Judas has been vilified ever since.*

---

* Too late for the hardback edition of this book, *Reading Judas* by Elaine Pagels and Karen L. King has now appeared. Based on Karen King's translation of the Gospel of Judas, it takes a sympathetic view of that alleged arch-traitor (who appears in the third person in the gospel itself).

I have described atonement, the central doctrine of Christianity, as vicious, sado-masochistic and repellent. We should also dismiss it as barking mad, but for its ubiquitous familiarity which has dulled our objectivity. If God wanted to forgive our sins, why not just forgive them, without having himself tortured and executed in payment – thereby, incidentally, condemning remote future generations of Jews to pogroms and persecution as 'Christ-killers': did that hereditary sin pass down in the semen too?

Paul, as the Jewish scholar Geza Vermes makes clear, was steeped in the old Jewish theological principle that without blood there is no atonement.[97] The author of the Epistle to the Hebrews (9: 22) said as much. Progressive ethicists today find it hard to defend any kind of retributive theory of punishment, let alone the scapegoat theory – executing an innocent to pay for the sins of the guilty. In any case (one can't help wondering), who was God trying to impress? Presumably himself – judge and jury as well as execution victim. To cap it all, Adam, the supposed perpetrator of the original sin, never existed in the first place: an awkward fact – excusably unknown to Paul but presumably known to an omniscient God (and Jesus, if you believe he was God?) – which fundamentally undermines the premise of the whole tortuously nasty theory. Oh, but of course, the story of Adam and Eve was only ever *symbolic*, wasn't it? *Symbolic?* So, in order to impress himself, Jesus had himself tortured and executed, in vicarious punishment for a *symbolic* sin committed by a *non-existent* individual? As I said, barking mad, as well as viciously unpleasant.

Before leaving the Bible, I need to call attention to one particularly unpalatable aspect of its ethical teaching. Christians seldom realize that much of the moral consideration for others which is apparently promoted by both the Old and New Testaments was originally intended to apply only to a narrowly defined in-group. 'Love thy neighbour' didn't mean what we now think it means. It meant only 'Love another Jew.' The point is devastatingly made by the American physician and evolutionary anthropologist John Hartung. He has written a remarkable paper on the evolution and biblical history of in-group morality, laying stress, too, on the flip side – out-group hostility.

LOVE THY NEIGHBOUR

John Hartung's black humour is evident from the outset,[98] where he tells of a Southern Baptist initiative to count the number of Alabamans in hell. As reported in the *New York Times* and *Newsday* the final total, 1.86 million, was estimated using a secret weighting formula whereby Methodists are more likely to be saved than Roman Catholics, while 'virtually everyone not belonging to a church congregation was counted among the lost'. The preternatural smugness of such people is reflected today in the various 'rapture' websites, where the author always takes it completely for granted that he will be among those who 'disappear' into heaven when the 'end times' come. Here is a typical example, from the author of 'Rapture Ready', one of the more odiously sanctimonious specimens of the genre: 'If the rapture should take place, resulting in my absence, it will become necessary for tribulation saints to mirror or financially support this site.'*

Hartung's interpretation of the Bible suggests that it offers no grounds for such smug complacency among Christians. Jesus limited his in-group of the saved strictly to Jews, in which respect he was following the Old Testament tradition, which was all he knew. Hartung clearly shows that 'Thou shalt not kill' was never intended to mean what we now think it means. It meant, very specifically, thou shalt not kill Jews. And all those commandments that make reference to 'thy neighbour' are equally exclusive. 'Neighbour' means fellow Jew. Moses Maimonides, the highly respected twelfth-century rabbi and physician, expounds the full meaning of 'Thou shalt not kill' as

* You may not know the meaning of 'tribulation saints' in this sentence. Don't bother: you have better things to do.

follows: 'If one slays a single Israelite, he transgresses a negative commandment, for Scripture says, Thou shalt not murder. If one murders wilfully in the presence of witnesses, he is put to death by the sword. Needless to say, one is not put to death if he kills a heathen.' Needless to say!

Hartung quotes the Sanhedrin (the Jewish Supreme Court, headed by the high priest) in similar vein, as exonerating a man who hypothetically killed an Israelite by mistake, while intending to kill an animal or a heathen. This teasing little moral conundrum raises a nice point. What if he were to throw a stone into a group of nine heathens and one Israelite and have the misfortune to kill the Israelite? Hm, difficult! But the answer is ready. 'Then his non-liability can be inferred from the fact that the majority were heathens.'

Hartung uses many of the same biblical quotations as I have used in this chapter, about the conquest of the Promised Land by Moses, Joshua and the Judges. I was careful to concede that religious people don't think in a biblical way any more. For me, this demonstrated that our morals, whether we are religious or not, come from another source; and that other source, whatever it is, is available to all of us, regardless of religion or lack of it. But Hartung tells of a horrifying study by the Israeli psychologist George Tamarin. Tamarin presented to more than a thousand Israeli schoolchildren, aged between eight and fourteen, the account of the battle of Jericho in the book of Joshua:

> Joshua said to the people, 'Shout; for the LORD has given you the city. And the city and all that is within it shall be devoted to the LORD for destruction . . . But all silver and gold, and vessels of bronze and iron, are sacred to the LORD; they shall go into the treasury of the LORD.' . . . Then they utterly destroyed all in the city, both men and women, young and old, oxen, sheep, and asses, with the edge of the sword . . . And they burned the city with fire,

> and all within it; only the silver and gold, and the vessels of bronze and of iron, they put into the treasury of the house of the LORD.

Tamarin then asked the children a simple moral question: 'Do you think Joshua and the Israelites acted rightly or not?' They had to choose between A (total approval), B (partial approval) and C (total disapproval). The results were polarized: 66 per cent gave total approval and 26 per cent total disapproval, with rather fewer (8 per cent) in the middle with partial approval. Here are three typical answers from the total approval (A) group:

> In my opinion Joshua and the Sons of Israel acted well, and here are the reasons: God promised them this land, and gave them permission to conquer. If they would not have acted in this manner or killed anyone, then there would be the danger that the Sons of Israel would have assimilated among the Goyim.

> In my opinion Joshua was right when he did it, one reason being that God commanded him to exterminate the people so that the tribes of Israel will not be able to assimilate amongst them and learn their bad ways.

> Joshua did good because the people who inhabited the land were of a different religion, and when Joshua killed them he wiped their religion from the earth.

The justification for the genocidal massacre by Joshua is religious in every case. Even those in category C, who gave total disapproval, did so, in some cases, for backhanded religious reasons. One girl, for example, disapproved of Joshua's conquering Jericho because, in order to do so, he had to enter it:

> I think it is bad, since the Arabs are impure and if one enters an impure land one will also become impure and share their curse.

Two others who totally disapproved did so because Joshua destroyed everything, including animals and property, instead of keeping some as spoil for the Israelites:

> I think Joshua did not act well, as they could have spared the animals for themselves.

> I think Joshua did not act well, as he could have left the property of Jericho; if he had not destroyed the property it would have belonged to the Israelites.

Once again the sage Maimonides, often cited for his scholarly wisdom, is in no doubt where he stands on this issue: 'It is a positive commandment to destroy the seven nations, as it is said: *Thou shalt utterly destroy them.* If one does not put to death any of them that falls into one's power, one transgresses a negative commandment, as it is said: *Thou shalt save alive nothing that breatheth*.'

Unlike Maimonides, the children in Tamarin's experiment were young enough to be innocent. Presumably the savage views they expressed were those of their parents, or the cultural group in which they were brought up. It is, I suppose, not unlikely that Palestinian children, brought up in the same war-torn country, would offer equivalent opinions in the opposite direction. These considerations fill me with despair. They seem to show the immense power of religion, and especially the religious upbringing of children, to divide people and foster historic enmities and hereditary vendettas. I cannot help remarking that two out of Tamarin's three representative quotations from group A mentioned the evils of assimilation, while the third one stressed the importance of killing people in order to stamp out their religion.

Tamarin ran a fascinating control group in his experiment. A different group of 168 Israeli children were given the same text from the book of Joshua, but with Joshua's own name replaced by 'General Lin' and 'Israel' replaced by 'a Chinese kingdom 3,000 years ago'. Now the experiment gave opposite results. Only 7 per cent approved of General Lin's behaviour, and 75 per cent disapproved. In other words, when their loyalty to Judaism was removed from the calculation, the majority of the children agreed with the moral judgements that most modern humans would share. Joshua's action was a deed of barbaric genocide. But it all looks different from a religious point of view. And the difference starts early in life. It was religion that made the difference between children condemning genocide and condoning it.

In the latter half of Hartung's paper, he moves on to the New Testament. To give a brief summary of his thesis, Jesus was a devotee of the same in-group morality – coupled with out-group hostility – that was taken for granted in the Old Testament. Jesus was a loyal Jew. It was Paul who invented the idea of taking the Jewish God to the Gentiles. Hartung puts it more bluntly than I dare: 'Jesus would have turned over in his grave if he had known that Paul would be taking his plan to the pigs.'

Hartung has some good fun with the book of Revelation, which is certainly one of the weirdest books in the Bible. It is supposed to have been written by St John and, as *Ken's Guide to the Bible* neatly put it, if his epistles can be seen as John on pot, then Revelation is John on acid.[99] Hartung draws attention to the two verses in Revelation where the number of those 'sealed' (which some sects, such as the Jehovah's Witnesses, interpret to mean 'saved') is limited to 144,000. Hartung's point is that they all had to be Jews: 12,000 from each of the 12 tribes. Ken Smith goes further, pointing out that the 144,000 elect 'did not defile themselves with women', which presumably means that none of them could *be* women. Well, that's the sort of thing we've come to expect.

There's a lot more in Hartung's entertaining paper. I shall simply recommend it once more, and summarize it in a quotation:

> The Bible is a blueprint of in-group morality, complete with instructions for genocide, enslavement of out-groups, and world domination. But the Bible is not evil by virtue of its objectives or even its glorification of murder, cruelty, and rape. Many ancient works do that – The Iliad, the Icelandic Sagas, the tales of the ancient Syrians and the inscriptions of the ancient Mayans, for example. But no one is selling the Iliad as a foundation for morality. Therein lies the problem. The Bible is sold, and bought, as a guide to how people should live their lives. And it is, by far, the world's all-time best seller.

Lest it be thought that the exclusiveness of traditional Judaism is unique among religions, look at the following confident verse from a hymn by Isaac Watts (1674–1748):

> Lord, I ascribe it to Thy Grace,
> And not to chance, as others do,
> That I was born of Christian Race
> And not a Heathen or a Jew.

What puzzles me about this verse is not the exclusiveness *per se* but the logic. Since plenty of others *were* born into religions other than Christianity, how did God decide which future people should receive such favoured birth? Why favour Isaac Watts and those individuals whom he visualized singing his hymn? In any case, before Isaac Watts was conceived, what was the nature of the entity being favoured? These are deep waters, but perhaps not too deep for a mind tuned to theology. Isaac Watts's hymn is reminiscent of three daily prayers that male Orthodox and Conservative (but not Reform) Jews are taught to recite:

'Blessed are You for not making me a Gentile. Blessed are You for not making me a woman. Blessed are You for not making me a slave.'

Religion is undoubtedly a divisive force, and this is one of the main accusations levelled against it. But it is frequently and rightly said that wars, and feuds between religious groups or sects, are seldom actually about theological disagreements. When an Ulster Protestant paramilitary murders a Catholic, he is not muttering to himself, 'Take that, transubstantiationist, mariolatrous, incense-reeking bastard!' He is much more likely to be avenging the death of another Protestant killed by another Catholic, perhaps in the course of a sustained transgenerational vendetta. Religion is a *label* of in-group/out-group enmity and vendetta, not necessarily worse than other labels such as skin colour, language or preferred football team, but often available when other labels are not.

Yes yes, of course the troubles in Northern Ireland are political. There really has been economic and political oppression of one group by another, and it goes back centuries. There really are genuine grievances and injustices, and these seem to have little to do with religion; except that – and this is important and widely overlooked – without religion there would be no labels by which to decide whom to oppress and whom to avenge. And the real problem in Northern Ireland is that the labels are inherited down many generations. Catholics, whose parents, grandparents and great-grandparents went to Catholic schools, send their children to Catholic schools. Protestants, whose parents, grandparents and great-grandparents went to Protestant schools, send their children to Protestant schools. The two sets of people have the same skin colour, they speak the same language, they enjoy the same things, but they might as well belong to different species, so deep is the historic divide. And without religion, and religiously segregated education, the divide simply would not be there. The warring tribes would have intermarried and long since dissolved into each other. From Kosovo to Palestine, from Iraq to Sudan, from Ulster to the Indian sub-continent, look carefully at any region of the world where you find intractable enmity and violence between rival groups today. I cannot guarantee that you'll find religions as the dominant labels for in-groups and out-groups. But it's a good bet.

In India at the time of partition, more than a million people were massacred in religious riots between Hindus and Muslims (and fifteen million displaced from their homes). There were no badges other than religious ones with which to label whom to kill. Ultimately, there was nothing to divide them but religion. Salman Rushdie was moved by a more recent bout of religious massacres in India to write an article called 'Religion, as ever, is the poison in India's blood'.[100] Here's his concluding paragraph:

> What is there to respect in any of this, or in any of the crimes now being committed almost daily around the world in religion's dreaded name? How well, with what fatal results, religion erects totems, and how willing we are to kill for them! And when we've done it often enough, the deadening of affect that results makes it easier to do it again.
>
> So India's problem turns out to be the world's problem. What happened in India has happened in God's name.
>
> The problem's name is God.

I do not deny that humanity's powerful tendencies towards in-group loyalties and out-group hostilities would exist even in the absence of religion. Fans of rival football teams are an example of the phenomenon writ small. Even football supporters sometimes divide along religious lines, as in the case of Glasgow Rangers and Glasgow Celtic. Languages (as in Belgium), races and tribes (especially in Africa) can be important divisive tokens. But religion amplifies and exacerbates the damage in at least three ways:

- Labelling of children. Children are described as 'Catholic children' or 'Protestant children' etc. from an early age, and certainly far too early for them to have made up their own minds on what they think about religion (I return to this abuse of childhood in Chapter 9).

- Segregated schools. Children are educated, again often from a very early age, with members of a religious in-group and separately from children whose families adhere to other religions. It is not an exaggeration to say that the troubles in Northern Ireland would disappear in a generation if segregated schooling were abolished.

- Taboos against 'marrying out'. This perpetuates hereditary feuds and vendettas by preventing the mingling of feuding groups. Intermarriage, if it were permitted, would naturally tend to mollify enmities.

The village of Glenarm in Northern Ireland is the seat of the Earls of Antrim. On one occasion within living memory, the then Earl did the unthinkable: he married a Catholic. Immediately, in houses throughout Glenarm, the blinds were drawn in mourning. A horror of 'marrying out' is also widespread among religious Jews. Several of the Israeli children quoted above mentioned the dire perils of 'assimilation' at the forefront of their defence of Joshua's Battle of Jericho. When people of different religions do marry, it is described with foreboding on both sides as a 'mixed marriage' and there are often prolonged battles over how the children are to be brought up. When I was a child and still carried a guttering torch for the Anglican Church, I remember being dumbfounded to be told of a rule that when a Roman Catholic married an Anglican, the children were always brought up Catholic. I could readily understand why a priest of either denomination would try to insist on this condition. What I couldn't understand (still can't) was the asymmetry. Why didn't the Anglican priests retaliate with the equivalent rule in reverse? Just less ruthless, I suppose. My old chaplain and Betjeman's 'Our Padre' were simply too nice.

Sociologists have done statistical surveys of religious homogamy (marrying somebody of the same religion) and heterogamy (marrying somebody of a different religion). Norval D. Glenn, of the University of Texas at Austin, gathered a number of such studies up to 1978 and analysed them together.[101] He concluded that there is a significant tendency towards religious homogamy in Christians (Protestants marry Protestants, and Catholics Catholics, and this goes beyond the ordinary 'boy next door effect'), but that it is especially marked among Jews. Out of a total sample of 6,021 married respondents to the questionnaire, 140 called themselves Jews and, of these, 85.7 per cent married Jews. This is hugely greater than the randomly expected percentage of homogamous marriages. And of course it will not come as news to anybody. Observant Jews are strongly discouraged from 'marrying out', and the taboo shows itself in Jewish jokes about mothers warning their boys about blonde shiksas lying in wait to entrap them. Here are typical statements by three American rabbis:

- 'I refuse to officiate at interfaith marriages.'
- 'I officiate when couples state their intention to raise children as Jews.'
- 'I officiate if couples agree to premarital counselling.'

Rabbis who will agree to officiate together with a Christian priest are rare, and much in demand.

Even if religion did no other harm in itself, its wanton and carefully nurtured divisiveness – its deliberate and cultivated pandering to humanity's natural tendency to favour in-groups and shun out-groups – would be enough to make it a significant force for evil in the world.

THE MORAL *ZEITGEIST*

This chapter began by showing that we do not – even the religious among us – ground our morality in holy books, no matter what we may fondly imagine. How, then, do we decide what is right and what is wrong? No matter how we answer that question, there is a consensus about what we do as a matter of fact consider right and wrong: a consensus that prevails surprisingly widely. The consensus has no obvious connection with religion. It extends, however, to most religious people, whether or not they *think* their morals come from scripture. With notable exceptions such as the Afghan Taliban and the American Christian equivalent, most people pay lip service to the same broad liberal consensus of ethical principles. The majority of us don't cause needless suffering; we believe in free speech and protect it even if we disagree with what is being said; we pay our taxes; we don't cheat, don't kill, don't commit incest, don't do things to others that we would not wish done to us. Some of these good principles can be found in holy books, but buried alongside much else that no decent person would wish to follow: and the holy books do not supply any rules for distinguishing the good principles from the bad.

One way to express our consensual ethics is as a 'New Ten Commandments'. Various individuals and institutions have attempted this. What is significant is that they tend to produce rather similar results to each other, and what they produce is characteristic of the times in which they happen to live. Here is one set of 'New Ten Commandments' from today, which I happened to find on an atheist website.[102]

- Do not do to others what you would not want them to do to you.
- In all things, strive to cause no harm.
- Treat your fellow human beings, your fellow living things,

and the world in general with love, honesty, faithfulness and respect.

- Do not overlook evil or shrink from administering justice, but always be ready to forgive wrongdoing freely admitted and honestly regretted.
- Live life with a sense of joy and wonder.
- Always seek to be learning something new.
- Test all things; always check your ideas against the facts, and be ready to discard even a cherished belief if it does not conform to them.
- Never seek to censor or cut yourself off from dissent; always respect the right of others to disagree with you.
- Form independent opinions on the basis of your own reason and experience; do not allow yourself to be led blindly by others.
- Question everything.

This little collection is not the work of a great sage or prophet or professional ethicist. It is just one ordinary web logger's rather endearing attempt to summarize the principles of the good life today, for comparison with the biblical Ten Commandments. It was the first list I found when I typed 'New Ten Commandments' into a search engine, and I deliberately didn't look any further. The whole point is that it is the sort of list that any ordinary, decent person today would come up with. Not everybody would home in on exactly the same list of ten. The philosopher John Rawls might include something like the following: 'Always devise your rules as if you didn't know whether you were going be at the top or the bottom of the pecking order.' An alleged Inuit system for sharing out food is a practical example of the Rawls principle: the individual who cuts up the food gets last pick.

In my own amended Ten Commandments, I would choose some of the above, but I would also try to find room for, among others:

- Enjoy your own sex life (so long as it damages nobody else) and leave others to enjoy theirs in private whatever their inclinations, which are none of your business.
- Do not discriminate or oppress on the basis of sex, race or (as far as possible) species.
- Do not indoctrinate your children. Teach them how to think for themselves, how to evaluate evidence, and how to disagree with you.
- Value the future on a timescale longer than your own.

But never mind these small differences of priority. The point is that we have almost all moved on, and in a big way, since biblical times. Slavery, which was taken for granted in the Bible and throughout most of history, was abolished in civilized countries in the nineteenth century. All civilized nations now accept what was widely denied up to the 1920s, that a woman's vote, in an election or on a jury, is the equal of a man's. In today's enlightened societies (a category that manifestly does not include, for example, Saudi Arabia), women are no longer regarded as property, as they clearly were in biblical times. Any modern legal system would have prosecuted Abraham for child abuse. And if he had actually carried through his plan to sacrifice Isaac, we would have convicted him of first-degree murder. Yet, according to the *mores* of his time, his conduct was entirely admirable, obeying God's commandment. Religious or not, we have all changed massively in our attitude to what is right and what is wrong. What is the nature of this change, and what drives it?

In any society there exists a somewhat mysterious consensus, which changes over the decades, and for which it is not pretentious to use the German loan-word *Zeitgeist* (spirit of the times). I said that female suffrage was now universal in the world's democracies, but this reform is in fact astonishingly recent. Here are some dates at which women were granted the vote:

New Zealand	1893
Australia	1902
Finland	1906
Norway	1913
United States	1920
Britain	1928
France	1945
Belgium	1946
Switzerland	1971
Kuwait	2006

This spread of dates through the twentieth century is a gauge of the shifting *Zeitgeist*. Another is our attitude to race. In the early part of the twentieth century, almost everybody in Britain (and many other countries too) would be judged racist by today's standards. Most white people believed that black people (in which category they would have lumped the very diverse Africans with unrelated groups from India, Australia and Melanesia) were inferior to white people in almost all respects except – patronizingly – sense of rhythm. The 1920s equivalent of James Bond was that cheerfully debonair boyhood hero, Bulldog Drummond. In one novel, *The Black Gang*, Drummond refers to 'Jews, foreigners, and other unwashed folk'. In the climax scene of *The Female of the Species*, Drummond is cleverly disguised as Pedro, black servant of the arch-villain. For his dramatic disclosure, to the reader as well as to the villain, that 'Pedro' is really Drummond himself, he could have said: 'You think I am Pedro. Little do you realize, I am your arch-enemy Drummond, blacked up.' Instead, he chose these words: 'Every beard is not false, but every nigger smells. That beard ain't false, dearie, and dis nigger don't smell. So I'm thinking, there's something wrong somewhere.' I read it in the 1950s, three decades after it was written, and it was (just) still possible for a boy to thrill to the drama and not notice the racism. Nowadays, it would be inconceivable.

Thomas Henry Huxley, by the standards of his times, was an enlightened and liberal progressive. But his times were not ours, and in 1871 he wrote the following:

> No rational man, cognizant of the facts, believes that the average negro is the equal, still less the superior, of the white man. And if this be true, it is simply incredible that, when all his disabilities are removed, and our prognathous relative has a fair field and no favor, as well as no oppressor, he will be able to compete successfully with his bigger-brained and smaller-jawed rival, in a contest which is to be carried on by thoughts and not by bites. The highest places in the hierarchy of civilization will assuredly not be within the reach of our dusky cousins.[103]

It is a commonplace that good historians don't judge statements from past times by the standards of their own. Abraham Lincoln, like Huxley, was ahead of his time, yet his views on matters of race also sound backwardly racist in ours. Here he is in a debate in 1858 with Stephen A. Douglas:

> I will say, then, that I am not, nor ever have been, in favor of bringing about in any way the social and political equality of the white and black races; that I am not, nor ever have been, in favor of making voters or jurors of negroes, nor of qualifying them to hold office, nor to intermarry with white people; and I will say, in addition to this, that there is a physical difference between the white and black races which I believe will forever forbid the two races living

> together on terms of social and political equality. And in as much as they cannot so live, while they do remain together there must be the position of superior and inferior, and I as much as any other man am in favor of having the superior position assigned to the white race.[104]

Had Huxley and Lincoln been born and educated in our time, they would have been the first to cringe with the rest of us at their own Victorian sentiments and unctuous tone. I quote them only to illustrate how the *Zeitgeist* moves on. If even Huxley, one of the great liberal minds of his age, and even Lincoln, who freed the slaves, could say such things, just think what the *average* Victorian must have thought. Going back to the eighteenth century it is, of course, well known that Washington, Jefferson and other men of the Enlightenment held slaves. The *Zeitgeist* moves on, so inexorably that we sometimes take it for granted and forget that the change is a real phenomenon in its own right.

There are numerous other examples. When the sailors first landed in Mauritius and saw the gentle dodos, it never occurred to them to do anything other than club them to death. They didn't even want to eat them (they were described as unpalatable). Presumably, hitting defenceless, tame, flightless birds over the head with a club was just something to do. Nowadays such behaviour would be unthinkable, and the extinction of a modern equivalent of the dodo, even by accident, let alone by deliberate human killing, is regarded as a tragedy.

Just such a tragedy, by the standards of today's cultural climate, was the more recent extinction of *Thylacinus*, the Tasmanian wolf. These now iconically lamented creatures had a bounty on their heads until as recently as 1909. In Victorian novels of Africa, 'elephant', 'lion' and 'antelope' (note the revealing singular) are 'game' and what you do to game, without a second thought, is shoot it. Not for food. Not for self-defence.

For 'sport'. But now the *Zeitgeist* has changed. Admittedly, rich, sedentary 'sportsmen' may shoot wild African animals from the safety of a Land-Rover and take the stuffed heads back home. But they have to pay through the nose to do so, and are widely despised for it. Wildlife conservation and the conservation of the environment have become accepted values with the same moral status as was once accorded to keeping the sabbath and shunning graven images.

The swinging sixties are legendary for their liberal modernity. But at the beginning of that decade a prosecuting barrister, in the trial for obscenity of *Lady Chatterley's Lover*, could still ask the jury: 'Would you approve of your young sons, young daughters – because girls can read as well as boys [can you *believe* he said that?] – reading this book? Is it a book you would have lying around in your own house? Is it a book you would even wish your wife or your servants to read?' This last rhetorical question is a particularly stunning illustration of the speed with which the *Zeitgeist* changes.

The American invasion of Iraq is widely condemned for its civilian casualties, yet these casualty figures are orders of magnitude lower than comparable numbers for the Second World War. There seems to be a steadily shifting standard of what is morally acceptable. Donald Rumsfeld, who sounds so callous and odious today, would have sounded like a bleeding-heart liberal if he had said the same things during the Second World War. Something has shifted in the intervening decades. It has shifted in all of us, and the shift has no connection with religion. If anything, it happens in spite of religion, not because of it.

The shift is in a recognizably consistent direction, which most of us would judge as improvement. Even Adolf Hitler, widely regarded as pushing the envelope of evil into uncharted territory, would not have stood out in the time of Caligula or of Genghis Khan. Hitler no doubt killed more people than Genghis, but he had twentieth-century technology at his disposal. And did even Hitler gain his greatest *pleasure*, as

Genghis avowedly did, from seeing his victims' 'near and dear bathed in tears'? We judge Hitler's degree of evil by the standards of today, and the moral *Zeitgeist* has moved on since Caligula's time, just as the technology has. Hitler seems especially evil only by the more benign standards of our time.

Within my lifetime, large numbers of people thoughtlessly bandied derogatory nicknames and national stereotypes: Frog, Wop, Dago, Hun, Yid, Coon, Nip, Wog. I won't claim that such words have disappeared, but they are now widely deplored in polite circles. The word 'negro', even though not intended to be insulting, can be used to date a piece of English prose. Prejudices are indeed revealing giveaways of the date of a piece of writing. In his own time, a respected Cambridge theologian, A. C. Bouquet, was able to begin the chapter on Islam of his *Comparative Religion* with these words: 'The Semite is not a natural monotheist, as was supposed about the middle of the nineteenth century. He is an animist.' The obsession with race (as opposed to culture) and the revealing use of the singular ('The Semite . . . He is an animist') to reduce an entire plurality of people to one 'type' are not heinous by any standards. But they are another tiny indicator of the changing *Zeitgeist.* No Cambridge professor of theology or any other subject would today use those words. Such subtle hints of changing *mores* tell us that Bouquet was writing no later than the middle of the twentieth century. It was in fact 1941.

Go back another four decades, and the changing standards become unmistakable. In a previous book I quoted H. G. Wells's utopian *New Republic,* and I shall do so again because it is such a shocking illustration of the point I am making.

> And how will the New Republic treat the inferior races? How will it deal with the black? . . . the yellow man? . . . the Jew? . . . those swarms of black, and brown, and dirty-white, and yellow people, who do not come into the new needs of efficiency? Well, the world is a world, and not a charitable institution,

> and I take it they will have to go . . . And the ethical system of these men of the New Republic, the ethical system which will dominate the world state, will be shaped primarily to favour the procreation of what is fine and efficient and beautiful in humanity – beautiful and strong bodies, clear and powerful minds . . . And the method that nature has followed hitherto in the shaping of the world, whereby weakness was prevented from propagating weakness . . . is death . . . The men of the New Republic . . . will have an ideal that will make the killing worth the while.

That was written in 1902, and Wells was regarded as a progressive in his own time. In 1902 such sentiments, while not widely agreed, would have made for an acceptable dinner-party argument. Modern readers, by contrast, literally gasp with horror when they see the words. We are forced to realize that Hitler, appalling though he was, was not quite as far outside the *Zeitgeist* of his time as he seems from our vantage-point today. How swiftly the *Zeitgeist* changes – and it moves in parallel, on a broad front, throughout the educated world.

Where, then, have these concerted and steady changes in social consciousness come from? The onus is not on me to answer. For my purposes it is sufficient that they certainly have not come from religion. If forced to advance a theory, I would approach it along the following lines. We need to explain why the changing moral *Zeitgeist* is so widely synchronized across large numbers of people; and we need to explain its relatively consistent direction.

First, how is it synchronized across so many people? It spreads itself from mind to mind through conversations in bars and at dinner parties, through books and book reviews, through newspapers and broadcasting, and nowadays through the Internet. Changes in the moral climate are signalled in editorials, on radio talk shows, in political speeches, in the patter of stand-up comedians and the scripts of soap operas, in the votes of parliaments making laws and the decisions of judges interpreting them. One way to put it would be in terms of changing meme frequencies in the meme pool, but I shall not pursue that.

Some of us lag behind the advancing wave of the changing moral *Zeitgeist* and some of us are slightly ahead. But most of us in the twenty-first century are bunched together and way ahead of our counterparts in the Middle Ages, or in the time of Abraham, or even as recently as the 1920s. The whole wave keeps moving, and even the vanguard of an earlier century (T. H. Huxley is the obvious example) would find itself way behind the laggers of a later century. Of course, the advance is not a smooth incline but a meandering sawtooth. There are local and temporary setbacks such as the United States is suffering from its government in the early 2000s. But over the longer timescale, the progressive trend is unmistakable and it will continue.

What impels it in its consistent direction? We mustn't neglect the driving role of individual leaders who, ahead of their time, stand up and persuade the rest of us to move on with them. In America, the ideals of racial equality were fostered by political leaders of the calibre of Martin Luther King, and entertainers, sportsmen and other public figures and role models such as Paul Robeson, Sidney Poitier, Jesse Owens and Jackie Robinson. The emancipations of slaves and of women owed much to charismatic leaders. Some of these leaders were religious; some were not. Some who were religious did their good deeds because they were religious. In other cases their religion was incidental. Although Martin Luther King was a Christian, he derived his philosophy of non-violent civil disobedience directly from Gandhi, who was not.

Then, too, there is improved education and, in particular, the increased understanding that each of us shares a common humanity with members of other races and with the other sex – both deeply unbiblical ideas that come from biological science, especially evolution. One reason black people and women and, in Nazi Germany, Jews and gypsies have been treated badly is that they were not perceived as fully human. The philosopher Peter Singer, in *Animal Liberation*, is the most eloquent advocate of the view that we should move to a post-speciesist condition in which humane treatment is meted out to all species that have the brainpower to appreciate it. Perhaps this hints at the direction in which the moral *Zeitgeist* might move in future centuries. It would be a natural extrapolation of earlier reforms like the abolition of slavery and the emancipation of women.

It is beyond my amateur psychology and sociology to go any further in explaining why the moral *Zeitgeist* moves in its broadly concerted way. For my purposes it is enough that, as a matter of observed fact, it *does* move, and it is not driven by religion – and certainly not by scripture. It is probably not a single force like gravity, but a complex interplay of disparate forces like the one that propels Moore's Law, describing the exponential increase in computer power. Whatever its cause, the manifest phenomenon of *Zeitgeist* progression is more than enough to undermine the claim that we need God in order to be good, or to decide what is good.

WHAT ABOUT HITLER AND STALIN? WEREN'T THEY ATHEISTS?

The *Zeitgeist* may move, and move in a generally progressive direction, but as I have said it is a sawtooth not a smooth improvement, and there have been some appalling reversals. Outstanding reversals, deep and terrible ones, are provided by the dictators of the twentieth century. It is important to separate the evil intentions of men like Hitler and Stalin from the vast power that they wielded in achieving them. I have already observed that Hitler's ideas and intentions were not self-evidently more evil than those of Caligula – or some of the Ottoman sultans, whose staggering feats of nastiness are described in Noel Barber's *Lords of the Golden Horn*. Hitler had twentieth-century weapons, and twentieth-century communications technology at his disposal. Nevertheless, Hitler and Stalin were, by any standards, spectacularly evil men.

'Hitler and Stalin were atheists. What have you got to say about that?' The question comes up after just about every public lecture that I ever give on the subject of religion, and in most of my radio interviews as well. It is put in a truculent way, indignantly freighted with two assumptions: not only (1) were Stalin and Hitler atheists, but (2) they did their terrible deeds *because* they were atheists. Assumption (1) is true for Stalin and dubious for Hitler. But assumption (1) is irrelevant anyway, because assumption (2) is false. It is certainly illogical if it is thought to follow from (1). Even if we accept that Hitler and Stalin shared atheism in common, they both also had moustaches, as does Saddam Hussein. So what? The interesting question is not whether evil (or good) individual human beings were religious or were atheists. We are not in the business of counting evil heads and compiling two rival roll calls of iniquity. The fact that Nazi belt buckles were inscribed with '*Gott mit uns*' doesn't prove anything, at least not without a lot more discussion. What matters is not whether Hitler and Stalin were atheists, but whether atheism systematically *influences* people to do bad things. There is not the smallest evidence that it does.

There seems no doubt that, as a matter of fact, Stalin was an atheist. He received his education at an Orthodox seminary, and his mother never lost her disappointment that he had not entered the priesthood as she intended – a fact that, according to Alan Bullock, caused Stalin much amusement.[105] Perhaps because of his training for the priesthood, the mature Stalin was scathing about the Russian Orthodox Church, and about Christianity and religion in general. But there is no evidence that his atheism motivated his brutality. His earlier religious training probably didn't either, unless it was through teaching him to revere absolutist faith, strong authority and a belief that ends justify means.

The legend that Hitler was an atheist has been assiduously cultivated, so much so that a great many people believe it without question, and it is regularly and defiantly trotted out by religious apologists. The truth of the matter is far from clear. Hitler was born into a Catholic family, and went to Catholic schools and churches as a child. Obviously that is not significant in itself: he could easily have given it up, as Stalin gave up his Russian Orthodoxy after leaving the Tiflis Theological Seminary. But Hitler never formally renounced his Catholicism, and there are indications throughout his life that he remained religious. If not Catholic, he seems to have retained a belief in some sort of divine providence. For example he stated in *Mein Kampf* that, when he heard the news of the declaration of the First World War, 'I sank down on my knees and thanked Heaven out of the fullness of my heart for the favour of having been permitted to live in such a time.'[106] But that was 1914, when he was still only twenty-five. Perhaps he changed after that?

In 1920, when Hitler was thirty-one, his close associate Rudolf Hess, later to be deputy Führer, wrote in a letter to the Prime Minister of Bavaria, 'I know Herr Hitler very well personally and am quite close to him. He has an unusually honourable character, full of profound kindness, is religious, a good Catholic.'[107] Of course, it could be said that, since Hess got the 'honourable character' and the 'profound kindness' so crashingly wrong, maybe he got the 'good Catholic' wrong too! Hitler could scarcely be described as a 'good' anything, which reminds me of the most comically audacious argument I have heard in favour of the proposition that Hitler must have been an atheist. Paraphrasing from many sources, Hitler was a bad man, Christianity teaches goodness, therefore Hitler can't have been a Christian! Goering's remark about Hitler, 'Only a Catholic could unite Germany,' might, I suppose, have meant somebody brought up Catholic rather than a believing Catholic.

In a speech of 1933 in Berlin, Hitler said, 'We were convinced that the people need and require this faith. We have therefore undertaken the fight against the atheistic movement, and that not merely with a few theoretical declarations: we have stamped it out.'[108] That might indicate only that, like many others, Hitler 'believed in belief'. But as late as 1941 he told his adjutant, General Gerhard Engel, 'I shall remain a Catholic for ever.'

Even if he didn't remain a sincerely believing Christian, Hitler would have to have been positively unusual not to have been influenced by the long Christian tradition of blaming Jews as Christ-killers. In a speech in Munich in 1923, Hitler said, 'The first thing to do is to rescue [Germany] from the Jew who is ruining our country . . . We want to prevent our Germany from suffering, as Another did, the death upon the Cross.'[109] In his *Adolf Hitler: The Definitive Biography*, John Toland wrote of Hitler's religious position at the time of the 'final solution':

> Still a member in good standing of the Church of Rome despite detestation of its hierarchy, he carried within him its teaching that the Jew was the killer of god. The extermination, therefore, could be done without a twinge of conscience since he was merely acting as the avenging hand of god – so long as it was done impersonally, without cruelty.

Christian hatred of Jews is not just a Catholic tradition. Martin Luther was a virulent anti-Semite. At the Diet of Worms he said that 'All Jews should be driven from Germany.' And he wrote a whole book, *On the Jews and their Lies*, which probably influenced Hitler. Luther described the Jews as a 'brood of vipers', and the same phrase was used by Hitler in a remarkable speech of 1922, in which he several times repeated that he was a Christian:

> My feeling as a Christian points me to my Lord and Saviour as a fighter. It points me to the man who

> once in loneliness, surrounded by a few followers, recognized these Jews for what they were and summoned men to fight against them and who, God's truth! was greatest not as a sufferer but as a fighter. In boundless love as a Christian and as a man I read through the passage which tells us how the Lord at last rose in His might and seized the scourge to drive out of the Temple the brood of vipers and adders. How terrific was His fight for the world against the Jewish poison. To-day, after two thousand years, with deepest emotion I recognize more profoundly than ever before the fact that it was for this that He had to shed His blood upon the Cross. As a Christian I have no duty to allow myself to be cheated, but I have the duty to be a fighter for truth and justice . . . And if there is anything which could demonstrate that we are acting rightly it is the distress that daily grows. For as a Christian I have also a duty to my own people.[110]

It is hard to know whether Hitler picked up the phrase 'brood of vipers' from Luther, or whether he got it directly from Matthew 3: 7, as Luther presumably did. As for the theme of Jewish persecution as part of God's will, Hitler returned to it in *Mein Kampf*: 'Hence today I believe that I am acting in accordance with the will of the Almighty Creator: *by defending myself against the Jew, I am fighting for the work of the Lord*.' That was 1925. He said it again in a speech in the Reichstag in 1938, and he said similar things throughout his career.

Quotations like those have to be balanced by others from his *Table Talk*, in which Hitler expressed virulently anti-Christian views, as recorded by his secretary. The following all date from 1941:

> The heaviest blow that ever struck humanity was the coming of Christianity. Bolshevism is Christianity's

> illegitimate child. Both are inventions of the Jew. The deliberate lie in the matter of religion was introduced into the world by Christianity . . .
>
> The reason why the ancient world was so pure, light and serene was that it knew nothing of the two great scourges: the pox and Christianity.
>
> When all is said, we have no reason to wish that the Italians and Spaniards should free themselves from the drug of Christianity. Let's be the only people who are immunised against the disease.

Hitler's *Table Talk* contains more quotations like those, often equating Christianity with Bolshevism, sometimes drawing an analogy between Karl Marx and St Paul and never forgetting that both were Jews (though Hitler, oddly, was always adamant that Jesus himself was not a Jew). It is possible that Hitler had by 1941 experienced some kind of deconversion or disillusionment with Christianity. Or is the resolution of the contradictions simply that he was an opportunistic liar whose words cannot be trusted, in either direction?

It could be argued that, despite his own words and those of his associates, Hitler was not really religious but just cynically exploiting the religiosity of his audience. He may have agreed with Napoleon, who said, 'Religion is excellent stuff for keeping common people quiet,' and with Seneca the Younger: 'Religion is regarded by the common people as true, by the wise as false, and by the rulers as useful.' Nobody could deny that Hitler was capable of such insincerity. If this was his real motive for pretending to be religious, it serves to remind us that Hitler didn't carry out his atrocities single-handed. The terrible deeds themselves were carried out by soldiers and their officers, most of whom were surely Christian. Indeed, the Christianity of the German people underlies the very hypothesis we are discussing – a hypothesis to explain the supposed insincerity of Hitler's religious professings! Or, perhaps Hitler felt that he had to display some token sympathy for Christianity, otherwise his regime would not have received the support it did from the Church. This support showed itself in various ways, including Pope Pius XII's persistent refusal to take a stand against the Nazis – a subject of considerable embarrassment to the modern Church. Either Hitler's professions of Christianity were sincere, or he faked his Christianity in order to win – successfully – co-operation from German Christians and the Catholic Church. In either case, the evils of Hitler's regime can hardly be held up as flowing from atheism.

Even when he was railing against Christianity, Hitler never ceased using the language of Providence: a mysterious agency which, he believed, had singled him out for a divine mission to lead Germany. He sometimes called it Providence, at other times God. After the *Anschluss*, when Hitler returned in triumph to Vienna in 1938, his exultant speech mentioned God in this providential guise: 'I believe it was God's will to send a boy from here into the Reich, to let him grow up and to raise him to be the leader of the nation so that he could lead back his homeland into the Reich.'[111]

When he narrowly escaped assassination in Munich in November 1939, Hitler credited Providence with intervening to save his life by causing him to alter his schedule: 'Now I am completely content. The fact that I left the Bürgerbräukeller earlier than usual is a corroboration of Providence's intention to let me reach my goal.'[112] After this failed assassination the Archbishop of Munich, Cardinal Michael Faulhaber, ordered that a *Te Deum* should be said in his cathedral, 'To thank Divine Providence in the name of the archdiocese for the Führer's fortunate escape.' Some of Hitler's followers, with the support of Goebbels, made no bones about building Nazism into a religion in its own right. The following, by the chief of the united trade unions, has the feel of a prayer, and even has the cadences of the Christian Lord's Prayer ('Our Father') or the Creed:

> Adolf Hitler! We are united with you alone! We want to renew our vow in this hour: On this earth we believe only in Adolf Hitler. We believe that National Socialism is the sole saving faith for our people. We believe that there is a Lord God in heaven, who created us, who leads us, who directs us and who blesses us visibly. And we believe that this Lord God sent Adolf Hitler to us, so that Germany might become a foundation for all eternity.[113]

Jonathan Glover, in his remarkable and chilling book, *Humanity: A Moral History of the Twentieth Century*, remarks that

> Many also accepted the religious cult of Stalin, expressed by a Lithuanian writer: 'I approached Stalin's portrait, took it off the wall, placed it on the table and, resting my head on my hands, I gazed and meditated. What should I do? The Leader's face, as always so serene, his eyes so clear-sighted, they penetrated into the distance. It seems that his penetrating look pierces my little room and goes out to embrace the entire globe . . . With my every fibre, every nerve, every drop of blood I feel that, at this moment, nothing exists in this entire world but this dear and beloved face.'

Such quasi-religious adulation is all the more repellent for coming, in Glover's book, immediately after his account of Stalin's shatteringly horrible cruelties.

Stalin was probably an atheist and Hitler probably wasn't; but even if they were both atheists, the bottom line of the Stalin/Hitler debating point is very simple. Individual atheists may do evil things but they don't do evil things in the name of atheism. Stalin and Hitler did extremely evil things, in the name of, respectively, dogmatic and doctrinaire Marxism, and an insane and unscientific eugenics theory tinged with sub-Wagnerian ravings. Religious wars really are fought in the name of religion, and they have been horribly frequent in history. I cannot think of any war that has been fought in the name of atheism. Why should it? A war might be motivated by economic greed, by political ambition, by ethnic or racial prejudice, by deep grievance or revenge, or by patriotic belief in the destiny of a nation. Even more plausible as a motive for war is an unshakeable faith that one's own religion is the only true one, reinforced by a holy book that explicitly condemns all heretics and followers of rival religions to death, and explicitly promises that the soldiers of God will go straight to a martyrs' heaven. Sam Harris, as so often, hits the bullseye, in *The End of Faith*:

> The danger of religious faith is that it allows otherwise normal human beings to reap the fruits of madness and consider them *holy*. Because each new generation of children is taught that religious propositions need not be justified in the way that all others must, civilization is still besieged by the armies of the preposterous. We are, even now, killing ourselves over ancient literature. Who would have thought something so tragically absurd could be possible?

By contrast, why would anyone go to war for the sake of an *absence* of belief?

CHAPTER 8

What's wrong with religion? Why be so hostile?

Religion has actually convinced people that there's an invisible man – living in the sky – who watches everything you do, every minute of every day. And the invisible man has a special list of ten things he does not want you to do. And if you do any of these ten things, he has a special place, full of fire and smoke and burning and torture and anguish, where he will send you to live and suffer and burn and choke and scream and cry forever and ever 'til the end of time . . . But He loves you!

GEORGE CARLIN

I do not, by nature, thrive on confrontation. I don't think the adversarial format is well designed to get at the truth, and I regularly refuse invitations to take part in formal debates. I was once invited to debate with the then Archbishop of York, in Edinburgh. I felt honoured by this, and accepted. After the debate, the religious physicist Russell Stannard reproduced in his book *Doing Away with God?* a letter that he wrote to the *Observer*:

> Sir, Under the gleeful headline 'God comes a poor Second before the Majesty of Science', your science correspondent reported (on Easter Sunday of all days) how Richard Dawkins 'inflicted grievous

> intellectual harm' on the Archbishop of York in a debate on science and religion. We were told of 'smugly smiling atheists' and 'Lions 10; Christians nil'.

Stannard went on to chide the *Observer* for failing to report a subsequent encounter between him and me, together with the Bishop of Birmingham and the distinguished cosmologist Sir Hermann Bondi, at the Royal Society, which had *not* been staged as an adversarial debate, and which had been a lot more constructive as a result. I can only agree with his implied condemnation of the adversarial debate format. In particular, for reasons explained in *A Devil's Chaplain*, I never take part in debates with creationists.*

Despite my dislike of gladiatorial contests, I seem somehow to have acquired a reputation for pugnacity towards religion. Colleagues who agree that there is no God, who agree that we do not need religion to be moral, and agree that we can explain the roots of religion and of morality in non-religious terms, nevertheless come back at me in gentle puzzlement. Why are you so hostile? What is actually wrong with religion? Does it really do so much harm that we should actively fight against it? Why not live and let live, as one does with Taurus and Scorpio, crystal energy and ley lines? Isn't it all just harmless nonsense?

I might retort that such hostility as I or other atheists occasionally voice towards religion is limited to words. I am not going to bomb anybody, behead them, stone them, burn them at the stake, crucify them, or fly planes into their skyscrapers, just because of a theological disagreement. But my interlocutor usually doesn't leave it at that. He may go on to say something like this: 'Doesn't your hostility mark you out as a funda-

* I do not have the *chutzpah* to refuse on the grounds offered by one of my most distinguished scientific colleagues, whenever a creationist tries to stage a formal debate with him (I shall not name him, but his words should be read in an Australian accent): 'That would look great on your CV; not so good on mine.'

mentalist atheist, just as fundamentalist in your own way as the wingnuts of the Bible Belt in theirs?' I need to dispose of this accusation of fundamentalism, for it is distressingly common.

FUNDAMENTALISM AND THE SUBVERSION OF SCIENCE

Fundamentalists know they are right because they have read the truth in a holy book and they know, in advance, that nothing will budge them from their belief. The truth of the holy book is an axiom, not the end product of a process of reasoning. The book is true, and if the evidence seems to contradict it, it is the evidence that must be thrown out, not the book. By contrast, what I, as a scientist, believe (for example, evolution) I believe not because of reading a holy book but because I have studied the evidence. It really is a very different matter. Books about evolution are believed not because they are holy. They are believed because they present overwhelming quantities of mutually buttressed evidence. In principle, any reader can go and check that evidence. When a science book is wrong, somebody eventually discovers the mistake and it is corrected in subsequent books. That conspicuously doesn't happen with holy books.

Philosophers, especially amateurs with a little philosophical learning, and even more especially those infected with 'cultural relativism', may raise a tiresome red herring at this point: a scientist's belief in *evidence* is itself a matter of fundamentalist faith. I have dealt with this elsewhere, and will only briefly repeat myself here. All of us believe in evidence in our own lives, whatever we may profess with our amateur philosophical hats on. If I am accused of murder, and prosecuting counsel sternly asks me whether it is true that I was in Chicago on the night of the crime, I cannot get away with a philosophical evasion: 'It depends what you mean by "true".' Nor with an anthropological, relativist plea: 'It is only in your Western scientific sense of "in" that I was in Chicago. The Bongolese have a completely different concept of "in", according to which you are only truly "in" a place if you are an anointed elder entitled to take snuff from the dried scrotum of a goat.'[114]

Maybe scientists are fundamentalist when it comes to defining in some abstract way what is meant by 'truth'. But so is everybody else. I am no more fundamentalist when I say evolution is true than when I say it is true that New Zealand is in the southern hemisphere. We believe in evolution because the evidence supports it, and we would abandon it overnight if new evidence arose to disprove it. No real fundamentalist would ever say anything like that.

It is all too easy to confuse fundamentalism with passion. I may well appear passionate when I defend evolution against a fundamentalist creationist, but this is not because of a rival fundamentalism of my own. It is because the evidence for evolution is overwhelmingly strong and I am passionately distressed that my opponent can't see it – or, more usually, refuses to look at it because it contradicts his holy book. My passion is increased when I think about how much the poor fundamentalists, and those whom they influence, are *missing*. The truths of evolution, along with many other scientific truths, are so engrossingly fascinating and beautiful; how truly tragic to die having missed out on all that! Of course that makes me passionate. How could it not? But my belief in evolution is not fundamentalism, and it is not faith, because I know what it would take to change my mind, and I would gladly do so if the necessary evidence were forthcoming.

It does happen. I have previously told the story of a respected elder statesman of the Zoology Department at Oxford when I was an undergraduate. For years he had passionately believed, and taught, that the Golgi Apparatus (a microscopic feature of the interior of cells) was not real: an artefact, an illusion. Every Monday afternoon it was the custom for the whole department to listen to a research talk by a visiting lecturer. One Monday, the visitor was an American cell biologist who presented completely convincing evidence that the Golgi Apparatus was real.

At the end of the lecture, the old man strode to the front of the hall, shook the American by the hand and said – with passion – 'My dear fellow, I wish to thank you. I have been wrong these fifteen years.' We clapped our hands red. No fundamentalist would ever say that. In practice, not all scientists would. But all scientists pay lip service to it as an ideal – unlike, say, politicians who would probably condemn it as flip-flopping. The memory of the incident I have described still brings a lump to my throat.

As a scientist, I am hostile to fundamentalist religion because it actively debauches the scientific enterprise. It teaches us not to change our minds, and not to want to know exciting things that are available to be known. It subverts science and saps the intellect. The saddest example I know is that of the American geologist Kurt Wise, who now directs the Center for Origins Research at Bryan College, Dayton, Tennessee. It is no accident that Bryan College is named after William Jennings Bryan, prosecutor of the science teacher John Scopes in the Dayton 'Monkey Trial' of 1925. Wise could have fulfilled his boyhood ambition to become a professor of geology at a real university, a university whose motto might have been 'Think critically' rather than the oxymoronic one displayed on the Bryan website: 'Think critically and biblically'. Indeed, he obtained a real degree in geology at the University of Chicago, followed by two higher degrees in geology and paleontology at Harvard (no less) where he studied under Stephen Jay Gould (no less). He was a highly qualified and genuinely promising young scientist, well on his way to achieving his dream of teaching science and doing research at a proper university.

Then tragedy struck. It came, not from outside but from within his own mind, a mind fatally subverted and weakened by a fundamentalist religious upbringing that required him to believe that the Earth – the subject of his Chicago and Harvard geological education – was less than ten thousand years old. He was too intelligent not to recognize the head-on collision between his religion and his science, and the conflict in his mind made him increasingly uneasy. One day, he could bear the strain no more, and he clinched the matter with a pair of scissors. He took a bible and went right through it, literally cutting out every verse that would have to go if the scientific world-view were true. At the end of this ruthlessly honest and labour-intensive exercise, there was so little left of his bible that,

> try as I might, and even with the benefit of intact margins throughout the pages of Scripture, I found it impossible to pick up the Bible without it being rent in two. I had to make a decision between evolution and Scripture. Either the Scripture was true and evolution was wrong or evolution was true and I must toss out the Bible . . . It was there that night that I accepted the Word of God and rejected all that would ever counter it, including evolution. With that, in great sorrow, I tossed into the fire all my dreams and hopes in science.

I find that terribly sad; but whereas the Golgi Apparatus story moved me to tears of admiration and exultation, the Kurt Wise story is just plain pathetic – pathetic and contemptible. The wound, to his career and his life's happiness, was self-inflicted, so unnecessary, so easy to escape. All he had to do was toss out the bible. Or interpret it symbolically, or allegorically, as the theologians do. Instead, he did the fundamentalist thing and tossed out science, evidence and reason, along with all his dreams and hopes.

Perhaps uniquely among fundamentalists, Kurt Wise is honest – devastatingly, painfully, shockingly honest. Give him the Templeton Prize; he might be the first really sincere recipient. Wise brings to the surface what is secretly going on underneath, in the minds of fundamentalists generally, when they encounter scientific evidence that contradicts their beliefs. Listen to his peroration:

> Although there are scientific reasons for accepting a young earth, I am a young-age creationist because that is my understanding of the Scripture. As I shared with my professors years ago when I was in college, if all the evidence in the universe turns against creationism, I would be the first to admit it, but I would still be a creationist because that is what the Word of God seems to indicate. Here I must stand.[115]

He seems to be quoting Luther as he nailed his theses to the door of the church in Wittenberg, but poor Kurt Wise reminds me more of Winston Smith in *1984* – struggling desperately to believe that two plus two equals five if Big Brother says it does. Winston, however, was being tortured. Wise's doublethink comes not from the imperative of physical torture but from the imperative – apparently just as undeniable to some people – of religious faith: arguably a form of mental torture. I am hostile to religion because of what it did to Kurt Wise. And if it did that to a Harvard-educated geologist, just think what it can do to others less gifted and less well armed.

Fundamentalist religion is hell-bent on ruining the scientific education of countless thousands of innocent, well-meaning, eager young minds. Non-fundamentalist, 'sensible' religion may not be doing that. But it is making the world safe for fundamentalism by teaching children, from their earliest years, that unquestioning faith is a virtue.

THE DARK SIDE OF ABSOLUTISM

In the previous chapter, when trying to explain the shifting moral *Zeitgeist*, I invoked a widespread consensus of liberal, enlightened, decent people. I made the rosy-spectacled assumption that 'we' all broadly agree with this consensus, some more than others, and I had in mind most of the people likely to read this book, whether they are religious or not. But of course, not everybody is of the consensus (and not everybody will have any desire to read my book). It has to be admitted that absolutism is far from dead. Indeed, it rules the minds of a great number of people in the world today, most dangerously so in the Muslim world and in the incipient American theocracy (see Kevin Phillips's book of that name). Such absolutism nearly always results from strong religious faith, and it constitutes a major reason for suggesting that religion can be a force for evil in the world.

One of the fiercest penalties in the Old Testament is the one exacted for blasphemy. It is still in force in certain countries. Section 295-C of the Pakistan penal code prescribes the death penalty for this 'crime'. On 18 August 2001, Dr Younis Shaikh, a medical doctor and lecturer, was sentenced to death for blasphemy. His particular crime was to tell students that the prophet Muhammad was not a Muslim before he invented the religion at the age of forty. Eleven of his students reported him to the authorities for this 'offence'. The blasphemy law in Pakistan is more usually invoked against Christians, such as Augustine Ashiq 'Kingri' Masih, who was sentenced to death in Faisalabad in 2000. Masih, as a Christian, was not allowed to marry his sweetheart because she was a Muslim and – incredibly – Pakistani (and Islamic) law does not allow a Muslim woman to marry a non-Muslim man. So he tried to convert to Islam and was then accused of doing so for base motives. It is not clear from the report I have read whether this in itself was the capital crime, or whether it was something he is alleged to have said about the prophet's own morals. Either way, it certainly was not the kind of offence that would warrant a death sentence in any country whose laws are free of religious bigotry.

In 2006 in Afghanistan, Abdul Rahman was sentenced to death for converting to Christianity. Did he kill anyone, hurt anybody, steal anything, damage anything? No. All he did was change his mind. Internally and privately, he changed his mind. He entertained certain *thoughts* which were not to the liking of the ruling party of his country. And this, remember, is not the Afghanistan of the Taliban but the 'liberated' Afghanistan of Hamid Karzai, set up by the American-led coalition. Mr Rahman finally escaped execution, but only on a plea of insanity, and only after intense international pressure. He has now sought asylum in Italy, to avoid being murdered by zealots eager to do their Islamic duty. It is still an article of the *constitution* of 'liberated' Afghanistan that the penalty for apostasy is death. Apostasy, remember, doesn't mean actual harm to persons or property. It is pure thoughtcrime, to use George Orwell's *1984* terminology, and the official punishment for it under Islamic law is death. On 3 September 1992, to take one example where it was actually carried out, Sadiq Abdul Karim Malallah was publicly beheaded in Saudi Arabia after being lawfully convicted of apostasy and blasphemy.[116]

I once had a televised encounter with Sir Iqbal Sacranie, mentioned in Chapter 1 as Britain's leading 'moderate' Muslim. I challenged him on the death penalty as punishment for apostasy. He wriggled and squirmed, but was unable either to deny or decry it. He kept trying to change the subject, saying it was an unimportant detail. This is a man who has been knighted by the British government for promoting good 'interfaith relations'.

But let's have no complacency in Christendom. As recently as 1922 in Britain, John William Gott was sentenced to nine months' hard labour for blasphemy: he compared Jesus to a clown. Almost unbelievably, the crime of blasphemy is still on the statute book in Britain,[117] and in 2005 a Christian group tried to bring a private prosecution for blasphemy against the BBC for broadcasting *Jerry Springer, the Opera*.

In the United States of recent years the phrase 'American Taliban' was begging to be coined, and a swift Google search nets more than a dozen websites that have done so. The quotations that they anthologize, from American religious leaders and faith-based politicians, chillingly recall the narrow bigotry, heartless cruelty and sheer nastiness of the Afghan

Taliban, the Ayatollah Khomeini and the Wahhabi authorities of Saudi Arabia. The web page called 'The American Taliban' is a particularly rich source of obnoxiously barmy quotations, beginning with a prize one from somebody called Ann Coulter who, American colleagues have persuaded me, is not a spoof, invented by *The Onion*: 'We should invade their countries, kill their leaders and convert them to Christianity.'[118] Other gems include Congressman Bob Dornan's 'Don't use the word "gay" unless it's an acronym for "Got Aids Yet?"' and General William G. Boykin's 'George Bush was not elected by a majority of the voters in the United States, he was appointed by God'. All the ingredients are there: slavish adherence to a misunderstood old text; hatred of women, modernity, rival religions, science and pleasure; love of punishment, bullying, narrow-minded, bossy interference in every aspect of life. The Afghan Taliban and the American Taliban are good examples of what happens when people take their scriptures literally and seriously. They provide a horrifying modern enactment of what life might have been like under the theocracy of the Old Testament. Kimberly Blaker's *The Fundamentals of Extremism: The Christian Right in America* is a book-length exposé of the menace of the Christian Taliban (not under that name).

FAITH AND HOMOSEXUALITY

In Afghanistan under the Taliban, the official punishment for homosexuality was execution, by the tasteful method of burial alive under a wall pushed over on top of the victim. The 'crime' itself being a private act, performed by consenting adults who were doing nobody else any harm, we again have here the classic hallmark of religious absolutism. My own country has no right to be smug. Private homosexuality was a criminal offence in Britain up until – astonishingly – 1967. In 1954 the British mathematician Alan Turing, a candidate along with John von Neumann for the title of father of the computer, committed suicide after being convicted of the criminal offence of homosexual behaviour in private. Admittedly Turing was not buried alive under a wall pushed over by a tank. He was offered a choice between two years in prison (you can imagine how the other prisoners would have treated him) and a course of hormone injections which could be said to amount to chemical castration, and would have caused him to grow breasts. His final, private choice was an apple that he had injected with cyanide.[119]

As the pivotal intellect in the breaking of the German Enigma codes, Turing arguably made a greater contribution to defeating the Nazis than Eisenhower or Churchill. Thanks to Turing and his 'Ultra' colleagues at Bletchley Park, Allied generals in the field were consistently, over long periods of the war, privy to detailed German plans before the German generals had time to implement them. After the war, when Turing's role was no longer top secret, he should have been knighted and fêted as a saviour of his nation. Instead, this gentle, stammering, eccentric genius was destroyed, for a 'crime', committed in private, which harmed nobody. Once again, the unmistakable trademark of the faith-based moralizer is to care passionately about what other people do (or even think) in *private.*

The attitude of the 'American Taliban' towards homosexuality epitomizes their religious absolutism. Listen to the Reverend Jerry Falwell, founder of Liberty University: 'AIDS is not just God's punishment for homosexuals; it is God's punishment for the society that tolerates homosexuals.'[120] The thing I notice first about such people is their wonderful Christian charity. What kind of an electorate could, term after term, vote in a man of such ill-informed bigotry as Senator Jesse Helms, Republican of North Carolina? A man who has sneered: '*The New York Times* and *Washington Post* are both infested with homosexuals themselves. Just about every person down there is a homosexual or lesbian.'[121] The answer, I suppose, is the kind of electorate that sees morality in narrowly religious terms and feels threatened by anybody who doesn't share the same absolutist faith.

I have already quoted Pat Robertson, founder of the Christian Coalition. He stood as a serious candidate for the Republican party nomination for President in 1988, and garnered more than three million volunteers to work in his campaign, plus a comparable quantity of money: a disquieting level of support, given that the following quotations are entirely typical of him: '[Homosexuals] want to come into churches and disrupt church services and throw blood all around and try to give people AIDS and spit in the face of ministers.' '[Planned Parenthood] is teaching kids to fornicate, teaching people to have adultery, every kind of bestiality, homosexuality, lesbianism – everything that the Bible condemns.' Robertson's attitude to women, too, would warm the black hearts of the Afghan Taliban: 'I know this is painful for the ladies to hear, but if you get married, you have accepted the headship of a man, your husband. Christ is the head of the household and the husband is the head of the wife, and that's the way it is, period.'

Gary Potter, President of Catholics for Christian Political Action, had this to say: 'When the Christian majority takes over this country, there will be no satanic churches, no more free distribution of pornography, no more talk of rights for homosexuals. After the Christian majority takes control, pluralism will be seen as immoral and evil and the state will not permit anybody the right to practice evil.' 'Evil', as is very clear from the quotation, doesn't mean doing things that have bad consequences for people. It means private thoughts and actions that are not to 'the Christian majority's' private liking.

Pastor Fred Phelps, of the Westboro Baptist Church, is another strong preacher with an obsessive dislike of homosexuals. When Martin Luther King's widow died, Pastor Fred organized a picket of her funeral, proclaiming: 'God Hates Fags & Fag-Enablers! Ergo, God hates Coretta Scott King and is now tormenting her with fire and brimstone where the worm never dies and the fire is never quenched, and the smoke of her torment ascendeth up for ever and ever.'[122] It is easy to write Fred Phelps off as a nut, but he has plenty of support from people and their money. According to his own website, Phelps has organized 22,000 anti-homosexual demonstrations since 1991 (that's an average of four per day) in the USA, Canada, Jordan and Iraq, displaying slogans such as 'THANK GOD FOR AIDS'. A particularly charming feature of his website is the automated tally of the number of days a particular, named, deceased homosexual has been burning in hell.

Attitudes to homosexuality reveal much about the sort of morality that is inspired by religious faith. An equally instructive example is abortion and the sanctity of human life.

FAITH AND THE SANCTITY OF HUMAN LIFE

Human embryos are examples of human life. Therefore, by absolutist religious lights, abortion is simply wrong: fully fledged murder. I am not sure what to make of my admittedly anecdotal observation that many of those who most ardently oppose the taking of embryonic life also seem to be more than usually enthusiastic about taking adult life. To be fair, this does not, as a rule, apply to Roman Catholics, who are among the most vociferous opponents of abortion. The born-again George W. Bush, however, is typical of today's religious ascendancy. He, and they, are stalwart defenders of human life, as long as it is embryonic life (or terminally ill life) – even to the point of preventing medical research that would certainly save many lives.[123] The obvious ground for opposing the death penalty is respect for human life. Since 1976, when the Supreme Court reversed the ban on the death penalty, Texas has been responsible for more than one-third of all executions in all fifty states of the Union. And Bush presided over more executions in Texas than any other governor in the state's history, averaging one death every nine days. Perhaps he was simply doing his duty and carrying out the laws of the state?[124] But then, what are we to make of the famous report by the CNN journalist Tucker Carlson? Carlson, who himself supports the death penalty, was shocked by Bush's 'humorous' imitation of a female prisoner on death row, pleading to the Governor for a stay of execution: ' "Please," Bush whimpers, his lips pursed in mock desperation, "Don't kill me." '[125] Perhaps this woman would have met with more sympathy if she had pointed out that she had once been an embryo. The contemplation of embryos really does seem to have the most extraordinary effect upon many people of faith. Mother Teresa of Calcutta actually said, in her speech accepting the Nobel Peace Prize, 'The greatest destroyer of peace is abortion.' *What?* How can a woman with such cock-eyed judgement be taken seriously on any topic, let alone be thought seriously worthy of a Nobel Prize? Anybody tempted to be taken in by the sanctimoniously hypocritical Mother Teresa should read Christopher Hitchens's book *The Missionary Position: Mother Teresa in Theory and Practice.*

Returning to the American Taliban, listen to Randall Terry, founder of Operation Rescue, an organization for intimidating abortion providers. 'When I, or people like me, are running the country, you'd better flee, because we will find you, we will try you, and we'll execute you. I mean every word of it. I will make it part of my mission to see to it that they are tried and executed.' Terry was here referring to doctors who provide abortions, and his Christian inspiration is clearly shown by other statements:

> I want you to just let a wave of intolerance wash over you. I want you to let a wave of hatred wash over you. Yes, hate is good . . . Our goal is a Christian nation. We have a Biblical duty, we are called by God, to conquer this country. We don't want equal time. We don't want pluralism.
>
> Our goal must be simple. We must have a Christian nation built on God's law, on the Ten Commandments. No apologies.[126]

This ambition to achieve what can only be called a Christian fascist state is entirely typical of the American Taliban. It is an almost exact mirror image of the Islamic fascist state so ardently sought by many people in other parts of the world. Randall Terry is not – yet – in political power. But no observer of the American political scene at the time of writing (2006) can afford to be sanguine.

A consequentialist or utilitarian is likely to approach the abortion question in a very different way, by trying to weigh up suffering. Does the embryo suffer? (Presumably not if it is aborted before it has a nervous system; and even if it is old enough to have a nervous system it surely suffers less than, say, an adult cow in a slaughterhouse.) Does the pregnant woman, or her family, suffer if she does not have an abortion? Very possibly so; and, in any case, given that the embryo lacks a nervous system, shouldn't the mother's well-developed nervous system have the choice?

This is not to deny that a consequentialist might have grounds to oppose abortion. 'Slippery slope' arguments can be framed by consequentialists (though I wouldn't in this case). Maybe embryos don't suffer, but a culture that tolerates the taking of human life risks going too far: where will it all end? In infanticide? The moment of birth provides a natural Rubicon for defining rules, and one could argue that it is hard to find another one earlier in embryonic development. Slippery slope arguments could therefore lead us to give the moment of birth more significance than utilitarianism, narrowly interpreted, would prefer.

Arguments against euthanasia, too, can be framed in slippery slope terms. Let's invent an imaginary quotation from a moral philosopher: 'If you allow doctors to put terminal patients out of their agony, the next thing you know everybody will be bumping off their granny to get her money. We philosophers may have grown out of absolutism, but society needs the discipline of absolute rules such as "Thou shalt not kill," otherwise it doesn't know where to stop. Under some circumstances absolutism might, for all the wrong reasons in a less than ideal world, have better *consequences* than naïve consequentialism!

We philosophers might have a hard time prohibiting the eating of people who were already dead and unmourned – say road-killed tramps. But, for slippery slope reasons, the absolutist taboo against cannibalism is too valuable to lose.'

Slippery slope arguments might be seen as a way in which consequentialists can reimport a form of indirect absolutism. But the religious foes of abortion don't bother with slippery slopes. For them, the issue is much simpler. An embryo is a 'baby', killing it is murder, and that's that: end of discussion. Much follows from this absolutist stance. For a start, embryonic stem-cell research must cease, despite its huge potential for medical science, because it entails the deaths of embryonic cells. The inconsistency is apparent when you reflect that society already accepts IVF (in vitro fertilization), in which doctors routinely stimulate women to produce surplus eggs, to be fertilized outside the body. As many as a dozen viable zygotes may be produced, of which two or three are then implanted in the uterus. The expectation is that, of these, only one or possibly two will survive. IVF, therefore, kills conceptuses at two stages of the procedure, and society in general has no problem with this. For twenty-five years, IVF has been a standard procedure for bringing joy into the lives of childless couples.

Religious absolutists, however, can have problems with IVF. The *Guardian* of 3 June 2005 carried a bizarre story under the headline 'Christian couples answer call to save embryos left by IVF'. The story is about an organization called Snowflakes which seeks to 'rescue' surplus embryos left over at IVF clinics. 'We really felt like the Lord was calling us to try to give one of these embryos – these children – a chance to live,' said a woman in Washington State, whose fourth child resulted from this 'unexpected alliance that conservative Christians have been forming with the world of test-tube babies'. Worried about that alliance, her husband had consulted a church elder, who advised, 'If you want to free the slaves, you sometimes have to make a deal with the slave trader.' I wonder what these people would say if they knew that the majority of conceived embryos spontaneously abort anyway. It is probably best seen as a kind of natural 'quality control'.

A certain kind of religious mind cannot see the moral difference between killing a microscopic cluster of cells on the one hand, and killing a full-grown doctor on the other. I have already quoted Randall Terry and 'Operation Rescue'. Mark Juergensmeyer, in his chilling book *Terror in the Mind of God*, prints a photograph of the Reverend Michael Bray with his friend the Reverend Paul Hill, holding a banner reading: 'Is it wrong to stop the murder of innocent babies?' Both look like nice, rather preppy young men, smiling engagingly, casually well-dressed, the very opposite of staring-eyed loonies. Yet they and their friends of the Army of God (AOG) made it their business to set fire to abortion clinics, and they have made no secret of their desire to kill doctors. On 29 July 1994, Paul Hill took a shotgun and murdered Dr John Britton and his bodyguard James Barrett outside Britton's clinic in Pensacola, Florida. He then gave himself up to the police, saying he had killed the doctor to prevent the future deaths of 'innocent babies'.

Michael Bray defends such actions articulately and with every appearance of high moral purpose, as I discovered when I interviewed him, in a public park in Colorado Springs, for my television documentary on religion.* Before coming on to the abortion question, I got the measure of Bray's Bible-based morality by asking him some preliminary questions. I pointed out that biblical law condemns adulterers to death by stoning. I expected him to disavow this particular example as obviously beyond the pale, but he surprised me. He was happy to agree that, after due process of law, adulterers should be executed. I then pointed out that Paul Hill, with Bray's full support, had not followed due process but had taken the law into his own hands and killed a doctor. Bray defended his fellow clergyman's

---

* The animal liberationists who threaten violence against scientists using animals for medical research would claim an equally high moral purpose.

action in the same terms as he had when Juergensmeyer interviewed him, making a distinction between retributive killing, say of a retired doctor, and killing a practising doctor as a means of preventing him from 'regularly killing babies'. I then put it to him that, sincere though Paul Hill's beliefs no doubt were, society would descend into a terrible anarchy if everybody invoked personal conviction in order to take the law into their own hands, rather than abiding by the law of the land. Wasn't the right course to try to get the law changed, democratically? Bray replied: 'Well, this is the problem when we don't have law that's really authentic law; when we have laws that are made up by people on the spot, capriciously, as we have seen in the case of the so-called law of abortion rights, that was imposed upon the people by judges . . .' We then got into an argument about the American constitution and where laws come from. Bray's attitude to such matters turned out to be very reminiscent of those militant Muslims living in Britain who openly announce themselves as bound only by Islamic law, not by the democratically enacted laws of their adopted country.

In 2003 Paul Hill was executed for the murder of Dr Britton and his bodyguard, saying he would do it again to save the unborn. Candidly looking forward to dying for his cause, he told a news conference, 'I believe the state, by executing me, will be making me a martyr.' Right-wing anti-abortionists protesting at his execution were joined in unholy alliance by left-wing opponents of the death penalty who urged the Governor of Florida, Jeb Bush, to 'stop the martyrdom of Paul Hill'. They plausibly argued that the judicial killing of Hill would actually encourage more murders, the precise opposite of the deterrent effect that the death penalty is supposed to have. Hill himself smiled all the way to the execution chamber, saying, 'I expect a great reward in heaven . . . I am looking forward to glory.'[127] And he suggested that others should take up his violent cause. Anticipating revenge attacks for the 'martyrdom' of Paul Hill, the police went on heightened alert as he was executed, and several individuals connected with the case received threatening letters accompanied by bullets.

This whole terrible business stems from a simple difference of perception. There are people who, because of their religious convictions, think abortion is murder and are prepared to kill in defence of embryos, which they choose to call 'babies'. On the other side are equally sincere supporters of abortion, who either have different religious convictions, or no religion, coupled with well-thought-out consequentialist morals. They too see themselves as idealists, providing a medical service for patients in need, who would otherwise go to dangerously incompetent back-street quacks. Both sides see the other side as murderers or advocates of murder. Both sides, by their own lights, are equally sincere.

A spokeswoman for another abortion clinic described Paul Hill as a dangerous psychopath. But people like him don't think of themselves as dangerous psychopaths; they think of themselves as good, moral people, guided by God. Indeed, I don't think Paul Hill was a psychopath. Just very religious. Dangerous, yes, but not a psychopath. Dangerously religious. By the lights of his religious faith, Hill was entirely right and moral to shoot Dr Britton. What was wrong with Hill was his religious faith itself. Michael Bray, too, when I met him, didn't strike me as a psychopath. I actually quite liked him. I thought he was an honest and sincere man, quietly spoken and thoughtful, but his mind had unfortunately been captured by poisonous religious nonsense.

Strong opponents of abortion are almost all deeply religious. The sincere supporters of abortion, whether personally religious or not, are likely to follow a non-religious, consequentialist moral philosophy, perhaps invoking Jeremy Bentham's question, 'Can they *suffer*?' Paul Hill and Michael Bray saw no moral difference between killing an embryo and killing a doctor except that the embryo was, to them, a blamelessly innocent 'baby'. The consequentialist sees all the difference in the world. An early embryo has the sentience, as well as the semblance, of a tadpole. A doctor is a grown-up conscious being with hopes, loves, aspirations, fears, a massive store of humane knowledge, the capacity for deep emotion, very probably a devastated widow and orphaned children, perhaps elderly parents who dote on him.

Paul Hill caused real, deep, lasting suffering, to beings with nervous systems capable of suffering. His doctor victim did no such thing. Early embryos that have no nervous system most certainly do not suffer. And if late-aborted embryos with nervous systems suffer – though all suffering is deplorable – it is not because they are *human* that they suffer. There is no general reason to suppose that human embryos at any age suffer more than cow or sheep embryos at the same developmental stage. And there is every reason to suppose that all embryos, whether human or not, suffer far less than adult cows or sheep in a slaughterhouse, especially a ritual slaughterhouse where, for religious reasons, they must be fully conscious when their throats are ceremonially cut.

Suffering is hard to measure,[128] and the details might be disputed. But that doesn't affect my main point, which concerns the difference between secular consequentialist and religiously absolute moral philosophies.* One school of thought cares about whether embryos can suffer. The other cares about whether they are human. Religious moralists can be heard debating questions like, 'When does the developing embryo become a person – a human being?' Secular moralists are more likely to ask, 'Never mind whether it is *human* (what does that even *mean* for a little cluster of cells?); at what age does any developing embryo, of any species, become capable of *suffering*?'

---

* This doesn't, of course, exhaust the possibilities. A substantial majority of American Christians do not take an absolutist attitude to abortion, and are pro-choice. See e.g. the Religious Coalition for Reproductive Choice, at www.rcrc.org/.

THE GREAT BEETHOVEN FALLACY

The anti-abortionist's next move in the verbal chess game usually goes something like this. The point is not whether a human embryo can or cannot suffer at present. The point lies in its *potential.* Abortion has deprived it of the opportunity for a full human life in the future. This notion is epitomized by a rhetorical argument whose extreme stupidity is its only defence against a charge of serious dishonesty. I am speaking of the Great Beethoven Fallacy, which exists in several forms. Peter and Jean Medawar,* in *The Life Science*, attribute the following version to Norman St John Stevas (now Lord St John), a British Member of Parliament and prominent Roman Catholic layman. He, in turn, got it from Maurice Baring (1874–1945), a noted Roman Catholic convert and close associate of those Catholic stalwarts G. K. Chesterton and Hilaire Belloc. He cast it in the form of a hypothetical dialogue between two doctors.

> 'About the terminating of pregnancy, I want your opinion. The father was syphilitic, the mother tuberculous. Of the four children born, the first was blind, the second died, the third was deaf and dumb, the fourth was also tuberculous. What would you have done?'
>
> 'I would have terminated the pregnancy.'
>
> 'Then you would have murdered Beethoven.'[129]

The Internet is riddled with so-called pro-life websites that repeat this ridiculous story, and incidentally change factual premises with wanton abandon. Here's another version. 'If you knew a woman who was pregnant, who had 8 kids already, three of whom were deaf, two who were blind, one mentally retarded (all because she had syphilis), would you recommend that she have an abortion? Then you would have killed

---

* Sir Peter Medawar won the Nobel Prize for Physiology and Medicine, 1960.

Beethoven.'[130] This rendering of the legend demotes the great composer from fifth to ninth in the birth order, raises the number born deaf to three and the number born blind to two, and gives syphilis to the mother instead of the father. Most of the forty-three websites I found when searching for versions of the story attribute it not to Maurice Baring but to a certain Professor L. R. Agnew at UCLA Medical School, who is said to have put the dilemma to his students and to have told them, 'Congratulations, you have just murdered Beethoven.' We might charitably give L. R. Agnew the benefit of doubting his existence – it is amazing how these urban legends sprout. I cannot discover whether it was Baring who originated the legend, or whether it was invented earlier.

For invented it certainly was. It is completely false. The truth is that Ludwig van Beethoven was neither the ninth child nor the fifth child of his parents. He was the eldest – strictly the number two, but his elder sibling died in infancy, as was common in those days, and was not, so far as is known, blind or deaf or dumb or mentally retarded. There is no evidence that either of his parents had syphilis, although it is true that his mother eventually died of tuberculosis. There was a lot of it about at the time.

This is, in fact, a fully fledged urban legend, a fabrication, deliberately disseminated by people with a vested interest in spreading it. But the fact that it is a lie is, in any case, completely beside the point. Even if it were not a lie, the argument derived from it is a very bad argument indeed. Peter and Jean Medawar had no need to doubt the truth of the story in order to point out the fallacy of the argument: 'The reasoning behind this odious little argument is breathtakingly fallacious, for unless it is being suggested that there is some causal connection between having a tubercular mother and a syphilitic father and giving birth to a musical genius the world is no more likely to be deprived of a Beethoven by abortion than by chaste abstinence from intercourse.' The Medawars' laconically scornful dismissal is unanswerable (to borrow the plot of one of Roald Dahl's dark short stories, an equally fortuitous decision *not* to have an abortion in 1888 gave us Adolf Hitler). But you do need a modicum of intelligence – or perhaps freedom from a certain kind of religious upbringing – to get the point. Of the forty-three 'pro-life' websites quoting a version of the Beethoven legend which my Google search turned up on the day of writing, not a single one spotted the illogic in the argument. Every one of them (they were all religious sites, by the way) fell for the fallacy, hook, line and sinker. One of them even acknowledged Medawar (spelled Medavvar) as the source. So eager were these people to believe a fallacy congenial to their faith, they didn't even notice that the Medawars had quoted the argument solely in order to blow it out of the water.

As the Medawars were entirely right to point out, the logical conclusion to the 'human potential' argument is that we potentially deprive a human soul of the gift of existence every time we fail to seize any opportunity for sexual intercourse. Every refusal of any offer of copulation by a fertile individual is, by this dopey 'pro-life' logic, tantamount to the murder of a potential child! Even resisting rape could be represented as murdering a potential baby (and, by the way, there are plenty of 'pro-life' campaigners who would deny abortion even to women who have been brutally raped). The Beethoven argument is, we can clearly see, very bad logic indeed. Its surreal idiocy is best summed up in that splendid song 'Every sperm is sacred' sung by Michael Palin, with a chorus of hundreds of children, in the Monty Python film *The Meaning of Life* (if you haven't seen it, please do). The Great Beethoven Fallacy is a typical example of the kind of logical mess we get into when our minds are befuddled by religiously inspired absolutism.

Notice now that 'pro-life' doesn't exactly mean pro-*life* at all. It means pro-*human*-life. The granting of uniquely special rights to cells of the species *Homo sapiens* is hard to reconcile with the fact of evolution. Admittedly, this will not worry those many anti-abortionists who don't understand that evolution *is* a fact! But let me briefly spell out the argument for the benefit of anti-abortion activists who may be less ignorant of science.

The evolutionary point is very simple. The *humanness* of an embryo's cells cannot confer upon it any absolutely discontinuous moral status. It cannot, because of our evolutionary continuity with chimpanzees and, more distantly, with every species on the planet. To see this, imagine that an intermediate species, say *Australopithecus afarensis*, had chanced to survive and was discovered in a remote part of Africa. Would these creatures 'count as human' or not? To a consequentialist like me, the question doesn't deserve an answer, for nothing turns on it. It is enough that we would be fascinated and honoured to meet a new 'Lucy'. The absolutist, on the other hand, must answer the question, in order to apply the moral principle of granting humans unique and special status *because they are human.* If it came to the crunch, they would presumably need to set up courts, like those of apartheid South Africa, to decide whether a particular individual should 'pass for human'.

Even if a clear answer might be attempted for *Australopithecus*, the gradual continuity that is an inescapable feature of biological evolution tells us that there must be *some* intermediate who would lie sufficiently close to the 'borderline' to blur the moral principle and destroy its absoluteness. A better way to say this is that there are no natural borderlines in evolution. The illusion of a borderline is created by the fact that the evolutionary intermediates happen to be extinct. Of course, it could be argued that humans are more capable of, for example, suffering than other species. This could well be true, and we might legitimately give humans special status by virtue of it. But evolutionary continuity shows that there is no *absolute* distinction. Absolutist moral discrimination is devastatingly undermined by the fact of evolution. An uneasy awareness of this fact might, indeed, underlie one of the main motives creationists have for opposing evolution: they fear what they believe to be its moral consequences. They are wrong to do so but, in any case, it is surely very odd to think that a truth about the real world can be reversed by considerations of what would be morally desirable.

HOW 'MODERATION' IN FAITH FOSTERS FANATICISM

In illustration of the dark side of absolutism, I mentioned the Christians in America who blow up abortion clinics, and the Taliban of Afghanistan, whose list of cruelties, especially to women, I find too painful to recount. I could have expanded upon Iran under the ayatollahs, or Saudi Arabia under the Saud princes, where women cannot drive, and are in trouble if they even leave their homes without a male relative (who may, as a generous concession, be a small male child). See Jan Goodwin's *Price of Honour* for a devastating exposé of the treatment of women in Saudi Arabia and other present-day theocracies. Johann Hari, one of the (London) *Independent*'s liveliest columnists, wrote an article whose title speaks for itself: 'The best way to undermine the jihadists is to trigger a rebellion of Muslim women.'[131]

Or, switching to Christianity, I could have cited those American 'rapture' Christians whose powerful influence on American Middle Eastern policy is governed by their biblical belief that Israel has a God-given right to all the lands of Palestine.[132] Some rapture Christians go further and actually yearn for nuclear war because they interpret it as the 'Armageddon' which, according to their bizarre but disturbingly popular interpretation of the book of Revelation, will hasten the Second Coming. I cannot improve on Sam Harris's chilling comment, in his *Letter to a Christian Nation*:

> It is, therefore, not an exaggeration to say that if the city of New York were suddenly replaced by a ball of fire, some significant percentage of the American population would see a silver-lining in the subsequent mushroom cloud, as it would suggest to them that the best thing that is ever going to happen was about to happen: the return of Christ. It should be blindingly obvious that beliefs of this sort will do little

> to help us create a durable future for ourselves – socially, economically, environmentally, or geopolitically. Imagine the consequences if any significant component of the U.S. government actually believed that the world was about to end and that its ending would be *glorious*. The fact that nearly half of the American population apparently believes this, purely on the basis of religious dogma, should be considered a moral and intellectual emergency.

There are, then, people whose religious faith takes them right outside the enlightened consensus of my 'moral *Zeitgeist*'. They represent what I have called the dark side of religious absolutism, and they are often called extremists. But my point in this section is that even mild and moderate religion helps to provide the climate of faith in which extremism naturally flourishes.

In July 2005, London was the victim of a concerted suicide bomb attack: three bombs in the subway and one in a bus. Not as bad as the 2001 attack on the World Trade Center, and certainly not as unexpected (indeed, London had been braced for just such an event ever since Blair volunteered us as unwilling side-kicks in Bush's invasion of Iraq), nevertheless the London explosions horrified Britain. The newspapers were filled with agonized appraisals of what drove four young men to blow themselves up and take a lot of innocent people with them. The murderers were British citizens, cricket-loving, well-mannered, just the sort of young men whose company one might have enjoyed.

Why did these cricket-loving young men do it? Unlike their Palestinian counterparts, or their kamikaze counterparts in Japan, or their Tamil Tiger counterparts in Sri Lanka, these human bombs had no expectation that their bereaved families would be lionized, looked after or supported on martyrs' pensions. On the contrary, their relatives in some cases had to go into hiding. One of the men wantonly widowed his pregnant wife and orphaned his toddler. The action of these four young men has been nothing short of a disaster not just for themselves and their victims, but for their families and for the whole Muslim community in Britain, which now faces a backlash. Only religious faith is a strong enough force to motivate such utter madness in otherwise sane and decent people. Once again, Sam Harris put the point with percipient bluntness, taking the example of the Al-Qaeda leader Osama bin Laden (who had nothing to do with the London bombings, by the way). Why would anyone want to destroy the World Trade Center and everybody in it? To call bin Laden 'evil' is to evade our responsibility to give a proper answer to such an important question.

> The answer to this question is obvious – if only because it has been patiently articulated ad nauseam by bin Laden himself. The answer is that men like bin Laden *actually* believe what they say they believe. They believe in the literal truth of the Koran. Why did nineteen well-educated middle-class men trade their lives in this world for the privilege of killing thousands of our neighbors? Because they believed that they would go straight to paradise for doing so. It is rare to find the behavior of humans so fully and satisfactorily explained. Why have we been so reluctant to accept this explanation?[133]

The respected journalist Muriel Gray, writing in the (Glasgow) *Herald* on 24 July 2005, made a similar point, in this case with reference to the London bombings.

> Everyone is being blamed, from the obvious villainous duo of George W. Bush and Tony Blair, to the inaction of Muslim 'communities'. But it has never been clearer that there is only one place to lay the blame and it has ever been thus. The cause of all this misery, mayhem, violence, terror and ignorance

> is of course religion itself, and if it seems ludicrous to have to state such an obvious reality, the fact is that the government and the media are doing a pretty good job of pretending that it isn't so.

Our Western politicians avoid mentioning the R word (religion), and instead characterize their battle as a war against 'terror', as though terror were a kind of spirit or force, with a will and a mind of its own. Or they characterize terrorists as motivated by pure 'evil'. But they are not motivated by evil. However misguided we may think them, they are motivated, like the Christian murderers of abortion doctors, by what they perceive to be righteousness, faithfully pursuing what their religion tells them. They are not psychotic; they are religious idealists who, by their own lights, are rational. They perceive their acts to be good, not because of some warped personal idiosyncrasy, and not because they have been possessed by Satan, but because they have been brought up, from the cradle, to have total and unquestioning *faith.* Sam Harris quotes a failed Palestinian suicide bomber who said that what drove him to kill Israelis was 'the love of martyrdom . . . I didn't want revenge for anything. I just wanted to be a martyr.' On 19 November 2001 *The New Yorker* carried an interview by Nasra Hassan of another failed suicide bomber, a polite young Palestinian aged twenty-seven known as 'S'. It is so poetically eloquent of the lure of paradise, as preached by moderate religious leaders and teachers, that I think it is worth giving at some length:

> 'What is the attraction of martyrdom?' I asked.
>
> 'The power of the spirit pulls us upward, while the power of material things pulls us downward,' he said. 'Someone bent on martyrdom becomes immune to the material pull. Our planner asked, "What if the operation fails?" We told him, "In any case, we get to meet the Prophet and his companions, inshallah."

> 'We were floating, swimming, in the feeling that we were about to enter eternity. We had no doubts. We made an oath on the Koran, in the presence of Allah – a pledge not to waver. This jihad pledge is called *bayt al-ridwan*, after the garden in Paradise that is reserved for the prophets and the martyrs. I know that there are other ways to do jihad. But this one is sweet – the sweetest. All martyrdom operations, if done for Allah's sake, hurt less than a gnat's bite!'
>
> S showed me a video that documented the final planning for the operation. In the grainy footage, I saw him and two other young men engaging in a ritualistic dialogue of questions and answers about the glory of martyrdom . . .
>
> The young men and the planner then knelt and placed their right hands on the Koran. The planner said: 'Are you ready? Tomorrow, you will be in Paradise.'[134]

If I had been 'S', I'd have been tempted to say to the planner, 'Well, in that case, why don't you put *your* neck where your mouth is? Why don't *you* do the suicide mission and take the fast track to Paradise?' But what is so hard for us to understand is that – to repeat the point because it is so important – *these people actually believe what they say they believe.* The take-home message is that we should blame religion itself, not religious *extremism* – as though that were some kind of terrible perversion of real, decent religion. Voltaire got it right long ago: 'Those who can make you believe absurdities can make you commit atrocities.' So did Bertrand Russell: 'Many people would sooner die than think. In fact they do.'

As long as we accept the principle that religious faith must be respected simply because it is religious faith, it is hard to withhold respect from the faith of Osama bin Laden and the suicide bombers. The alternative, one so transparent that it should need no urging, is to abandon the principle of automatic respect for religious faith. This is one reason why I do everything in my power to warn people against faith itself, not just against so-called 'extremist' faith. The teachings of 'moderate' religion, though not extremist in themselves, are an open invitation to extremism.

It might be said that there is nothing special about religious faith here. Patriotic love of country or ethnic group can also make the world safe for its own version of extremism, can't it? Yes it can, as with the kamikazes in Japan and the Tamil Tigers in Sri Lanka. But religious faith is an especially potent silencer of rational calculation, which usually seems to trump all others. This is mostly, I suspect, because of the easy and beguiling promise that death is not the end, and that a martyr's heaven is especially glorious. But it is also partly because it discourages questioning, by its very nature.

Christianity, just as much as Islam, teaches children that unquestioned faith is a virtue. You don't have to make the case for what you believe. If somebody announces that it is part of his *faith*, the rest of society, whether of the same faith, or another, or of none, is obliged, by ingrained custom, to 'respect' it without question; respect it until the day it manifests itself in a horrible massacre like the destruction of the World Trade Center, or the London or Madrid bombings. Then there is a great chorus of disownings, as clerics and 'community leaders' (who elected *them*, by the way?) line up to explain that this extremism is a perversion of the 'true' faith. But how can there be a perversion of faith, if faith, lacking objective justification, doesn't have any demonstrable standard to pervert?

Ten years ago, Ibn Warraq, in his excellent book *Why I Am Not a Muslim*, made a similar point from the standpoint of a deeply knowledgeable scholar of Islam. Indeed, a good alternative title for Warraq's book might have been *The Myth of Moderate Islam*, which is the actual title of a more recent article in the (London) *Spectator* (30 July 2005) by another scholar, Patrick Sookhdeo, director of the Institute for the Study of

Islam and Christianity. 'By far the majority of Muslims today live their lives without recourse to violence, for the Koran is like a pick-and-mix selection. If you want peace, you can find peaceable verses. If you want war, you can find bellicose verses.'

Sookhdeo goes on to explain how Islamic scholars, in order to cope with the many contradictions that they found in the Qur'an, developed the principle of abrogation, whereby later texts trump earlier ones. Unfortunately, the peaceable passages in the Qur'an are mostly early, dating from Muhammad's time in Mecca. The more belligerent verses tend to date from later, after his flight to Medina. The result is that

> the mantra 'Islam is peace' is almost 1,400 years out of date. It was only for about 13 years that Islam was peace and nothing but peace .. For today's radical Muslims – just as for the mediaeval jurists who developed classical Islam – it would be truer to say 'Islam is war'. One of the most radical Islamic groups in Britain, al-Ghurabaa, stated in the wake of the two London bombings, 'Any Muslim that denies that terror is a part of Islam is kafir.' A kafir is an unbeliever (i.e. a non-Muslim), a term of gross insult . . .
>
> Could it be that the young men who committed suicide were neither on the fringes of Muslim society in Britain, nor following an eccentric and extremist interpretation of their faith, but rather that they came from the very core of the Muslim community and were motivated by a mainstream interpretation of Islam?

More generally (and this applies to Christianity no less than to Islam), what is really pernicious is the practice of teaching children that faith itself is a virtue. Faith is an evil precisely because it requires no justification and brooks no argument. Teaching children that unquestioned faith is a virtue primes them – given certain other ingredients that are not hard to come by – to grow up into potentially lethal weapons for future jihads or crusades. Immunized against fear by the promise of a martyr's paradise, the authentic faith-head deserves a high place in the history of armaments, alongside the longbow, the warhorse, the tank and the cluster bomb. If children were taught to question and think through their beliefs, instead of being taught the superior virtue of faith without question, it is a good bet that there would be no suicide bombers. Suicide bombers do what they do because they really believe what they were taught in their religious schools: that duty to God exceeds all other priorities, and that martyrdom in his service will be rewarded in the gardens of Paradise. And they were taught *that* lesson not necessarily by extremist fanatics but by decent, gentle, mainstream religious instructors, who lined them up in their madrasas, sitting in rows, rhythmically nodding their innocent little heads up and down while they learned every word of the holy book like demented parrots. Faith can be very very dangerous, and deliberately to implant it into the vulnerable mind of an innocent child is a grievous wrong. It is to childhood itself, and the violation of childhood by religion, that we turn in the next chapter.

CHAPTER 9

Childhood, abuse and the escape from religion

There is in every village a torch – the teacher: and an extinguisher – the clergyman.

VICTOR HUGO

I begin with an anecdote of nineteenth-century Italy. I am not implying that anything like this awful story could happen today. But the attitudes of mind that it betrays are lamentably current, even though the practical details are not. This nineteenth-century human tragedy sheds a pitiless light on present-day religious attitudes to children.

In 1858 Edgardo Mortara, a six-year-old child of Jewish parents living in Bologna, was legally seized by the papal police acting under orders from the Inquisition. Edgardo was forcibly dragged away from his weeping mother and distraught father to the Catechumens (house for the conversion of Jews and Muslims) in Rome, and thereafter brought up as a Roman Catholic. Aside from occasional brief visits under close priestly supervision, his parents never saw him again. The story is told by David I. Kertzer in his remarkable book, *The Kidnapping of Edgardo Mortara.*

Edgardo's story was by no means unusual in Italy at the time, and the reason for these priestly abductions was always the same. In every case, the child had been secretly baptized at some earlier date, usually by a Catholic nursemaid, and the Inquisition later came to hear of the baptism. It was a central part of the Roman Catholic belief-system that, once a child had been baptized, however informally and clandestinely, that child was irrevocably transformed into a Christian. In their mental world, to allow a 'Christian child' to stay with his Jewish parents was not an option, and they maintained this bizarre and cruel stance steadfastly, and with the utmost sincerity, in the face of worldwide outrage. That widespread outrage, by the way, was dismissed by the Catholic newspaper *Civiltà Cattolica* as due to the international power of rich Jews – sounds familiar, doesn't it?

Apart from the publicity it aroused, Edgardo Mortara's history was entirely typical of many others. He had once been looked after by Anna Morisi, an illiterate Catholic girl who was then fourteen. He fell ill and she panicked lest he might die. Brought up in a stupor of belief that a child who died unbaptized would suffer forever in hell, she asked advice from a Catholic neighbour who told her how to do a baptism. She went back into the house, threw some water from a bucket on little Edgardo's head and said, 'I baptize you in the name of the Father and of the Son and of the Holy Ghost.' And that was it. From that moment on, Edgardo was legally a Christian. When the priests of the Inquisition learned of the incident years later, they acted promptly and decisively, giving no thought to the sorrowful consequences of their action.

Amazingly for a rite that could have such monumental significance for a whole extended family, the Catholic Church allowed (and still allows) anybody to baptize anybody else. The baptizer doesn't have to be a priest. Neither the child, nor the parents, nor anybody else has to consent to the baptism. Nothing need be signed. Nothing need be officially witnessed. All that is necessary is a splash of water, a few words, a helpless child, and a superstitious and catechistically brainwashed babysitter. Actually, only the last of these is needed because, assuming the child is too young to be a witness, who is even to know? An American colleague who was brought up Catholic writes to me as follows: 'We used to baptize our dolls. I don't remember any of us baptizing our little Protestant friends but no doubt that has happened and happens today. We made little Catholics of our dolls, taking them to church, giving them Holy Communion etc. We were brainwashed to be good Catholic mothers early on.'

If nineteenth-century girls were anything like my modern correspondent, it is surprising that cases like Edgardo Mortara's were not more common than they were. As it was, such stories were distressingly frequent in nineteenth-century Italy, which leaves one asking the obvious question. Why did the Jews of the Papal States employ Catholic servants at all, given the appalling risk that could flow from doing so? Why didn't they take good care to engage Jewish servants? The answer, yet again, has nothing to do with sense and everything to do with religion. The Jews needed servants whose religion didn't forbid them to work on the sabbath. A Jewish maid could indeed be relied upon not to baptize your child into a spiritual orphanage. But she couldn't light the fire or clean the house on a Saturday. This was why, of the Bolognese Jewish families at the time who could afford servants, most hired Catholics.

In this book, I have deliberately refrained from detailing the horrors of the Crusades, the *conquistadores* or the Spanish Inquisition. Cruel and evil people can be found in every century and of every persuasion. But this story of the Italian Inquisition and its attitude to children is particularly revealing of the religious mind, and the evils that arise specifically *because* it is religious. First is the remarkable perception by the religious mind that a sprinkle of water and a brief verbal incantation can totally change a child's life, taking precedence over parental consent, the child's own consent, the child's own happiness and psychological well-being . . . over everything that ordinary common sense and human feeling would see as important. Cardinal Antonelli spelled it out at the time in a letter to Lionel Rothschild, Britain's first Jewish Member of Parliament, who had written to protest about Edgardo's abduction. The cardinal replied that he was powerless to intervene, and added, 'Here it may be opportune to observe that, if the voice of nature is powerful, even more powerful are the sacred duties of religion.' Yes, well, that just about says it all, doesn't it?

Second is the extraordinary fact that the priests, cardinals and Pope seem genuinely not to have understood what a terrible thing they were doing to poor Edgardo Mortara. It passes all sensible understanding, but they sincerely believed they were doing him a good turn by taking him away from his parents and giving him a Christian upbringing. They felt a duty of *protection*! A Catholic newspaper in the United States defended the Pope's stance on the Mortara case, arguing that it was unthinkable that a Christian government 'could leave a Christian child to be brought up by a Jew' and invoking the principle of religious liberty, 'the liberty of a child to be a Christian and not forced compulsorily to be a Jew . . . The Holy Father's protection of the child, in the face of all the ferocious fanaticism of infidelity and bigotry, is the grandest moral spectacle which the world has seen for ages.' Has there ever been a more flagrant misdirection of words like 'forced', 'compulsorily', 'ferocious', 'fanaticism' and 'bigotry'? Yet all the indications are that Catholic apologists, from the Pope down, sincerely believed that what they were doing was right: absolutely right morally, and right for the welfare of the child. Such is the power of (mainstream, 'moderate') religion to warp judgement and pervert ordinary human decency. The newspaper *Il Cattolico* was frankly bewildered at the widespread failure to see what a magnanimous favour the Church had done Edgardo Mortara when it rescued him from his Jewish family:

> Whoever among us gives a little serious thought to the matter, compares the condition of a Jew – without a true Church, without a King, and without a country, dispersed and always a foreigner wherever he lives on the face of the earth, and moreover,

> infamous for the ugly stain with which the killers of Christ are marked . . . will immediately understand how great is this temporal advantage that the Pope is obtaining for the Mortara boy.

Third is the presumptuousness whereby religious people *know*, without evidence, that the faith of their birth is the one true faith, all others being aberrations or downright false. The above quotations give vivid examples of this attitude on the Christian side. It would be grossly unjust to equate the two sides in this case, but this is as good a place as any to note that the Mortaras could at a stroke have had Edgardo back, if only they had accepted the priests' entreaties and agreed to be baptized themselves. Edgardo had been stolen in the first place because of a splash of water and a dozen meaningless words. Such is the fatuousness of the religiously indoctrinated mind, another pair of splashes is all it would have taken to reverse the process. To some of us, the parents' refusal indicates wanton stubbornness. To others, their principled stand elevates them into the long list of martyrs for all religions down the ages.

'Be of good comfort Master Ridley and play the man: we shall this day by God's grace light such a candle in England, as I trust shall never be put out.' No doubt there are causes for which to die is noble. But how could the martyrs Ridley, Latimer and Cranmer let themselves be burned rather than forsake their Protestant Little-endianism in favour of Catholic Big-endianism – does it really matter all that much from which end you open a boiled egg? Such is the stubborn – or admirable, if that is your view – conviction of the religious mind, that the Mortaras could not bring themselves to seize the opportunity offered by the meaningless rite of baptism. Couldn't they cross their fingers, or whisper 'not' under their breath while being baptized? No, they couldn't, because they had been brought up in a (moderate) religion, and therefore took the whole ridiculous charade seriously. As for me, I think only of poor little Edgardo – unwittingly born into a world dominated by the religious mind, hapless in the crossfire, all but orphaned in an act of well-meaning but, to a young child, shattering cruelty.

Fourth, to pursue the same theme, is the assumption that a six-year-old child can properly be said to have a religion at all, whether it is Jewish or Christian or anything else. To put it another way, the idea that baptizing an unknowing, uncomprehending child can change him from one religion to another at a stroke seems absurd – but it is surely not more absurd than labelling a tiny child as belonging to any particular religion in the first place. What mattered to Edgardo was not 'his' religion (he was too young to possess thought-out religious opinions) but the love and care of his parents and family, and he was deprived of those by celibate priests whose grotesque cruelty was mitigated only by their crass insensitivity to normal human feelings – an insensitivity that comes all too easily to a mind hijacked by religious faith.

Even without physical abduction, isn't it always a form of child abuse to label children as possessors of beliefs that they are too young to have thought about? Yet the practice persists to this day, almost entirely unquestioned. To question it is my main purpose in this chapter.

PHYSICAL AND MENTAL ABUSE

Priestly abuse of children is nowadays taken to mean sexual abuse, and I feel obliged, at the outset, to get the whole matter of sexual abuse into proportion and out of the way. Others have noted that we live in a time of hysteria about pedophilia, a mob psychology that calls to mind the Salem witch-hunts of 1692. In July 2000 the *News of the World*, widely acclaimed in the face of stiff competition as Britain's most disgusting newspaper, organized a 'name and shame' campaign, barely stopping short of inciting vigilantes to take direct violent action against pedophiles. The house of a hospital pediatrician was attacked by zealots unacquainted with the difference between a pediatrician and a pedophile.[135] The mob hysteria over pedophiles has reached epidemic proportions and driven parents to panic. Today's Just Williams, today's Huck Finns, today's Swallows and Amazons are deprived of the freedom to roam that was one of the delights of childhood in earlier times (when the actual, as opposed to perceived, risk of molestation was probably no less).

In fairness to the *News of the World*, at the time of its campaign passions had been aroused by a truly horrifying murder, sexually motivated, of an eight-year-old girl kidnapped in Sussex. Nevertheless, it is clearly unjust to visit upon all pedophiles a vengeance appropriate to the tiny minority who are also murderers. All three of the boarding schools I attended employed teachers whose affection for small boys overstepped the bounds of propriety. That was indeed reprehensible. Nevertheless if, fifty years on, they had been hounded by vigilantes or lawyers as no better than child murderers, I should have felt obliged to come to their defence, even as the victim of one of them (an embarrassing but otherwise harmless experience).

The Roman Catholic Church has borne a heavy share of such retrospective opprobrium. For all sorts of reasons I dislike the Roman Catholic Church. But I dislike unfairness even more, and I can't help wondering whether this one institution has been unfairly demonized over the issue, especially in Ireland and America. I suppose some additional public resentment flows from the hypocrisy of priests whose professional life is largely devoted to arousing guilt about 'sin'. Then there is the abuse of trust by a figure in authority, whom the child has been trained from the cradle to revere. Such additional resentments should make us all the more careful not to rush to judgement. We should be aware of the remarkable power of the mind to concoct false memories, especially when abetted by unscrupulous therapists and mercenary lawyers. The psychologist Elizabeth Loftus has shown great courage, in the face of spiteful vested interests, in demonstrating how easy it is for people to concoct memories that are entirely false but which seem, to the victim, every bit as real as true memories.[136] This is so counter-intuitive that juries are easily swayed by sincere but false testimony from witnesses.

In the particular case of Ireland, even without the sexual abuse, the brutality of the Christian Brothers,[137] responsible for the education of a significant proportion of the male population of the country, is legendary. And the same could be said of the often sadistically cruel nuns who ran many of Ireland's girls' schools. The infamous Magdalene Asylums, subject of Peter Mullan's film *The Magdalene Sisters*, continued in existence until as late as 1996. Forty years on, it is harder to get redress for floggings than for sexual fondlings, and there is no shortage of lawyers actively soliciting custom from victims who might not otherwise have raked over the distant past. There's gold in them thar long-gone fumbles in the vestry – some of them, indeed, so long gone that the alleged offender is likely to be dead and unable to present his side of the story. The Catholic Church worldwide has paid out more than a billion dollars in compensation.[138] You might almost sympathize with them, until you remember where their money came from in the first place.

Once, in the question time after a lecture in Dublin, I was asked what I thought about the widely publicized cases of sexual abuse by Catholic priests in Ireland. I replied that, horrible as sexual abuse no doubt was, the damage was arguably less than the long-term psychological damage inflicted by bringing the child up Catholic in the first place. It was an off-the-cuff remark made in the heat of the moment, and I was surprised that it earned a round of enthusiastic applause from that Irish audience (composed, admittedly, of Dublin intellectuals and presumably not representative of the country at large). But I was reminded of the incident later when I received a letter from an American woman in her forties who had been brought up Roman Catholic. At the age of seven, she told me, two unpleasant things had happened to her. She was sexually abused by her parish priest in his car. And, around the same time, a little schoolfriend of hers, who had tragically died, went to hell because she was a Protestant. Or so my correspondent had been led to believe by the then official doctrine of her parents' church. Her view as a mature adult was that, of these two examples of Roman Catholic child abuse, the one physical and the other mental, the second was by far the worst. She wrote:

> Being fondled by the priest simply left the impression (from the mind of a 7 year old) as 'yucky' while the memory of my friend going to hell was one of cold, immeasurable fear. I never lost sleep because of the priest – but I spent many a night being terrified that the people I loved would go to Hell. It gave me nightmares.

Admittedly, the sexual fondling she suffered in the priest's car was relatively mild compared with, say, the pain and disgust of a sodomized altar boy. And nowadays the Catholic Church is said not to make so much of hell as it once did. But the example shows that it is at least possible for psychological abuse of children to outclass physical. It is said that Alfred Hitchcock, the great cinematic specialist in the art of frightening people, was once driving through Switzerland when he suddenly pointed out of the car window and said, 'That is the most frightening sight I have ever seen.' It was a priest in conversation with a little boy, his hand on the boy's shoulder. Hitchcock leaned out of the car window and shouted, 'Run, little boy! Run for your life!'

'Sticks and stones may break my bones, but words can never hurt me.' The adage is true as long as you don't really *believe* the words. But if your whole upbringing, and everything you have ever been told by parents, teachers and priests, has led you to believe, *really believe*, utterly and completely, that sinners burn in hell (or some other obnoxious article of doctrine such as that a woman is the property of her husband), it is entirely plausible that words could have a more long-lasting and damaging effect than deeds. I am persuaded that the phrase 'child abuse' is no exaggeration when used to describe what teachers and priests are doing to children whom they encourage to believe in something like the punishment of unshriven mortal sins in an eternal hell.

In the television documentary *Root of All Evil?* to which I have already referred, I interviewed a number of religious leaders and was criticized for picking on American extremists rather than respectable mainstreamers like archbishops.* It sounds like a fair criticism – except that, in early 21st-century America, what *seems* extreme to the outside world is actually mainstream. One of my interviewees who most appalled the British television audience, for example, was Pastor Ted Haggard of Colorado Springs. But, far from being extreme in Bush's America, 'Pastor Ted' is president of the thirty-million-strong National Association of Evangelicals, and he claims to be favoured with a telephone consultation with President Bush every Monday. If I had wanted to interview real extremists by modern American standards, I'd have gone for 'Reconstructionists' whose 'Dominion Theology' openly advocates a Christian theocracy in America. As a concerned American colleague writes to me:

> Europeans need to know there is a traveling theo-freak show which actually advocates reinstatement of Old Testament law – killing of homosexuals etc. – and the right to hold office, or even to vote, for Christians only. Middle class crowds cheer to this rhetoric. If secularists are not vigilant, Dominionists

---

* The Archbishop of Canterbury, the Cardinal Archbishop of Westminster and the Chief Rabbi of Britain were all invited to be interviewed by me. All declined, doubtless for good reasons. The Bishop of Oxford agreed, and he was as delightful, and as far from being extremist, as they surely would have been.

> and Reconstructionists will soon be mainstream in a true American theocracy.*

Another of my television interviewees was Pastor Keenan Roberts, from the same state of Colorado as Pastor Ted. Pastor Roberts's particular brand of nuttiness takes the form of what he calls Hell Houses. A Hell House is a place where children are brought, by their parents or their Christian schools, to be scared witless over what might happen to them after they die. Actors play out fearsome tableaux of particular 'sins' like abortion and homosexuality, with a scarlet-clad devil in gloating attendance. These are a prelude to the *pièce de résistance*, Hell Itself, complete with realistic sulphurous smell of burning brimstone and the agonized screams of the forever damned.

After watching a rehearsal, in which the devil was suitably diabolical in the hammed-up style of a villain of Victorian melodrama, I interviewed Pastor Roberts in the presence of his cast. He told me that the optimum age for a child to visit a Hell House is twelve. This shocked me somewhat, and I asked him whether it would worry him if a twelve-year-old child had nightmares after one of his performances. He replied, presumably honestly:

> I would rather for them to understand that Hell is a place that they absolutely do not want to go. I would

---

* The following seems to be real, although I at first suspected a satirical hoax by *The Onion*: www.talk2action.org/story/2006/5/29/195855/959. It is a computer game called Left Behind: Eternal Forces. P. Z. Myers sums it up on his excellent Pharyngula website. 'Imagine: you are a foot soldier in a paramilitary group whose purpose is to remake America as a Christian theocracy and establish its worldly vision of the dominion of Christ over all aspects of life . . . You are on a mission – both a religious mission and a military mission – to convert or kill Catholics, Jews, Muslims, Buddhists, gays, and anyone who advocates the separation of church and state . . .' See http://scienceblogs.com/pharyngula/2006/05/30/gta-meet-lbef/; for a review, see http://select.nytimes.com/gst/abstract.html?res=F1071FFD3C550C718CDDAA0894DE404482.

> rather reach them with that message at twelve than to not reach them with that message and have them live a life of sin and to never find the Lord Jesus Christ. And if they end up having nightmares, as a result of experiencing this, I think there's a higher good that would ultimately be achieved and accomplished in their life than simply having nightmares.

I suppose that, if you really and truly believed what Pastor Roberts says he believes, you would feel it right to intimidate children too.

We cannot write off Pastor Roberts as an extremist wingnut. Like Ted Haggard, he is mainstream in today's America. I'd be surprised if even they would buy into the belief of some of their co-religionists that you can hear the screams of the damned if you listen in on volcanoes,[139] and that the giant tube worms found in hot deep-ocean vents are fulfilments of Mark 9: 43–4: 'And if thy hand offend thee, cut it off: it is better for thee to enter into life maimed, than having two hands to go into hell, into the fire that never shall be quenched: where their worm dieth not, and the fire is not quenched.' Whatever they believe hell is actually like, all these hell-fire enthusiasts seem to share the gloating *Schadenfreude* and complacency of those who know they are among the saved, well conveyed by that foremost among theologians, St Thomas Aquinas, in *Summa Theologica*: 'That the saints may enjoy their beatitude and the grace of God more abundantly they are permitted to see the punishment of the damned in hell.' Nice man.*

The fear of hell-fire can be very real, even among otherwise rational people. After my television documentary on religion, among the many letters I received was this, from an obviously bright and honest woman:

---

* Compare Ann Coulter's charming Christian charity: 'I defy any of my co-religionists to tell me they do not laugh at the idea of Dawkins burning in hell' (Coulter 2006: 268).

> I went to a Catholic school from the age of five, and was indoctrinated by nuns who wielded straps, sticks and canes. During my teens I read Darwin, and what he said about evolution made such a lot of sense to the logical part of my mind. However, I've gone through life suffering much conflict and a deep down fear of hell fire which gets triggered quite frequently. I've had some psychotherapy which has enabled me to work through some of my earlier problems but can't seem to overcome this deep fear.
>
> So, the reason I'm writing to you is would you send me please the name and address of the therapist you interviewed on this week's programme who deals with this particular fear.

I was moved by her letter, and (suppressing a momentary and ignoble regret that there is no hell for those nuns to go to) replied that she should trust in her reason as a great gift which she – unlike less fortunate people – obviously possessed. I suggested that the extreme horribleness of hell, as portrayed by priests and nuns, is inflated to compensate for its implausibility. If hell were plausible, it would only have to be moderately unpleasant in order to deter. Given that it is so unlikely to be true, it has to be advertised as very very scary indeed, to balance its implausibility and retain some deterrence value. I also put her in touch with the therapist she mentioned, Jill Mytton, a delightful and deeply sincere woman whom I had interviewed on camera. Jill had herself been raised in a more than usually odious sect called the Exclusive Brethren: so unpleasant that there is even a website, http://wikipeebia.com/,* entirely devoted to caring for those who have escaped from it.

Jill Mytton was brought up to be terrified of hell, escaped from Christianity as an adult, and now counsels and helps

---

* See also http://culthelp.info/index.php?option=com_content&task=view&id:221<emid=8

others similarly traumatized in childhood: 'If I think back to my childhood, it's one dominated by fear. And it was the fear of disapproval while in the present, but also of eternal damnation. And for a child, images of hell-fire and gnashing of teeth are actually very real. They are not metaphorical at all.' I then asked her to spell out what she had actually been told about hell, as a child, and her eventual reply was as moving as her expressive face during the long hesitation before she answered: 'It's strange, isn't it? After all this time it still has the power to . . . affect me . . . when you . . . when you ask me that question. Hell is a fearful place. It's complete rejection by God. It's complete judgement, there is real fire, there is real torment, real torture, and it goes on for ever so there is no respite from it.'

She went on to tell me of the support group she runs for escapees from a childhood similar to her own, and she dwelt on how difficult it is for many of them to leave: 'The process of leaving is extraordinarily difficult. Ah, you are leaving behind a whole social network, a whole system that you've practically been brought up in, you are leaving behind a belief-system that you have held for years. Very often you leave families and friends . . . You don't really exist any more for them.' I was able to chime in with my own experience of letters from people in America saying they have read my books and have given up their religion as a consequence. Disconcertingly many go on to say that they daren't tell their families, or that they have told their families with terrible results. The following is typical. The writer is a young American medical student.

> I felt the urge to write you an email because I share your view on religion, a view that is, as I'm sure you're aware, isolating in America. I grew up in a Christian family and even though the idea of religion never sat well with me I only recently got up the nerve to tell someone. That someone was my girlfriend who was . . . horrified. I realize that a declaration of atheism could be shocking but now it's as if she views me as

> a completely different person. She can't trust me, she says, because my morals don't come from God. I don't know if we'll get past this, and I don't particularly want to share my belief with other people who are close to me because I fear the same reaction of distaste . . . I don't expect a response. I only write to you because I hoped you'd sympathize and share in my frustration. Imagine losing someone you loved, and who loved you, on the basis of religion. Aside from her view that I'm now a Godless heathen we were perfect for each other. It reminds me of your observation that people do insane things in the name of their faith. Thanks for listening.

I replied to this unfortunate young man, pointing out to him that, while his girlfriend had discovered something about him, he too had discovered something about her. Was she really good enough for him? I doubted it.

I have already mentioned the American comic actor Julia Sweeney and her dogged and endearingly humorous struggle to find some redeeming features in religion and to rescue the God of her childhood from her growing adult doubts. Eventually her quest ended happily, and she is now an admirable role model for young atheists everywhere. The *dénouement* is perhaps the most moving scene of her show *Letting Go of God.* She had tried everything. And then . . .

> . . . as I was walking from my office in my backyard into my house, I realized there was this little teeny-weenie voice whispering in my head. I'm not sure how long it had been there, but it suddenly got just one decibel louder. It whispered, 'There is no god.'
>
> And I tried to ignore it. But it got a teeny bit louder. 'There is no god. There is no god. *Oh my god, there is no god.*' . . .
>
> And I shuddered. I felt I was slipping off the raft.

> And then I thought, 'But I can't. I don't know if I *can* not believe in God. I need God. I mean, we have a history' . . .
>
> 'But I don't know how to not believe in God. I don't know how you do it. How do you get up, how do you get through the day?' I felt unbalanced . . .
>
> I thought, 'Okay, calm down. Let's just try on the not-believing-in-God glasses for a moment, just for a second. Just put on the no-God glasses and take a quick look around and then immediately throw them off.' And I put them on and I looked around.
>
> I'm embarrassed to report that I initially felt dizzy. I actually had the thought, 'Well, how does the Earth stay up in the sky? You mean, we're just hurtling through space? That's so vulnerable!' I wanted to run out and catch the Earth as it fell out of space into my hands.
>
> And then I remembered, 'Oh yeah, gravity and angular momentum is gonna keep us revolving around the sun for probably a long, long time.'

When I saw *Letting Go of God* in a Los Angeles theatre I was deeply moved by this scene. Especially when Julia went on to tell us of her parents' reaction to a press report of her cure:

> My first call from my mother was more of a scream. 'Atheist? ATHEIST?!?!'
>
> My dad called and said, 'You have betrayed your family, your school, your city.' It was like I had sold secrets to the Russians. They both said they weren't going to talk to me any more. My dad said, 'I don't even want you to come to my funeral.' After I hung up, I thought, 'Just try and stop me.'

Part of Julia Sweeney's gift is to make you cry and laugh at the same time:

> I think that my parents had been mildly disappointed when I'd said I didn't believe in God any more, but being an *atheist* was another thing altogether.

Dan Barker's *Losing Faith in Faith: From Preacher to Atheist* is the story of his gradual conversion from devout fundamentalist minister and zealous travelling preacher to the strong and confident atheist he is today. Significantly, Barker continued to go through the motions of preaching Christianity for a while after he had become an atheist, because it was the only career he knew and he felt locked into a web of social obligations. He now knows many other American clergymen who are in the same position as he was but have confided only in him, having read his book. They dare not admit their atheism even to their own families, so terrible is the anticipated reaction. Barker's own story had a happier conclusion. To begin with, his parents were deeply and agonizingly shocked. But they listened to his quiet reasoning, and eventually became atheists themselves.

Two professors from one university in America wrote to me independently about their parents. One said that his mother suffers permanent grief because she fears for his immortal soul. The other one said that his father wishes he had never been born, so convinced is he that his son is going to spend eternity in hell. These are highly educated university professors, confident in their scholarship and their maturity, who have presumably left their parents behind in all matters of the intellect, not just religion. Just think what the ordeal must be like for less intellectually robust people, less equipped by education and rhetorical skill than they are, or than Julia Sweeney is, to argue their corner in the face of obdurate family members. As it was for many of Jill Mytton's patients, perhaps.

Earlier in our televised conversation, Jill had described this kind of religious upbringing as a form of mental abuse, and I returned to the point, as follows: 'You use the words religious abuse. If you were to compare the abuse of bringing up a child really to believe in hell . . . how do you think that would compare in trauma terms with sexual abuse?' She replied: 'That's a very difficult question . . . I think there are a lot of similarities actually, because it is about abuse of trust; it is about denying the child the right to feel free and open and able to relate to the world in the normal way . . . it's a form of denigration; it's a form of denial of the true self in both cases.'

IN DEFENCE OF CHILDREN

My colleague the psychologist Nicholas Humphrey used the 'sticks and stones' proverb in introducing his Amnesty Lecture in Oxford in 1997.[140] Humphrey began his lecture by arguing that the proverb is not always true, citing the case of Haitian Voodoo believers who die, apparently from some psychosomatic effect of terror, within days of having a malign 'spell' cast upon them. He then asked whether Amnesty International, the beneficiary of the lecture series to which he was contributing, should campaign against hurtful or damaging speeches or publications. His answer was a resounding no to such censorship in general: 'Freedom of speech is too precious a freedom to be meddled with.' But he then went on to shock his liberal self by advocating one important exception: to argue in favour of censorship for the special case of children . . .

> . . . moral and religious education, and especially the education a child receives at home, where parents are allowed – even expected – to determine for their children what counts as truth and falsehood, right and wrong. Children, I'll argue, have a human right not to have their minds crippled by exposure to other people's bad ideas – no matter who these other people are. Parents, correspondingly, have no God-given licence to enculturate their children in whatever ways they personally choose: no right to limit the horizons of their children's knowledge, to

> bring them up in an atmosphere of dogma and superstition, or to insist they follow the straight and narrow paths of their own faith.
>
> In short, children have a right not to have their minds addled by nonsense, and we as a society have a duty to protect them from it. So we should no more allow parents to teach their children to believe, for example, in the literal truth of the Bible or that the planets rule their lives, than we should allow parents to knock their children's teeth out or lock them in a dungeon.

Of course, such a strong statement needs, and received, much qualification. Isn't it a matter of opinion what is nonsense? Hasn't the applecart of orthodox science been upset often enough to chasten us into caution? Scientists may think it is nonsense to teach astrology and the literal truth of the Bible, but there are others who think the opposite, and aren't they entitled to teach it to their children? Isn't it just as arrogant to insist that children should be taught science?

I thank my own parents for taking the view that children should be taught not so much *what* to think as *how* to think. If, having been fairly and properly exposed to all the scientific evidence, they grow up and decide that the Bible is literally true or that the movements of the planets rule their lives, that is their privilege. The important point is that it is *their* privilege to decide what they shall think, and not their parents' privilege to impose it by *force majeure.* And this, of course, is especially important when we reflect that children become the parents of the next generation, in a position to pass on whatever indoctrination may have moulded them.

Humphrey suggests that, as long as children are young, vulnerable and in need of protection, truly moral guardianship shows itself in an honest attempt to second-guess what they *would* choose for themselves if they were old enough to do so. He movingly quotes the example of a young Inca girl whose

500-year-old remains were found frozen in the mountains of Peru in 1995. The anthropologist who discovered her wrote that she had been the victim of a ritual sacrifice. By Humphrey's account, a documentary film about this young 'ice maiden' was shown on American television. Viewers were invited

> to marvel at the spiritual commitment of the Inca priests and to share with the girl on her last journey her pride and excitement at having been selected for the signal honour of being sacrificed. The message of the television programme was in effect that the practice of human sacrifice was in its own way a glorious cultural invention – another jewel in the crown of multiculturalism, if you like.

Humphrey is scandalized, and so am I.

> Yet, how dare anyone even suggest this? How dare they invite us – in our sitting rooms, watching television – to feel uplifted by contemplating an act of ritual murder: the murder of a dependent child by a group of stupid, puffed up, superstitious, ignorant old men? How dare they invite us to find good for ourselves in contemplating an immoral action against someone else?

Again, the decent liberal reader may feel a twinge of unease. Immoral by our standards, certainly, and stupid, but what about Inca standards? Surely, to the Incas, the sacrifice was a moral act and far from stupid, sanctioned by all that they held sacred? The little girl was, no doubt, a loyal believer in the religion in which she was brought up. Who are we to use a word like 'murder', judging Inca priests by our own standards rather than theirs? Perhaps this girl was rapturously happy with her fate: perhaps she really believed she was going straight to everlasting paradise, warmed by the radiant company of the Sun God.

Or perhaps – as seems far more likely – she screamed in terror.

Humphrey's point – and mine – is that, regardless of whether she was a willing victim or not, there is strong reason to suppose that she would not have been willing if she had been in full possession of the facts. For example, suppose she had known that the sun is really a ball of hydrogen, hotter than a million degrees Kelvin, converting itself into helium by nuclear fusion, and that it originally formed from a disc of gas out of which the rest of the solar system, including Earth, also condensed . . . Presumably, then, she would not have worshipped it as a god, and this would have altered her perspective on being sacrificed to propitiate it.

The Inca priests cannot be blamed for their ignorance, and it could perhaps be thought harsh to judge them stupid and puffed up. But they can be blamed for foisting their own beliefs on a child too young to decide whether to worship the sun or not. Humphrey's additional point is that today's documentary film makers, and we their audience, can be blamed for seeing beauty in that little girl's death – 'something that enriches *our* collective culture'. The same tendency to glory in the quaintness of ethnic religious habits, and to justify cruelties in their name, crops up again and again. It is the source of squirming internal conflict in the minds of nice liberal people who, on the one hand, cannot bear suffering and cruelty, but on the other hand have been trained by postmodernists and relativists to respect other cultures no less than their own. Female genital mutilation (sometimes called circumcision) is undoubtedly hideously painful, it sabotages sexual pleasure in women (indeed, this is probably its underlying purpose), and one half of the decent liberal mind wants to abolish the practice. The other half, however, 'respects' ethnic cultures and feels that we should not interfere if 'they' want to mutilate 'their' girls.* The point, of

* It is a regular practice in Britain today. A senior Schools Inspector told me of London girls in 2006 being sent to an 'uncle' in Bradford to be circumcised. Authorities turn a blind eye, for fear of being thought racist in 'the community'.

course, is that 'their' girls are actually the girls' *own* girls, and their wishes should not be ignored. Trickier to answer, what if a girl says she wants to be circumcised? But *would* she, with the hindsight of a fully informed adult, wish that it had never happened? Humphrey makes the point that no adult woman who has somehow missed out on circumcision as a child volunteers for the operation later in life.

After a discussion of the Amish, and their right to bring up 'their own' children in 'their own' way, Humphrey is scathing about our enthusiasm as a society for

> maintaining cultural diversity. All right, you may want to say, so it's tough on a child of the Amish, or the Hasidim, or the gypsies to be shaped up by their parents in the ways they are – but at least the result is that these fascinating cultural traditions continue. Would not our whole civilization be impoverished if they were to go? It's a shame, maybe, when individuals have to be sacrificed to maintain such diversity. But there it is: it's the price we pay as a society. Except, I would feel bound to remind you, we do not pay it, *they* do.

The issue came to public attention in 1972 when the US Supreme Court ruled on a test case, Wisconsin versus Yoder, which concerned the right of parents to withdraw their children from school on religious grounds. The Amish people live in closed communities in various parts of the United States, mostly speaking an archaic dialect of German called Pennsylvania Dutch and eschewing, to varying extents, electricity, internal combustion engines, zippers and other manifestations of modern life. There is, indeed, something attractively quaint about an island of seventeenth-century life as a spectacle for today's eyes. Isn't it worth preserving, for the sake of the enrichment of human diversity? And the only way to preserve it is to allow the Amish to educate their own children in their own way, and protect them from the corrupting influence of modernity. But, we surely want to ask, shouldn't the children themselves have some say in the matter?

The Supreme Court was asked to rule in 1972, when some Amish parents in Wisconsin withdrew their children from high school. The very idea of education beyond a certain age was contrary to Amish religious values, and scientific education especially so. The State of Wisconsin took the parents to court, claiming that the children were being deprived of their right to an education. After passing up through the courts, the case eventually reached the United States Supreme Court, which handed down a split (6:1) decision in favour of the parents.[141] The majority opinion, written by Chief Justice Warren Burger, included the following: 'As the record shows, compulsory school attendance to age 16 for Amish children carries with it a very real threat of undermining the Amish community and religious practice as they exist today; they must either abandon belief and be assimilated into society at large, or be forced to migrate to some other and more tolerant region.'

Justice William O. Douglas's minority opinion was that the children themselves should have been consulted. Did they really want to cut short their education? Did they, indeed, really want to stay in the Amish religion? Nicholas Humphrey would have gone further. Even if the children had been asked and had expressed a preference for the Amish religion, can we suppose that they would have done so if they had been educated and informed about the available alternatives? For this to be plausible, shouldn't there be examples of young people from the outside world voting with their feet and volunteering to join the Amish? Justice Douglas went further in a slightly different direction. He saw no particular reason to give the *religious* views of parents special status in deciding how far they should be allowed to deprive their children of education. If religion is grounds for exemption, might there not be secular beliefs that also qualify?

The majority of the Supreme Court drew a parallel with

some of the positive values of monastic orders, whose presence in our society arguably enriches it. But, as Humphrey points out, there is a crucial difference. Monks volunteer for the monastic life of their own free will. Amish children never volunteered to be Amish; they were born into it and they had no choice.

There is something breathtakingly condescending, as well as inhumane, about the sacrificing of anyone, especially children, on the altar of 'diversity' and the virtue of preserving a variety of religious traditions. The rest of us are happy with our cars and computers, our vaccines and antibiotics. But you quaint little people with your bonnets and breeches, your horse buggies, your archaic dialect and your earth-closet privies, you enrich our lives. Of course you must be allowed to trap your children with you in your seventeenth-century time warp, otherwise something irretrievable would be lost to us: a part of the wonderful diversity of human culture. A small part of me can see something in this. But the larger part is made to feel very queasy indeed.

AN EDUCATIONAL SCANDAL

The Prime Minister of my country, Tony Blair, invoked 'diversity' when challenged in the House of Commons by Jenny Tonge MP to justify government subsidy of a school in the north-east of England that (almost uniquely in Britain) teaches literal biblical creationism. Mr Blair replied that it would be unfortunate if concerns about that issue were to interfere with our getting 'as diverse a school system as we properly can'.[142] The school in question, Emmanuel College in Gateshead, is one of the 'city academies' set up in a proud initiative of the Blair government. Rich benefactors are encouraged to put up a relatively small sum of money (£2 million in the case of Emmanuel), which buys a much larger sum of government money (£20 million for the school, plus running costs and salaries in perpetuity), and also buys the benefactor the right to control the ethos of the school, the appointment of a majority of the school governors, the policy for exclusion or inclusion of pupils, and much else.

Emmanuel's 10 per cent benefactor is Sir Peter Vardy, a wealthy car salesman with a creditable desire to give today's children the education he wishes he had had, and a less creditable desire to imprint his personal religious convictions upon them.* Vardy has unfortunately become embroiled with a clique of American-inspired fundamentalist teachers, led by Nigel McQuoid, sometime headmaster of Emmanuel and now director of a whole consortium of Vardy schools. The level of McQuoid's scientific understanding can be judged from his belief that the world is less than ten thousand years old, and also from the following quotation: 'But to think that we just evolved from a bang, that we used to be monkeys, that seems unbelievable when you look at the complexity of the human body . . . If you tell children there is no purpose to their life – that they are just a chemical mutation – that doesn't build self-esteem.'[143]

No scientist has ever suggested that a child is a 'chemical mutation'. The use of the phrase in such a context is illiterate nonsense, on a par with the declarations of 'Bishop' Wayne Malcolm, leader of the Christian Life City church in Hackney, east London, who, according to the *Guardian* of 18 April 2006, 'disputes the scientific evidence for evolution'. Malcolm's understanding of the evidence he disputes can be gauged from his statement that 'There is clearly an absence in the fossil record for intermediate levels of development. If a frog turned into a monkey, shouldn't you have lots of fronkies?'

Well, science is not Mr McQuoid's subject either, so we should, in fairness, turn to his head of science, Stephen Layfield, instead. On 21 September 2001, Mr Layfield gave a lecture at Emmanuel College on 'The Teaching of Science: A

---

* H. L. Mencken was prophetic when he wrote: 'Deep within the heart of every evangelist lies the wreck of a car salesman.'

Biblical Perspective'. The text of the lecture was posted on a Christian website (www. christian.org.uk). But you won't find it there now. The Christian Institute removed the lecture the very day after I had called attention to it in an article in the *Daily Telegraph* on 18 March 2002, where I subjected it to a critical dissection.[144] It is hard, however, to delete something permanently from the World Wide Web. Search engines achieve their speed partly by keeping caches of information, and these inevitably persist for a while even after the originals have been deleted. An alert British journalist, Andrew Brown, the *Independent*'s first religious affairs correspondent, promptly located the Layfield lecture, downloaded it from the Google cache and posted it, safe from deletion, on his own website, http://www.darwinwars.com/lunatic/liars/layfield.html. You will notice that the words chosen by Brown for the URL make entertaining reading in themselves. They lose their power to amuse, however, when we look at the content of the lecture itself.

Incidentally, when a curious reader wrote to Emmanuel College to ask why the lecture had been removed from the website, he received the following disingenuous reply from the school, again recorded by Andrew Brown:

> Emmanuel College has been at the centre of a debate regarding the teaching of creation in schools. At a practical level Emmanuel College has had a huge number of press calls. This has involved a considerable amount of time for the Principal and senior Directors of the College. All of these people have other jobs to do. In order to assist we have temporarily removed a lecture by Stephen Layfield from our website.

Of course, the school officials may well have been too busy to explain to journalists their stance on teaching creationism. But why, then, remove from their website the text of a lecture that does precisely that, and to which they could have referred the journalists, thereby saving themselves a great deal of time? No, they removed their head of science's lecture because they recognized that they had something to hide. The following paragraph is from the beginning of his lecture:

> Let us state then right from the start that we reject the notion popularised, perhaps inadvertently, by Francis Bacon in the 17th century that there are 'Two Books' (i.e. the Book of nature & the Scriptures) which may be mined independently for truth. Rather, we stand firm upon the bare proposition that God has spoken authoritatively and inerrantly in the pages of holy Scripture. However fragile, old-fashioned or naive this assertion may ostensibly appear, especially to an unbelieving, TV-drunk modern culture, we can be sure that it is as robust a foundation as it is possible to lay down and build upon.

You have to keep pinching yourself. You are not dreaming. This is not some preacher in a tent in Alabama but the head of *science* at a school into which the British government is pouring money, and which is Tony Blair's pride and joy. A devout Christian himself, Mr Blair in 2004 performed the ceremonial opening of one of the later additions to the Vardy fleet of schools.[145] Diversity may be a virtue, but this is diversity gone mad.

Layfield proceeds to itemize the comparison between science and scripture, concluding, in every case where there seems to be a conflict, that scripture is to be preferred. Noting that earth science is now included in the national curriculum, Layfield says, 'It would seem particularly prudent for all who deliver this aspect of the course to familiarise themselves with the Flood geology papers of Whitcomb & Morris.' Yes, 'Flood geology' means what you think it means. We're talking Noah's Ark here.

Noah's Ark! – when the children could be learning the spine-tingling fact that Africa and South America were once joined, and have drawn apart at the speed with which fingernails grow. Here's more from Layfield (the head of science) on Noah's flood as the recent and rapid explanation for phenomena which, according to real geological evidence, took hundreds of millions of years to grind out:

> We must acknowledge within our grand geophysical paradigm the historicity of a world-wide flood as outlined in Gen 6–10. If the Biblical narrative is secure and the listed genealogies (e.g. Gen 5; 1 Chro 1; Matt 1 & Lu 3) are substantially full, we must reckon that this global catastrophe took place in the relatively recent past. Its effects are everywhere abundantly apparent. Principal evidence is found in the fossil-laden sedimentary rocks, the extensive reserves of hydrocarbon fuels (coal, oil and gas) and the 'legendary' accounts of just such a great flood common to various population groups world-wide. The feasibility of maintaining an ark full of representative creatures for a year until the waters had sufficiently receded has been well documented by, among others, John Woodmorrappe.

In a way this is even worse than the utterances of know-nothings like Nigel McQuoid or Bishop Wayne Malcolm quoted above, because Layfield is educated in science. Here's another astonishing passage:

> As we stated at the beginning, Christians, with very good reason, reckon the Scriptures of the Old & New Testaments a reliable guide concerning just what we are to believe. They are not merely religious documents. They provide us with a true account of Earth history which we ignore at our peril.

The implication that the scriptures provide a literal account of geological history would make any reputable theologian wince. My friend Richard Harries, Bishop of Oxford, and I wrote a joint letter to Tony Blair, and we got it signed by eight bishops and nine senior scientists.[146] The nine scientists included the then President of the Royal Society (previously Tony Blair's chief scientific adviser), both the biological and physical secretaries of the Royal Society, the Astronomer Royal (now President of the Royal Society), the director of the Natural History Museum, and Sir David Attenborough, perhaps the most respected man in England. The bishops included one Roman Catholic and seven Anglican bishops – senior religious leaders from all around England. We received a perfunctory and inadequate reply from the Prime Minister's office, referring to the school's good examination results and its good report from the official schools inspection agency, OFSTED. It apparently didn't occur to Mr Blair that, if the OFSTED inspectors give a rave report to a school whose head of science teaches that the entire universe began after the domestication of the dog, there just might be something a teeny weeny bit wrong with the standards of the inspectorate.

Perhaps the most disturbing section of Stephen Layfield's lecture is his concluding 'What can be done?', where he considers the tactics to be employed by those teachers wishing to introduce fundamentalist Christianity into the science classroom. For example, he urges science teachers to

> note every occasion when an evolutionary/old-earth paradigm (millions or billions of years) is explicitly mentioned or implied by a text-book, examination question or visitor and courteously point out the fallibility of the statement. Wherever possible, we must give the alternative (always better) Biblical explanation of the same data. We shall look at a few examples from each of Physics, Chemistry & Biology in due course.

The rest of Layfield's lecture is nothing less than a propaganda manual, a resource for religious teachers of biology, chemistry and physics who wish, while remaining just inside the guidelines of the national curriculum, to subvert evidence-based science education and replace it with biblical scripture.

On 15 April 2006, James Naughtie, one of the BBC's most experienced anchormen, interviewed Sir Peter Vardy on radio. The main subject of the interview was a police investigation of allegations, denied by Vardy, that bribes – knighthoods and peerages – had been offered by the Blair government to rich men, in an attempt to get them to subscribe to the city academies scheme. Naughtie also asked Vardy about the creationism issue, and Vardy categorically denied that Emmanuel promotes young-Earth creationism to its pupils. One of Emmanuel's alumni, Peter French, has equally categorically stated,[147] 'We were taught that the earth was 6000 years old.'* Who is telling the truth here? Well, we don't know, but Stephen Layfield's lecture lays out his policy for teaching science pretty candidly. Has Vardy never read Layfield's very explicit manifesto? Does he really not know what his head of science has been up to? Peter Vardy made his money selling used cars. Would you buy one from him? And would you, like Tony Blair, sell him a school for 10 per cent of its price – throwing in an offer to pay all his running costs into the bargain? Let's be charitable to Blair and assume that he, at least, has not read the Layfield lecture. I suppose it is too much to hope that his attention may now be drawn to it.

Headmaster McQuoid offered a defence of what he clearly saw as his school's open-mindedness, which is remarkable for its patronizing complacency:

> the best example I can give of what it is like here is a sixth-form philosophy lecture I was giving. Shaquille

---

* To get an idea of the scale of this error, it is equivalent to believing that the distance from New York to San Francisco is 7.8 yards.

> was sitting there and he says, 'The Koran is correct and true.' And Clare, over here, says, 'No, the Bible is true.' So we talked about the similarities between what they say and the places where they disagree. And we agreed that they could not both be true. And eventually I said, 'Sorry Shaquille, you are wrong, it is the Bible that is true.' And he said, 'Sorry Mr McQuoid, you are wrong, it is the Koran.' And they went on to lunch and carried on discussing it there. That's what we want. We want children to know why it is they believe what they believe and to defend it.[148]

What a charming picture! Shaquille and Clare went to lunch together, vigorously arguing their cases and defending their incompatible beliefs. But is it really so charming? Isn't it actually rather a deplorable picture that Mr McQuoid has painted? Upon what, after all, did Shaquille and Clare base their argument? What cogent evidence was each one able to bring to bear, in their vigorous and constructive debate? Clare and Shaquille each simply asserted that her or his holy book was superior, and that was that. That is apparently all they said, and that, indeed, is all you can say when you have been taught that truth comes from scripture rather than from evidence. Clare and Shaquille and their fellows were not being educated. They were being let down by their school, and their school principal was abusing, not their bodies, but their minds.

CONSCIOUSNESS-RAISING AGAIN

And now, here's another charming picture. At Christmas-time one year my daily newspaper, the *Independent*, was looking for a seasonal image and found a heart-warmingly ecumenical one at a school nativity play. The Three Wise Men were played by, as the caption glowingly said, Shadbreet (a Sikh), Musharaff (a Muslim) and Adele (a Christian), all aged four.

Charming? Heart-warming? No, it is not, it is neither; it is grotesque. How could any decent person think it right to label four-year-old children with the cosmic and theological opinions of their parents? To see this, imagine an identical photograph, with the caption changed as follows: 'Shadbreet (a Keynesian), Musharaff (a Monetarist) and Adele (a Marxist), all aged four.' Wouldn't this be a candidate for irate letters of protest? It certainly should be. Yet, because of the weirdly privileged status of religion, not a squeak was heard, nor is it ever heard on any similar occasion. Just imagine the outcry if the caption had read, 'Shadbreet (an Atheist), Musharaff (an Agnostic) and Adele (a Secular Humanist), all aged four.' Mightn't the parents actually be investigated to see if they were fit to bring up children? In Britain, where we lack a constitutional separation between church and state, atheist parents usually go with the flow and let schools teach their children whatever religion prevails in the culture. 'The-Brights.net' (an American initiative to rebrand atheists as 'Brights' in the same way as homosexuals successfully rebranded themselves as 'gays') is scrupulous in setting out the rules for children to sign up: 'The decision to be a Bright must be the child's. Any youngster who is told he or she must, or should, be a Bright can NOT be a Bright.' Can you even begin to imagine a church or mosque issuing such a self-denying ordinance? But shouldn't they be compelled to do so? Incidentally, I signed up to the Brights, partly because I was genuinely curious whether such a word could be memetically engineered into the language. I don't know, and would like to, whether the transmutation of 'gay' was deliberately engineered or whether it just happened.[149] The Brights campaign got off to a shaky start when it was furiously denounced by some atheists, petrified of being branded 'arrogant'. The Gay Pride movement, fortunately, suffers from no such false modesty, which may be why it succeeded.

In an earlier chapter, I generalized the theme of 'consciousness-raising', starting with the achievement of feminists in making us flinch when we hear a phrase like 'men of goodwill' instead of 'people of goodwill'. Here I want to raise consciousness in another way. I think we should all wince when we hear a small child being labelled as belonging to some particular religion or another. Small children are too young to decide their views on the origins of the cosmos, of life and of morals. The very sound of the phrase 'Christian child' or 'Muslim child' should grate like fingernails on a blackboard.

Here is a report, dated 3 September 2001, from the Irish *Aires* show on the American radio station KPFT-FM.

> Catholic schoolgirls faced protests from Loyalists as they attempted to enter the Holy Cross Girls' Primary School on the Ardoyne Road in north Belfast. Royal Ulster Constabulary (RUC) officers and British Army (BA) soldiers had to clear the protestors who were attempting to blockade the school. Crash barriers were erected to allow the children to get through the protest to the school. Loyalists jeered and shouted sectarian abuse as the children, some as young as four years of age, were escorted by the parents into the school. As children and parents entered the front gate of the school Loyalists threw bottles and stones.

Naturally, any decent person will wince at the ordeal of these unfortunate schoolgirls. I am trying to encourage us to wince, too, at the very idea of labelling them 'Catholic schoolgirls' at all. ('Loyalists', as I pointed out in Chapter 1, is the mealy-mouthed Northern Ireland euphemism for Protestants, just as 'Nationalists' is the euphemism for Catholics. People who do not hesitate to brand children 'Catholics' or 'Protestants' stop short of applying those same religious labels – far more appropriately – to adult terrorists and mobs.)

Our society, including the non-religious sector, has accepted the preposterous idea that it is normal and right to indoctrinate tiny children in the religion of their parents, and to slap religious labels on them – 'Catholic child', 'Protestant child', 'Jewish child', 'Muslim child', etc. – although no other comparable labels: no conservative children, no liberal children, no Republican children, no Democrat children. Please, please raise your consciousness about this, and raise the roof whenever you hear it happening. A child is not a Christian child, not a Muslim child, but a child of Christian parents or a child of Muslim parents. This latter nomenclature, by the way, would be an excellent piece of consciousness-raising for the children themselves. A child who is told she is a 'child of Muslim parents' will immediately realize that religion is something for her to choose – or reject – when she becomes old enough to do so.

A good case can indeed be made for the educational benefits of teaching comparative religion. Certainly my own doubts were first aroused, at the age of about nine, by the lesson (which came not from school but from my parents) that the Christian religion in which I was brought up was only one of many mutually incompatible belief-systems. Religious apologists themselves realize this and it often frightens them. After that nativity play story in the *Independent*, not a single letter to the Editor complained of the religious labelling of the four-year-olds. The only negative letter came from 'The Campaign for Real Education', whose spokesman, Nick Seaton, said multi-faith religious education was extremely dangerous because 'Children these days are taught that all religions are of equal worth, which means that their own has no special value.' Yes indeed; that is exactly what it means. Well might this spokesman worry. On another occasion, the same individual said, 'To present all faiths as equally valid is wrong. Everybody is entitled to think their faith is superior to others, be they Hindus, Jews, Muslims or Christians – otherwise what's the point in having faith?'[150]

What indeed? And what transparent nonsense this is! These faiths are mutually incompatible. Otherwise what is the point of thinking your faith superior? Most of them, therefore, cannot be 'superior to others'. Let children learn about different faiths, let them notice their incompatibility, and let them draw their own conclusions about the consequences of that incompatibility. As for whether any are 'valid', let them make up their own minds when they are old enough to do so.

RELIGIOUS EDUCATION AS A PART OF LITERARY CULTURE

I must admit that even I am a little taken aback at the biblical ignorance commonly displayed by people educated in more recent decades than I was. Or maybe it isn't a decade thing. As long ago as 1954, according to Robert Hinde in his thoughtful book *Why Gods Persist*, a Gallup poll in the United States of America found the following. Three-quarters of Catholics and Protestants could not name a single Old Testament prophet. More than two-thirds didn't know who preached the Sermon on the Mount. A substantial number thought that Moses was one of Jesus's twelve apostles. That, to repeat, was in the United States, which is dramatically more religious than other parts of the developed world.

The King James Bible of 1611 – the Authorized Version – includes passages of outstanding literary merit in its own right, for example the Song of Songs, and the sublime Ecclesiastes (which I am told is pretty good in the original Hebrew too). But the main reason the English Bible needs to be part of our education is that it is a major source book for literary culture. The same applies to the legends of the Greek and Roman gods, and we learn about them without being asked to believe in them. Here is a quick list of biblical, or Bible-inspired, phrases and sentences that occur commonly in literary or conversational English, from great poetry to hackneyed cliché, from proverb to gossip.

> Be fruitful and multiply • East of Eden • Adam's Rib • Am I my brother's keeper? • The mark of Cain • As old as Methuselah • A mess of pottage • Sold his

birthright • Jacob's ladder • Coat of many colours • Amid the alien corn • Eyeless in Gaza • The fat of the land • The fatted calf • Stranger in a strange land • Burning bush • A land flowing with milk and honey • Let my people go • Flesh pots • An eye for an eye and a tooth for a tooth • Be sure your sin will find you out • The apple of his eye • The stars in their courses • Butter in a lordly dish • The hosts of Midian • Shibboleth • Out of the strong came forth sweetness • He smote them hip and thigh • Philistine • A man after his own heart • Like David and Jonathan • Passing the love of women • How are the mighty fallen? • Ewe lamb • Man of Belial • Jezebel • Queen of Sheba • Wisdom of Solomon • The half was not told me • Girded up his loins • Drew a bow at a venture • Job's comforters • The patience of Job • I am escaped with the skin of my teeth • The price of wisdom is above rubies • Leviathan • Go to the ant thou sluggard; consider her ways, and be wise • Spare the rod and spoil the child • A word in season • Vanity of vanities • To everything there is a season, and a time to every purpose • The race is not to the swift, nor the battle to the strong • Of making many books there is no end • I am the rose of Sharon • A garden inclosed • The little foxes • Many waters cannot quench love • Beat their swords into plowshares • Grind the faces of the poor • The wolf also shall dwell with the lamb, and the leopard shall lie down with the kid • Let us eat and drink; for tomorrow we shall die • Set thine house in order • A voice crying in the wilderness • No peace for the wicked • See eye to eye • Cut off out of the land of the living • Balm in Gilead • Can the leopard change his spots? • The parting of the ways • A Daniel in the lions' den • They have sown the wind, and they shall reap the whirlwind • Sodom and Gomorrah • Man shall not live by bread alone • Get thee behind me Satan • The salt of the earth • Hide your light under a bushel • Turn the other cheek • Go the extra mile • Moth and rust doth corrupt • Cast your pearls before swine • Wolf in sheep's clothing • Weeping and gnashing of teeth • Gadarene swine • New wine in old bottles • Shake off the dust of your feet • He that is not with me is against me • Judgement of Solomon • Fell upon stony ground • A prophet is not without honour, save in his own country • The crumbs from the table • Sign of the times • Den of thieves • Pharisee • Whited sepulchre • Wars and rumours of wars • Good and faithful servant • Separate the sheep from the goats • I wash my hands of it • The sabbath was made for man, and not man for the sabbath • Suffer the little children • The widow's mite • Physician heal thyself • Good Samaritan • Passed by on the other side • Grapes of wrath • Lost sheep • Prodigal son • A great gulf fixed • Whose shoe latchet I am not worthy to unloose • Cast the first stone • Jesus wept • Greater love hath no man than this • Doubting Thomas • Road to Damascus • A law unto himself • Through a glass darkly • Death, where is thy sting? • A thorn in the flesh • Fallen from grace • Filthy lucre • The root of all evil • Fight the good fight • All flesh is as grass • The weaker vessel • I am Alpha and Omega • Armageddon • De profundis • Quo vadis • Rain on the just and on the unjust

Every one of these idioms, phrases or clichés comes directly from the King James Authorized Version of the Bible. Surely ignorance of the Bible is bound to impoverish one's appreciation of English literature? And not just solemn and serious literature. The following rhyme by Lord Justice Bowen is ingeniously witty:

> The rain it raineth on the just,
> And also on the unjust fella.
> But chiefly on the just, because
> The unjust hath the just's umbrella.

But the enjoyment is muffled if you can't take the allusion to Matthew 5: 45 ('For he maketh his sun to rise on the evil and on the good, and sendeth rain on the just and on the unjust'). And the fine point of Eliza Dolittle's fantasy in *My Fair Lady* would escape anybody ignorant of John the Baptist's end:

> 'Thanks a lot, King,' says I in a manner well bred,
> 'But all I want is 'Enry 'Iggins' 'ead.'

P. G. Wodehouse is, for my money, the greatest writer of light comedy in English, and I bet fully half my list of biblical phrases will be found as allusions within his pages. (A Google search will not find all of them, however. It will miss the derivation of the short-story title 'The Aunt and the Sluggard', from Proverbs 6: 6.) The Wodehouse canon is rich in other biblical phrases, not in my list above and not incorporated into the language as idioms or proverbs. Listen to Bertie Wooster's evocation of what it is like to wake up with a bad hangover: 'I had been dreaming that some bounder was driving spikes through my head – not just ordinary spikes, as used by Jael the wife of Heber, but red-hot ones.' Bertie himself was immensely proud of his only scholastic achievement, the prize he once earned for scripture knowledge.

What is true of comic writing in English is more obviously true of serious literature. Naseeb Shaheen's tally of more than thirteen hundred biblical references in Shakespeare's works is widely cited and very believable.[151] The *Bible Literacy Report* published in Fairfax, Virginia (admittedly financed by the infamous Templeton Foundation) provides many examples, and cites overwhelming agreement by teachers of English literature that biblical literacy is essential to full appreciation of their subject.[152] Doubtless the equivalent is true of French, German, Russian, Italian, Spanish and other great European literatures. And, for speakers of Arabic and Indian languages, knowledge of the Qur'an or the Bhagavad Gita is presumably just as essential for full appreciation of their literary heritage. Finally, to round off the list, you can't appreciate Wagner (whose music, as has been wittily said, is better than it sounds) without knowing your way around the Norse gods.

Let me not labour the point. I have probably said enough to convince at least my older readers that an atheistic world-view provides no justification for cutting the Bible, and other sacred books, out of our education. And of course we can retain a sentimental loyalty to the cultural and literary traditions of, say, Judaism, Anglicanism or Islam, and even participate in religious rituals such as marriages and funerals, without buying into the supernatural beliefs that historically went along with those traditions. We can give up belief in God while not losing touch with a treasured heritage.

CHAPTER 10

A much needed gap?

What can be more soul shaking than peering through a 100-inch telescope at a distant galaxy, holding a 100-million-year-old fossil or a 500,000-year-old stone tool in one's hand, standing before the immense chasm of space and time that is the Grand Canyon, or listening to a scientist who gazed upon the face of the universe's creation and did not blink? That is deep and sacred science.

MICHAEL SHERMER

'This book fills a much needed gap.' The jest works because we simultaneously understand the two opposite meanings. Incidentally, I thought it was an invented witticism but, to my surprise, I find that it has actually been used, in all innocence, by publishers. See www.amazon.co.uk/Tel-Quel-Reader-Patrick-Ffrench/dp/0415157145 for a book that 'fills a much needed gap in the literature available on the poststructuralist movement'. It seems deliciously appropriate that this avowedly superfluous book is all about Michel Foucault, Roland Barthes, Julia Kristeva and other icons of haute francophonyism.

Does religion fill a much needed gap? It is often said that there is a God-shaped gap in the brain which needs to be filled: we have a psychological need for God – imaginary friend, father, big brother, confessor, confidant – and the need has to be satisfied whether God really exists or not. But could it be that God clutters up a gap that we'd be better off filling with something else? Science, perhaps? Art? Human friendship? Humanism? Love of this life in the real world, giving no credence to other lives beyond the grave? A love of nature, or what the great entomologist E. O. Wilson has called *Biophilia*?

Religion has at one time or another been thought to fill four main roles in human life: explanation, exhortation, consolation and inspiration. Historically, religion aspired to *explain* our own existence and the nature of the universe in which we find ourselves. In this role it is now completely superseded by science, and I have dealt with it in Chapter 4. By *exhortation* I mean moral instruction on how we ought to behave, and I covered that in Chapters 6 and 7. I have not so far done justice to *consolation* and *inspiration*, and this final chapter will briefly deal with them. As a preliminary to consolation itself, I want to begin with the childhood phenomenon of the 'imaginary friend', which I believe has affinities with religious belief.

BINKER

Christopher Robin, I presume, did not believe that Piglet and Winnie the Pooh really spoke to him. But was Binker different?

> Binker – what I call him – is a secret of my own,
> And Binker is the reason why I never feel alone.
> Playing in the nursery, sitting on the stair,
> Whatever I am busy at, Binker will be there.
> Oh, Daddy is clever, he's a clever sort of man,
> And Mummy is the best since the world began,
> And Nanny is Nanny, and I call her Nan –
> But they can't See Binker.
> Binker's always talking, 'cos I'm teaching him to speak
> He sometimes likes to do it in a funny sort of squeak,
> And he sometimes likes to do it in a hoodling sort of roar . . .
> And I have to do it for him 'cos his throat is rather sore.
> Oh, Daddy is clever, he's a clever sort of man,
> And Mummy knows all that anybody can,
> And Nanny is Nanny, and I call her Nan –
> But they don't Know Binker.

Binker's brave as lions when we're running in the park;
Binker's brave as tigers when we're lying in the dark;
Binker's brave as elephants. He never, never cries . . .
Except (like other people) when the soap gets in his eyes.
Oh, Daddy is Daddy, he's a Daddy sort of man,
And Mummy is as Mummy as anybody can,
And Nanny is Nanny, and I call her Nan . . .
But they're not Like Binker.
Binker isn't greedy, but he does like things to eat,
So I have to say to people when they're giving me a sweet,
'Oh, Binker wants a chocolate, so could you give me two?'
And then I eat it for him, 'cos his teeth are rather new.
Well, I'm very fond of Daddy, but he hasn't time to play,
And I'm very fond of Mummy, but she sometimes goes away,
And I'm often cross with Nanny when she wants to brush my
hair . . .
But Binker's always Binker, and is certain to be there.

A. A. MILNE, *Now We Are Six**

Is the imaginary-friend phenomenon a higher illusion, in a different category from ordinary childhood make-believe? My own experience is not much help here. Like many parents, my mother kept a notebook of my childish sayings. In addition to simple pretendings (now I'm the man in the moon . . . an accelerator . . . a Babylonian) I was evidently fond of second-order pretendings (now I'm an owl pretending to be a waterwheel) which might be reflexive (now I'm a little boy pretending to be Richard). I never once believed I really was any of those things, and I think that is normally true of childhood make-believe games. But I didn't have a Binker. If the testimony of their adult selves is to be believed, at least some of those normal children who have imaginary friends really do believe they exist, and, in some cases, see them as clear and vivid hallucinations. I suspect that the Binker phenomenon of

* Reproduced by permission of the A. A. Milne Estate.

childhood may be a good model for understanding theistic belief in adults. I do not know whether psychologists have studied it from this point of view, but it would be a worthwhile piece of research. Companion and confidant, a Binker for life: that is surely one role that God plays – one gap that might be left if God were to go.

Another child, a girl, had a 'little purple man', who seemed to her a real and visible presence, and who would manifest himself, sparkling out of the air, with a gentle tinkling sound. He visited her regularly, especially when she felt lonely, but with decreasing frequency as she grew older. On a particular day just before she went to kindergarten, the little purple man came to her, heralded by his usual tinkling fanfare, and announced that he would not be visiting her any more. This saddened her, but the little purple man told her that she was getting bigger now and wouldn't need him in the future. He must leave her now, so that he could look after other children. He promised her that he would come back to her if ever she *really* needed him. He did return to her, many years later in a dream, when she had a personal crisis and was trying to decide what to do with her life. The door of her bedroom opened and a cartload of books appeared, pushed into the room by . . . the little purple man. She interpreted this as advice that she should go to university – advice that she took and later judged to be good. The story makes me almost tearful, and it brings me as close as I shall probably come to understanding the consoling and counselling role of imaginary gods in people's lives. A being may exist only in the imagination, yet still seem completely real to the child, and still give real comfort and good advice. Perhaps even better: imaginary friends – and imaginary gods – have the time and patience to devote all their attention to the sufferer. And they are much cheaper than psychiatrists or professional counsellors.

Did gods, in their role as consolers and counsellors, evolve from binkers, by a sort of psychological 'pedomorphosis'? Pedomorphosis is the retention into adulthood of childhood characteristics. Pekinese dogs have pedomorphic faces: the adults look like puppies. It is a well-known pattern in evolution, widely accepted as important for the development of such human characteristics as our bulbous forehead and short jaws. Evolutionists have described us as juvenile apes, and it is certainly true that juvenile chimpanzees and gorillas look more like humans than adult ones do. Could religions have evolved originally by gradual postponement, over generations, of the moment in life when children gave up their binkers – just as we slowed down, during evolution, the flattening of our foreheads and the protrusion of our jaws?

I suppose, for completeness, we should consider the reverse possibility. Rather than gods evolving from ancestral binkers, could binkers have evolved from ancestral gods? This seems to me less likely. I was led to think about it while reading the American psychologist Julian Jaynes's *The Origin of Consciousness in the Breakdown of the Bicameral Mind*, a book that is as strange as its title suggests. It is one of those books that is either complete rubbish or a work of consummate genius, nothing in between! Probably the former, but I'm hedging my bets.

Jaynes notes that many people perceive their own thought processes as a kind of dialogue between the 'self' and another internal protagonist inside the head. Nowadays we understand that both 'voices' are our own – or if we don't we are treated as mentally ill. This happened, briefly, to Evelyn Waugh. Never one to mince words, Waugh remarked to a friend: 'I haven't seen you for a long time, but then I've seen so few people because – did you know? – I went mad.' After his recovery, Waugh wrote a novel, *The Ordeal of Gilbert Pinfold*, which described his hallucinatory period, and the voices that he heard.

Jaynes's suggestion is that some time before 1000 BC people in general were unaware that the second voice – the Gilbert Pinfold voice – came from within themselves. They thought the Pinfold voice was a god: Apollo, say, or Astarte or Yahweh or, more probably, a minor household god, offering them advice or orders. Jaynes even located the voices of the gods in the opposite hemisphere of the brain from the one that controls audible speech. The 'breakdown of the bicameral' mind was, for Jaynes, a historical transition. It was the moment in history when it dawned on people that the external voices that they seemed to be hearing were really internal. Jaynes even goes so far as to define this historical transition as the dawning of human consciousness.

There is an ancient Egyptian inscription about the creator god Ptah, which describes the various other gods as variations of Ptah's 'voice' or 'tongue'. Modern translations reject the literal 'voice' and interpret the other gods as 'objectified conceptions of [Ptah's] mind'. Jaynes dismisses such educated readings, preferring to take the literal meaning seriously. The gods were hallucinated voices, speaking inside people's heads. Jaynes further suggests that such gods evolved from memories of dead kings, who still, in a manner of speaking, retained control over their subjects via imagined voices in their heads. Whether or not you find his thesis plausible, Jaynes's book is intriguing enough to earn its mention in a book on religion.

Now, to the possibility I raised of borrowing from Jaynes to construct a theory that gods and binkers are developmentally related, but the opposite way around from the paedomorphosis theory. It amounts to the suggestion that the breakdown of the bicameral mind didn't happen suddenly in history, but was a progressive pulling back into childhood of the moment when hallucinated voices and apparitions were rumbled as not real. In a kind of reversal of the paedomorphosis hypothesis, the hallucinated gods disappeared from adult minds first, then were pulled back earlier and earlier into childhood, until today they survive only in the Binker or little purple man phenomenon. The problem with this version of the theory is that it doesn't explain the persistence of gods into adulthood today.

It might be better not to treat gods as ancestral to binkers, or vice versa, but rather to see both as by-products of the same psychological predisposition. Gods and binkers have in common the power to comfort, and provide a vivid sounding board for trying out ideas. We have not moved far from Chapter 5's psychological by-product theory of the evolution of religion.

CONSOLATION

It is time to face up to the important role that God plays in consoling us; and the humanitarian challenge, if he does not exist, to put something in his place. Many people who concede that God probably doesn't exist, and that he is not necessary for morality, still come back with what they often regard as a trump card: the alleged psychological or emotional *need* for a god. If you take religion away, people truculently ask, what are you going to put in its place? What have you to offer the dying patients, the weeping bereaved, the lonely Eleanor Rigbys for whom God is their only friend?

The first thing to say in response to this is something that should need no saying. Religion's power to console doesn't make it true. Even if we make a huge concession; even if it were conclusively demonstrated that belief in God's existence is completely essential to human psychological and emotional well-being; even if all atheists were despairing neurotics driven to suicide by relentless cosmic angst – none of this would contribute the tiniest jot or tittle of evidence that religious belief is true. It might be evidence in favour of the desirability of convincing yourself that God exists, even if he doesn't. As I've already mentioned, Dennett, in *Breaking the Spell*, makes the distinction between belief in God and belief in belief: the belief that it is desirable to believe, even if the belief itself is false: 'Lord, I believe; help thou mine unbelief' (Mark 9: 24). The faithful are encouraged to *profess* belief, whether they are convinced by it or not. Maybe if you repeat something often enough, you will succeed in convincing yourself of its truth. I think we all know people who enjoy the idea of religious faith, and resent attacks on it, while reluctantly admitting that they don't have it themselves. I was slightly shocked to discover a first-class example on page 96 of my hero Peter Medawar's book *The Limits of Science*: 'I regret my disbelief in God and religious answers generally, for I believe it would give satisfaction and comfort to many in need of it if it were possible to discover good scientific and philosophic reasons to believe in God.'

Since reading of Dennett's distinction, I have found occasion to use it again and again. It is scarcely an exaggeration to say that the majority of atheists I know disguise their atheism behind a pious façade. They do not believe in anything supernatural themselves, but retain a vague soft spot for irrational belief. They believe in belief. It is amazing how many people seemingly cannot tell the difference between 'X is true' and 'It is desirable that people should believe that X is true'. Or maybe they don't really fall for this logical error, but simply rate truth as unimportant compared with human feelings. I don't want to decry human feelings. But let's be clear, in any particular conversation, what we are talking about: feelings, or truth. Both may be important, but they are not the same thing.

In any case, my hypothetical concession was extravagant and wrong. I know of no evidence that atheists have any general tendency towards unhappy, angst-ridden despond. Some atheists are happy. Others are miserable. Similarly, some Christians, Jews, Muslims, Hindus and Buddhists are miserable, while others are happy. There may be statistical evidence bearing on the relationship between happiness and belief (or unbelief), but I doubt if it is a strong effect, one way or the other. I find it more interesting to ask whether there is any good *reason* to feel depressed if we live without God. I shall end this book by arguing, on the contrary, that it is an understatement to say that one can lead a happy and fulfilled life without supernatural religion. First, though, I must examine the claims of religion to offer consolation.

Consolation, according to the *Shorter Oxford Dictionary*, is the alleviation of sorrow or mental distress. I shall divide consolation into two types.

1. ***Direct physical consolation.*** A man stuck for the night on a bare mountain may find comfort in a large, warm St Bernard dog, not forgetting, of course, the brandy barrel around its neck. A weeping child may be consoled by the embrace of strong arms wrapped around her and reassuring words whispered in her ear.

2. ***Consolation by discovery of a previously unappreciated fact, or a previously undiscovered way of looking at existing facts.*** A woman whose husband has been killed in war may be consoled by the discovery that she is pregnant by him, or that he died a hero. We can also get consolation through discovering a new way of thinking about a situation. A philosopher points out that there is nothing special about the moment when an old man dies. The child that he once was 'died' long ago, not by suddenly ceasing to live but by growing up. Each of Shakespeare's seven ages of man 'dies' by slowly morphing into the next. From this point of view, the moment when the old man finally expires is no different from the slow 'deaths' throughout his life.[153] A man who does not relish the prospect of his own death may find this changed perspective consoling. Or maybe not, but it is a potential example of consolation through reflection. Mark Twain's dismissal of the fear of death is another: 'I do not fear death. I had been dead for billions and billions of years before I was born, and had not suffered the slightest inconvenience from it.' The *aperçu* changes nothing about the fact of our inevitable death. But we have been offered a different way of looking at that inevitability and we may find it consoling. Thomas Jefferson, too, had no fear of death and he seems to have believed in no kind of afterlife. By Christopher Hitchens's account, 'As his days began to wane, Jefferson more than once wrote to friends that he

> faced the approaching end without either hope or fear. This was as much as to say, in the most unmistakable terms, that he was not a Christian.'

Robust intellects may be ready for the strong meat of Bertrand Russell's declaration, in his 1925 essay 'What I Believe':

> I believe that when I die I shall rot, and nothing of my ego will survive. I am not young and I love life. But I should scorn to shiver with terror at the thought of annihilation. Happiness is nonetheless true happiness because it must come to an end, nor do thought and love lose their value because they are not everlasting. Many a man has borne himself proudly on the scaffold; surely the same pride should teach us to think truly about man's place in the world. Even if the open windows of science at first make us shiver after the cosy indoor warmth of traditional humanizing myths, in the end the fresh air brings vigour, and the great spaces have a splendour of their own.

I was inspired by this essay of Russell's when I read it in my school library at the age of about sixteen, but I had forgotten it. It is possible that I was paying unconscious homage to Russell (as well as conscious homage to Darwin) when I wrote, in *A Devil's Chaplain* in 2003,

> There is more than just grandeur in this view of life, bleak and cold though it can seem from under the security blanket of ignorance. There is deep refreshment to be had from standing up and facing straight into the strong keen wind of understanding: Yeats's 'Winds that blow through the starry ways'.

How does religion compare with, say, science in providing these two types of consolation? Looking at Type 1 consolation first, it is entirely plausible that the strong arms of God, even if they are purely imaginary, could console in just the same kind of way as the real arms of a friend, or a St Bernard dog with a brandy cask around its neck. But of course scientific medicine can also offer comfort – usually more effectively than brandy.

Turning now to Type 2 consolation, it is easy to believe that religion could be extremely effective. People caught up in a terrible disaster, such as an earthquake, frequently report that they derive consolation from the reflection that it is all part of God's inscrutable plan: no doubt good shall come of it in the fullness of time. If someone fears death, sincere belief that he has an immortal soul can be consoling – unless, of course, he thinks he is going to hell or purgatory. False beliefs can be every bit as consoling as true ones, right up until the moment of disillusionment. This applies to non-religious beliefs too. A man with terminal cancer may be consoled by a doctor who lies to him that he is cured, just as effectively as another man who is told truthfully that he is cured. Sincere and wholehearted belief in life after death is even more immune to disillusionment than belief in a lying doctor. The doctor's lie remains effective only until the symptoms become unmistakable. A believer in life after death can never be ultimately disillusioned.

Polls suggest that approximately 95 per cent of the population of the United States believe they will survive their own death. Aspiring martyrs aside, I can't help wondering how many moderate religious people who claim such belief really hold it, in their heart of hearts. If they were truly sincere, shouldn't they all behave like the Abbot of Ampleforth? When Cardinal Basil Hume told him that he was dying, the abbot was delighted for him: 'Congratulations! That's brilliant news. I wish I was coming with you.'[154] The abbot, it seems, really was a sincere believer. But it is precisely because it is so rare and unexpected that his story catches our attention, almost provokes our amusement – in a fashion reminiscent of the cartoon of a young woman carrying a 'Make love not war' banner, stark naked, and with a bystander exclaiming, 'Now that's what I call sincerity!' Why don't all Christians and Muslims say something like the abbot when they hear that a friend is dying? When a devout woman is told by the doctor that she has only months to live, why doesn't she beam with excited anticipation, as if she has just won a holiday in the Seychelles? 'I can't wait!' Why don't faithful visitors at her bedside shower her with messages for those that have gone before? 'Do give my love to Uncle Robert when you see him . . .'

Why don't religious people talk like that when in the presence of the dying? Could it be that they don't really believe all that stuff they pretend to believe? Or perhaps they do believe it but fear the *process* of dying. With good reason, given that our species is the only one not allowed to go to the vet to be painlessly put out of our misery. But in that case, why does the most vocal opposition to euthanasia and assisted suicide come from the religious? On the 'Abbot of Ampleforth' or 'Holiday in the Seychelles' model of death, wouldn't you expect that religious people would be the least likely to cling unbecomingly to earthly life? Yet it is a striking fact that, if you meet somebody who is passionately opposed to mercy killing, or passionately against assisted suicide, you can bet a good sum that they will turn out to be religious. The official reason may be that all killing is a sin. But why deem it to be a sin if you sincerely believe you are accelerating a journey to heaven?

My attitude to assisted suicide, by contrast, takes off from Mark Twain's observation, already quoted. Being dead will be no different from being unborn – I shall be just as I was in the time of William the Conqueror or the dinosaurs or the trilobites. There is nothing to fear in that. But the process of dying could well be, depending on our luck, painful and unpleasant – the sort of experience from which we have become accustomed to being protected by a general anaesthetic, like having your appendix out. If your pet is dying in pain, you will be condemned for cruelty if you do not summon the vet to give him a general anaesthetic from which he will not come round. But if your doctor performs exactly the same merciful service for you when you are dying in pain, he runs the risk of being prosecuted for murder. When I am dying, I should like my life to be taken out under a general anaesthetic, exactly as if it were a diseased appendix. But I shall not be allowed that privilege, because I have the ill-luck to be born a member of *Homo sapiens* rather than, for example, *Canis familiaris* or *Felis catus*. At least, that will be the case unless I move to a more enlightened place like Switzerland, the Netherlands or Oregon. Why are such enlightened places so rare? Mostly because of the influence of religion.

But, it might be said, isn't there an important difference between having your appendix removed and having your life removed? Not really; not if you are about to die anyway. And not if you have a sincere religious belief in life after death. If you have that belief, dying is just a transition from one life to another. If the transition is painful, you should no more wish to undergo it without anaesthetic than you would wish to have your appendix removed without anaesthetic. It is those of us who see death as terminal rather than transitional who might naïvely be expected to resist euthanasia or assisted suicide. Yet we are the ones who support it.*

In the same vein, what are we to make of the observation of a senior nurse of my acquaintance, with a lifetime's experience in running a home for old people, where death is a regular occurrence? She has noticed over the years that the individuals who are most afraid of death are the religious ones. Her observation would need to be substantiated statistically but, assuming she is right, what is going on here? Whatever it is, it

---

* One study of attitudes to death among American atheists found the following: 50 per cent wanted a memorial celebration of their life; 99 per cent supported physician-assisted suicide for those who want it, and 75 per cent wanted it for themselves; 100 per cent wanted no contact with hospital staff who promote religion. See http://nursestoner.com/myresearch.html [link no longer active, 2016]. See Marilyn Smith-Stoner, *Journal of Palliative Medicine*: http://online.liebertpub.com/doi/abs/10.1089/jpm.2006.0197?src=recsys&journalCode=jpm

doesn't, on the face of it, speak strongly of religion's power to comfort the dying.* In the case of Catholics, maybe they are afraid of purgatory? The saintly Cardinal Hume said farewell to a friend in these words: 'Well, goodbye then. See you in purgatory, I suppose.' What *I* suppose is that there was a sceptical twinkle in those kind old eyes.

The doctrine of purgatory offers a preposterous revelation of the way the theological mind works. Purgatory is a sort of divine Ellis Island, a Hadean waiting room where dead souls go if their sins aren't bad enough to send them to hell, but they still need a bit of remedial checking out and purifying before they can be admitted to the sin-free-zone of heaven.† In medieval times, the Church used to sell 'indulgences' for money. This amounted to paying for some number of days' remission from purgatory, and the Church literally (and with breathtaking presumption) issued signed certificates specifying the number of days off that had been purchased. The Roman Catholic Church is an institution for whose gains the phrase 'ill-gotten' might have been specially invented. And of all its money-making rip-offs, the selling of indulgences must surely rank among the greatest con tricks in history, the medieval equivalent of the Nigerian Internet scam but far more successful.

As recently as 1903, Pope Pius X was still able to tabulate the number of days' remission from purgatory that each rank in the hierarchy was entitled to grant: cardinals two hundred days, archbishops a hundred days, bishops a mere fifty days. By his

---

* An Australian friend coined a wonderful phrase to describe the tendency for religiosity to increase in old age. Say it with an Australian intonation, going up at the end like a question: 'Cramming for the final?'

† Purgatory is not to be confused with Limbo, where babies who died unbaptized were supposed to go. And aborted foetuses? Blastocysts? Now, with characteristically presumptuous aplomb, Pope Benedict XVI has just abolished Limbo. Does that mean that all the babies who have been languishing there all these centuries will now suddenly float off to heaven? Or do they stay there and only the newly dead escape Limbo? Or have earlier popes been wrong all along, in spite of their infallibility? This is the kind of thing we are all supposed to 'respect'.

time, however, indulgences were no longer sold directly for money. Even in the Middle Ages, money was not the only currency in which you could buy parole from purgatory. You could pay in prayers too, either your own before death or the prayers of others on your behalf, after your death. And money could buy prayers. If you were rich, you could lay down provision for your soul in perpetuity. My own Oxford College, New College, was founded in 1379 (it was new then) by one of that century's great philanthropists, William of Wykeham, Bishop of Winchester. A medieval bishop could become the Bill Gates of the age, controlling the equivalent of the information highway (to God), and amassing huge riches. His diocese was exceptionally large, and Wykeham used his wealth and influence to found two great educational establishments, one in Winchester and one in Oxford. Education was important to Wykeham, but, in the words of the official New College history, published in 1979 to mark the sixth centenary, the fundamental purpose of the college was 'as a great chantry to make intercession for the repose of his soul. He provided for the service of the chapel by ten chaplains, three clerks and sixteen choristers, and he ordered that they alone were to be retained if the college's income failed.' Wykeham left New College in the hands of the Fellowship, a self-electing body which has been continuously in existence like a single organism for more than six hundred years. Presumably he trusted us to continue to pray for his soul through the centuries.

Today the college has only one chaplain* and no clerks, and the steady century-by-century torrent of prayers for Wykeham in purgatory has dwindled to a trickle of two prayers per year. The choristers alone go from strength to strength and their music is, indeed, magical. Even I feel a twinge of guilt, as a member of that Fellowship, for a trust betrayed. In the understanding of his own time, Wykeham was doing the equivalent of a rich man today making a large down payment to a

---

* Female – what would Bishop William have made of that?

cryogenics company which guarantees to freeze your body and keep it insulated from earthquakes, civil disorder, nuclear war and other hazards, until some future time when medical science has learned how to unfreeze it and cure whatever disease it was dying of. Are we later Fellows of New College reneging on a contract with our Founder? If so, we are in good company. Hundreds of medieval benefactors died trusting that their heirs, well paid to do so, would pray for them in purgatory. I can't help wondering what proportion of Europe's medieval treasures of art and architecture started out as down payments on eternity, in trusts now betrayed.

But what really fascinates me about the doctrine of purgatory is the *evidence* that theologians have advanced for it: evidence so spectacularly weak that it renders even more comical the airy confidence with which it is asserted. The entry on purgatory in the *Catholic Encyclopedia* has a section called 'proofs'. The essential evidence for the existence of purgatory is this. If the dead simply went to heaven or hell on the basis of their sins while on Earth, there would be no point in praying for them. 'For why pray for the dead, if there be no belief in the power of prayer to afford solace to those who as yet are excluded from the sight of God.' And we do pray for the dead, don't we? Therefore purgatory must exist, otherwise our prayers would be pointless! Q.E.D. This seriously is an example of what passes for reasoning in the theological mind.

That remarkable *non sequitur* is mirrored, on a larger scale, in another common deployment of the Argument from Consolation. There must be a God, the argument goes, because, if there were not, life would be empty, pointless, futile, a desert of meaninglessness and insignificance. How can it be necessary to point out that the logic falls at the first fence? Maybe life *is* empty. Maybe our prayers for the dead really *are* pointless. To presume the opposite is to presume the truth of the very conclusion we seek to prove. The alleged syllogism is transparently circular. Life without your wife may very well be intolerable, barren and empty, but this unfortunately doesn't stop her being dead. There is something infantile in the presumption that somebody else (parents in the case of children, God in the case of adults) has a responsibility to give your life meaning and point. It is all of a piece with the infantilism of those who, the moment they twist their ankle, look around for someone to sue. Somebody else must be responsible for my well-being, and somebody else must be to blame if I am hurt. Is it a similar infantilism that really lies behind the 'need' for a God? Are we back to Binker again?

The truly adult view, by contrast, is that our life is as meaningful, as full and as wonderful as we choose to make it. And we can make it very wonderful indeed. If science gives consolation of a non-material kind, it merges into my final topic, inspiration.

INSPIRATION

This is a matter of taste or private judgement, which has the slightly unfortunate effect that the method of argument I must employ is rhetoric rather than logic. I've done it before, and so have many others including, to name only recent examples, Carl Sagan in *Pale Blue Dot*, E. O. Wilson in *Biophilia*, Michael Shermer in *The Soul of Science* and Paul Kurtz in *Affirmations*. In *Unweaving the Rainbow* I tried to convey how lucky we are to be alive, given that the vast majority of people who could potentially be thrown up by the combinatorial lottery of DNA will in fact never be born. For those of us lucky enough to be here, I pictured the relative brevity of life by imagining a laser-thin spotlight creeping along a gigantic ruler of time. Everything before or after the spotlight is shrouded in the darkness of the dead past, or the darkness of the unknown future. We are staggeringly lucky to find ourselves in the spotlight. However brief our time in the sun, if we waste a second of it, or complain that it is dull or barren or (like a child) boring, couldn't this be seen as a callous insult to those unborn trillions who will never even be offered life in the first place? As many atheists have said better than me, the knowledge that we have only one life should make it all the more precious. The atheist view is correspondingly life-affirming and life-enhancing, while at the same time never being tainted with self-delusion, wishful thinking, or the whingeing self-pity of those who feel that life owes them something. Emily Dickinson said,

> That it will never come again
> Is what makes life so sweet.

If the demise of God will leave a gap, different people will fill it in different ways. My way includes a good dose of science, the honest and systematic endeavour to find out the truth about the real world. I see the human effort to understand the universe as a model-building enterprise. Each of us builds, inside our head, a model of the world in which we find ourselves. The minimal model of the world is the model our ancestors needed in order to survive in it. The simulation software was constructed and debugged by natural selection, and it is most adept in the world familiar to our ancestors on the African savannah: a three-dimensional world of medium-sized material objects, moving at medium speeds relative to one another. As an unexpected bonus, our brains turn out to be powerful enough to accommodate a much richer world model than the mediocre utilitarian one that our ancestors needed in order to survive. Art and science are runaway manifestations of this bonus. Let me paint one final picture, to convey the power of science to open the mind and satisfy the psyche.

THE MOTHER OF ALL BURKAS

One of the unhappiest spectacles to be seen on our streets today is the image of a woman swathed in shapeless black from head to toe, peering out at the world through a tiny slit. The burka is not just an instrument of oppression of women and claustral repression of their liberty and their beauty; not just a token of egregious male cruelty and tragically cowed female submission. I want to use the narrow slit in the veil as a symbol of something else.

Our eyes see the world through a narrow slit in the electromagnetic spectrum. Visible light is a chink of brightness in the vast dark spectrum, from radio waves at the long end to gamma rays at the short end. Quite *how* narrow is hard to appreciate and a challenge to convey. Imagine a gigantic black burka, with a vision slit of approximately the standard width, say about one inch. If the length of black cloth above the slit represents the short-wave end of the invisible spectrum, and if the length of black cloth below the slit represents the long-wave portion of the invisible spectrum, how long would the burka have to be in order to accommodate a one-inch slit to the same scale? It is hard to represent it sensibly without invoking logarithmic scales, so huge are the lengths we are dealing with. The last chapter of a book like this is no place to start tossing logarithms around, but you can take it from me that it would be the mother of all burkas. The one-inch window of visible light is derisorily tiny compared with the miles and miles of black cloth representing the invisible part of the spectrum, from radio waves at the hem of the skirt to gamma rays at the top of the head. What science does for us is widen the window. It opens up so wide that the imprisoning black garment drops away almost completely, exposing our senses to airy and exhilarating freedom.

Optical telescopes use glass lenses and mirrors to scan the heavens, and what they see is stars that happen to be radiating in the narrow band of wavelengths that we call visible light. But other telescopes 'see' in the X-ray or radio wavelengths, and present to us a cornucopia of alternative night skies. On a smaller scale, cameras with appropriate filters can 'see' in the ultraviolet and take photographs of flowers that show an alien range of stripes and spots that are visible to, and seemingly 'designed' for, insect eyes but which our unaided eyes can't see at all. Insect eyes have a spectral window of similar width to ours, but slightly shifted up the burka: they are blind to red and they see further into the ultraviolet than we do – into the 'ultraviolet garden'.*

The metaphor of the narrow window of light, broadening out into a spectacularly wide spectrum, serves us in other areas of science. We live near the centre of a cavernous museum of magnitudes, viewing the world with sense organs and nervous systems that are equipped to perceive and understand only a small middle range of sizes, moving at a middle range of speeds. We are at home with objects ranging in size from a few kilometres (the view from a mountaintop) to about a tenth of a millimetre (the point of a pin). Outside this range even our imagination is handicapped, and we need the help of instruments and of mathematics – which, fortunately, we can learn to deploy. The range of sizes, distances or speeds with which our imaginations are comfortable is a tiny band, set in the midst of a gigantic range of the possible, from the scale of quantum strangeness at the smaller end to the scale of Einsteinian cosmology at the larger.

Our imaginations are forlornly under-equipped to cope with distances outside the narrow middle range of the ancestrally familiar. We try to visualize an electron as a tiny ball, in orbit around a larger cluster of balls representing protons and neutrons. That isn't what it is like at all. Electrons are not like little balls. They are not like anything we recognize. It isn't clear that 'like' even means anything when we try to fly too close to reality's further horizons. Our imaginations are not yet tooled-up to penetrate the neighbourhood of the quantum. Nothing at that scale behaves in the way matter – as we are evolved to think – ought to behave. Nor can we cope with the behaviour of objects that move at some appreciable fraction of the speed of light. Common sense lets us down, because common sense

---

* 'The Ultraviolet Garden' was the title of one of my five Royal Institution Christmas Lectures, originally televised by the BBC under the general title 'Growing Up in the Universe'. The whole series of five lectures is available on DVD from www.richarddawkins.net/home.

evolved in a world where nothing moves very fast, and nothing is very small or very large.

At the end of a famous essay on 'Possible Worlds', the great biologist J. B. S. Haldane wrote, 'Now, my own suspicion is that the universe is not only queerer than we suppose, but queerer than we can suppose . . . I suspect that there are more things in heaven and earth than are dreamed of, or can be dreamed of, in any philosophy.' By the way, I am intrigued by the suggestion that the famous Hamlet speech invoked by Haldane is conventionally mis-spoken. The normal stress is on 'your':

> There are more things in heaven and earth, Horatio,
> Than are dreamt of in *your* philosophy.

Indeed, the line is often plonkingly quoted with the implication that Horatio stands for shallow rationalists and sceptics everywhere. But some scholars place the stress on 'philosophy', with 'your' almost vanishing: '. . . than are dreamt of inya *philosophy*.' The difference doesn't really matter for present purposes, except that the second interpretation already takes care of Haldane's 'any' philosophy.

The dedicatee of this book made a living from the strangeness of science, pushing it to the point of comedy. The following is taken from the same extempore speech in Cambridge in 1998 that I quoted in Chapter 1: 'The fact that we live at the bottom of a deep gravity well, on the surface of a gas-covered planet going around a nuclear fireball ninety million miles away and think this to be *normal* is obviously some indication of how skewed our perspective tends to be.' Where other science-fiction writers played on the oddness of science to arouse our sense of the mysterious, Douglas Adams used it to make us laugh (those who have read *The Hitchhiker's Guide to the Galaxy* might think of the 'infinite improbability drive', for instance). Laughter is arguably the best response to some of the stranger paradoxes of modern physics. The alternative, I sometimes think, is to cry.

Quantum mechanics, that rarefied pinnacle of twentieth-century scientific achievement, makes brilliantly successful predictions about the real world. Richard Feynman compared its precision to predicting a distance as great as the width of North America to an accuracy of one human hair's breadth. This predictive success seems to mean that quantum theory has got to be true in some sense; as true as anything we know, even including the most down-to-earth common-sense facts. Yet the *assumptions* that quantum theory needs to make, in order to deliver those predictions, are so mysterious that even the great Feynman himself was moved to remark (there are various versions of this quotation, of which the following seems to me the neatest): 'If you think you understand quantum theory . . . you don't understand quantum theory.'*

Quantum theory is so queer that physicists resort to one or another paradoxical 'interpretation' of it. Resort is the right word. David Deutsch, in *The Fabric of Reality*, embraces the 'many worlds' interpretation of quantum theory, perhaps because the worst that you can say of it is that it is preposterously *wasteful*. It postulates a vast and rapidly growing number of universes, existing in parallel and mutually undetectable except through the narrow porthole of quantum-mechanical experiments. In some of these universes I am already dead. In a small minority of them, you have a green moustache. And so on.

The alternative 'Copenhagen interpretation' is equally preposterous – not wasteful, just shatteringly paradoxical. Erwin Schrödinger satirized it with his parable of the cat. Schrödinger's cat is shut up in a box with a killing mechanism triggered by a quantum-mechanical event. Before we open the lid of the box, we don't know whether the cat is dead. Common sense tells us that, nevertheless, the cat must be either alive or dead inside the box. The Copenhagen interpretation

---

* A similar remark is attributed to Niels Bohr: 'Anyone who is not shocked by quantum theory has not understood it.'

contradicts common sense: all that exists before we open the box is a probability. As soon as we open the box, the wave function collapses and we are left with the single event: the cat is dead, or the cat is alive. Until we opened the box, it was neither dead nor alive.

The 'many worlds' interpretation of the same events is that in some universes the cat is dead; in other universes the cat is alive. Neither interpretation satisfies human common sense or intuition. The more macho physicists don't care. What matters is that the mathematics work, and the predictions are experimentally fulfilled. Most of us are too wimpish to follow them. We seem to *need* some sort of visualization of what is 'really' going on. I understand, by the way, that Schrödinger originally proposed his cat thought-experiment in order to expose what he saw as the absurdity of the Copenhagen interpretation.

The biologist Lewis Wolpert believes that the queerness of modern physics is just the tip of the iceberg. Science in general, as opposed to technology, does violence to common sense.[155] Wolpert calculates, for example, 'that there are many more molecules in a glass of water than there are glasses of water in the sea'. Since all the water on the planet cycles through the sea, it would seem to follow that every time you drink a glass of water, the odds are good that something of what you are drinking has passed through the bladder of Oliver Cromwell. There is, of course, nothing special about Cromwell, or bladders. Haven't you just breathed in a nitrogen atom that was once breathed out by the third iguanodon to the left of the tall cycad tree? Aren't you glad to be alive in a world where not only is such a conjecture possible but you are privileged to understand why? And publicly explain it to somebody else, not as your opinion or belief but as something that they, when they have understood your reasoning, will feel compelled to accept? Maybe this is an aspect of what Carl Sagan meant when he explained his motive in writing *The Demon-Haunted World: Science as a Candle in the Dark*: '*Not* explaining science seems to me perverse. When you're in love, you want to tell the world.

This book is a personal statement, reflecting my lifelong love affair with science.'

The evolution of complex life, indeed its very existence in a universe obeying physical laws, is wonderfully surprising – or would be but for the fact that surprise is an emotion that can exist only in a brain which is the product of that very surprising process. There is an anthropic sense, then, in which our existence should not be surprising. I'd like to think that I speak for my fellow humans in insisting, nevertheless, that it is desperately surprising.

Think about it. On one planet, and possibly only one planet in the entire universe, molecules that would normally make nothing more complicated than a chunk of rock, gather themselves together into chunks of rock-sized matter of such staggering complexity that they are capable of running, jumping, swimming, flying, seeing, hearing, capturing and eating other such animated chunks of complexity; capable in some cases of thinking and feeling, and falling in love with yet other chunks of complex matter. We now understand essentially how the trick is done, but only since 1859. Before 1859 it would have seemed very very odd indeed. Now, thanks to Darwin, it is merely very odd. Darwin seized the window of the burka and wrenched it open, letting in a flood of understanding whose dazzling novelty, and power to uplift the human spirit, perhaps had no precedent – unless it was the Copernican realization that the Earth was not the centre of the universe.

'Tell me,' the great twentieth-century philosopher Ludwig Wittgenstein once asked a friend, 'why do people always say it was natural for man to assume that the sun went round the Earth rather than that the Earth was rotating?' His friend replied, 'Well, obviously because it just *looks* as though the Sun is going round the Earth.' Wittgenstein responded, 'Well, what would it have looked like if it had looked as though the Earth was rotating?' I sometimes quote this remark of Wittgenstein in lectures, expecting the audience to laugh. Instead, they seem stunned into silence.

In the limited world in which our brains evolved, small objects are more likely to move than large ones, which are seen as the background to movement. As the world rotates, objects that seem large because they are near – mountains, trees and buildings, the ground itself – all move in exact synchrony with each other and with the observer, relative to heavenly bodies such as the sun and stars. Our evolved brains project an illusion of movement onto them rather than the mountains and trees in the foreground.

I now want to pursue the point mentioned above, that the way we see the world, and the reason why we find some things intuitively easy to grasp and others hard, is that *our brains are themselves evolved organs*: on-board computers, evolved to help us survive in a world – I shall use the name Middle World – where the objects that mattered to our survival were neither very large nor very small; a world where things either stood still or moved slowly compared with the speed of light; and where the very improbable could safely be treated as impossible. Our mental burka window is narrow because it didn't *need* to be any wider in order to assist our ancestors to survive.

Science has taught us, against all evolved intuition, that apparently solid things like crystals and rocks are really composed almost entirely of empty space. The familiar illustration represents the nucleus of an atom as a fly in the middle of a sports stadium. The next atom is right outside the stadium. The hardest, solidest, densest rock, then, is 'really' almost entirely empty space, broken only by tiny particles so far apart that they shouldn't count. So why do rocks look and feel solid and hard and impenetrable?

I won't try to imagine how Wittgenstein might have answered that question. But, as an evolutionary biologist, I would answer it like this. Our brains have evolved to help our bodies find their way around the world on the scale at which those bodies operate. We never evolved to navigate the world of atoms. If we had, our brains probably *would* perceive rocks as full of empty space. Rocks feel hard and impenetrable to our hands because our hands can't penetrate them. The reason they can't penetrate them is unconnected with the sizes and separations of the particles that constitute matter. Instead, it has to do with the force fields that are associated with those widely spaced particles in 'solid' matter. It is useful for our brains to *construct* notions like solidity and impenetrability, because such notions help us to navigate our bodies through a world in which objects – which we call solid – cannot occupy the same space as each other.

Having evolved in Middle World, we find it intuitively easy to grasp ideas like: 'When a human being moves, at the sort of medium velocity at which human beings and other Middle World objects do move, and hits another solid Middle World object like a wall, his progress is painfully arrested.' Our brains are not equipped to imagine what it would be like to be a neutrino passing through a wall, in the vast interstices of which that wall 'really' consists. Nor can our understanding cope with what happens when things move at close to the speed of light.

Unaided human intuition, evolved and schooled in Middle World, even finds it hard to believe Galileo when he tells us that a cannonball and a feather, given no air friction, would hit the ground at the same instant when dropped from a leaning tower. That is because, in Middle World, air friction is always there. If we had evolved in a vacuum, we would *expect* a feather and a cannonball to hit the ground simultaneously. We are evolved denizens of Middle World, and that limits what we are capable of imagining. The narrow window of our burka permits us, unless we are especially gifted or peculiarly well educated, to see only Middle World.

There is a sense in which we animals have to survive not just in Middle World but in the micro-world of atoms and electrons too. The very nerve impulses with which we do our thinking and our imagining depend upon activities in Micro World. But no action that our wild ancestors ever had to perform, no decision that they ever had to take, would have been assisted by an understanding of Micro World. If we were bacteria, constantly buffeted by thermal movements of molecules, it would be different. But we Middle Worlders are too cumbersomely massive to notice Brownian motion. Similarly, our lives are dominated by gravity but are almost oblivious to the delicate force of surface tension. A small insect would reverse that priority and would find surface tension anything but delicate.

Steve Grand, in *Creation: Life and How to Make It*, is almost scathing about our preoccupation with matter itself. We have this tendency to think that only solid, material 'things' are 'really' things at all. 'Waves' of electromagnetic fluctuation in a vacuum seem 'unreal'. Victorians thought that waves had to be waves 'in' some material medium. No such medium was known, so they invented one and named it the luminiferous ether. But we find 'real' matter comfortable to our understanding only because our ancestors evolved to survive in Middle World, where matter is a useful construct.

On the other hand, even we Middle Worlders can see that a whirlpool is a 'thing' with something like the reality of a rock, even though the matter in the whirlpool is constantly changing. In a desert plain in Tanzania, in the shadow of Ol Donyo Lengai, sacred volcano of the Masai, there is a large dune made of ash from an eruption in 1969. It is carved into shape by the wind. But the beautiful thing is that it *moves* bodily. It is what is technically known as a barchan (pronounced bahkahn). The entire dune walks across the desert in a westerly direction at a speed of about 17 metres per year. It retains its crescent shape and creeps along in the direction of the horns. The wind blows sand up the shallower slope. Then, as each sand grain hits the top of the ridge, it cascades down the steeper slope on the inside of the crescent.

Actually, even a barchan is more of a 'thing' than a wave. A wave *seems* to move horizontally across the open sea, but the molecules of water move vertically. Similarly, sound waves may travel from speaker to listener, but molecules of air don't: that would be a wind, not a sound. Steve Grand points out that you and I are more like waves than permanent 'things'. He invites his reader to think . . .

> . . . of an experience from your childhood. Something you remember clearly, something you can see, feel, maybe even smell, as if you were really there. After all, you really were there at the time, weren't you? How else would you remember it? But here is the bombshell: you *weren't* there. Not a single atom that is in your body today was there when that event took place . . . Matter flows from place to place and momentarily comes together to be you. Whatever you are, therefore, you are not the stuff of which you are made. If that doesn't make the hair stand up on the back of your neck, read it again until it does, because it is important.*

'Really' isn't a word we should use with simple confidence. If a neutrino had a brain which had evolved in neutrino-sized ancestors, it would say that rocks 'really' do consist mostly of empty space. We have brains that evolved in medium-sized ancestors, who couldn't walk through rocks, so our 'really' is a 'really' in which rocks are solid. 'Really', for an animal, is whatever its brain needs it to be, in order to assist its survival. And because different species live in such different worlds, there will be a troubling variety of 'reallys'.

What we see of the real world is not the unvarnished real world but a *model* of the real world, regulated and adjusted by sense data – a model that is constructed so that it is useful for dealing with the real world. The nature of that model depends on the kind of animal we are. A flying animal needs a different kind of world model from a walking, a climbing or a swimming animal. Predators need a different kind of model from prey,

* Some might dispute the literal truth of Grand's statement, for example in the case of bone molecules. But the spirit of it is surely valid. You are more like a wave than a static material 'thing'.

even though their worlds necessarily overlap. A monkey's brain must have software capable of simulating a three-dimensional maze of branches and trunks. A water boatman's brain doesn't need 3D software, since it lives on the surface of the pond in an Edwin Abbott Flatland. A mole's software for constructing models of the world will be customized for underground use. A naked mole rat probably has world-representing software similar to a mole's. But a squirrel, although it is a rodent like the mole rat, probably has world-rendering software much more like a monkey's.

I've speculated, in *The Blind Watchmaker* and elsewhere, that bats may 'see' colour with their ears. The world-model that a bat needs, in order to navigate through three dimensions catching insects, must surely be similar to the model that a swallow needs in order to perform much the same task. The fact that the bat uses echoes to update the variables in its model, while the swallow uses light, is incidental. Bats, I suggest, use perceived hues such as 'red' and 'blue' as internal labels for some useful aspect of echoes, perhaps the acoustic texture of surfaces; just as swallows use the same perceived hues to label long and short wavelengths of light. The point is that the nature of the model is governed by how it is to be *used* rather than by the sensory modality involved. The lesson of the bats is this. The general form of the mind model – as opposed to the variables that are constantly being inputted by sensory nerves – is an adaptation to the animal's way of life, no less than its wings, legs and tail are.

J. B. S. Haldane, in the article on 'possible worlds' that I quoted above, had something relevant to say about animals whose world is dominated by smell. He noted that dogs can distinguish two very similar volatile fatty acids – caprylic acid and caproic acid – each diluted to one part in a million. The only difference is that caprylic acid's main molecular chain is two carbon atoms longer than the main chain of caproic acid. A dog, Haldane guessed, would probably be able to place the acids 'in the order of their molecular weights by their smells, just as a man could place a number of piano wires in the order of their lengths by means of their notes'.

There is another fatty acid, capric acid, which is just like the other two except that it has yet two more carbon atoms in its main chain. A dog that had never met capric acid would perhaps have no more trouble imagining its smell than we would have trouble imagining a trumpet playing one note higher than we have heard a trumpet play before. It seems to me entirely reasonable to guess that a dog, or a rhinoceros, might treat mixtures of smells as harmonious chords. Perhaps there are discords. Probably not melodies, for melodies are built up of notes that start or stop abruptly with accurate timing, unlike smells. Or perhaps dogs and rhinos smell in colour. The argument would be the same as for the bats.

Once again, the perceptions that we call colours are tools used by our brains to label important distinctions in the outside world. Perceived hues – what philosophers call qualia – have no intrinsic connection with lights of particular wavelengths. They are internal labels that are *available* to the brain, when it constructs its model of external reality, to make distinctions that are especially salient to the animal concerned. In our case, or that of a bird, that means light of different wavelengths. In a bat's case, I have speculated, it might be surfaces of different echoic properties or textures, perhaps red for shiny, blue for velvety, green for abrasive. And in a dog's or a rhino's case, why should it not be smells? The power to imagine the alien world of a bat or a rhino, a pond skater or a mole, a bacterium or a bark beetle, is one of the privileges science grants us when it tugs at the black cloth of our burka and shows us the wider range of what is out there for our delight.

The metaphor of Middle World – of the intermediate range of phenomena that the narrow slit in our burka permits us to see – applies to yet other scales or 'spectrums'. We can construct a scale of improbabilities, with a similarly narrow window through which our intuition and imagination are capable of going. At one extreme of the spectrum of improbabilities are those would-be events that we call impossible. Miracles are events that are extremely improbable. A statue of a madonna could wave its hand at us. The atoms that make up its crystalline structure are all vibrating back and forth. Because there are so many of them, and because there is no agreed preference in their direction of motion, the hand, as we see it in Middle World, stays rock steady. But the jiggling atoms in the hand *could* all just *happen* to move in the same direction at the same time. And again. And again . . . In this case the hand would move, and we'd see it waving at us. It could happen, but the odds against are so great that, if you had set out writing the number at the origin of the universe, you still would not have written enough zeroes to this day. The power to calculate such odds – the power to quantify the near-impossible rather than just throw up our hands in despair – is another example of the liberating benefactions of science to the human spirit.

Evolution in Middle World has ill equipped us to handle very improbable events. But in the vastness of astronomical space, or geological time, events that seem impossible in Middle World turn out to be inevitable. Science flings open the narrow window through which we are accustomed to viewing the spectrum of possibilities. We are liberated by calculation and reason to visit regions of possibility that had once seemed out of bounds or inhabited by dragons. We have already made use of this widening of the window in Chapter 4, where we considered the improbability of the origin of life and how even a near-impossible chemical event must come to pass given enough planet years to play with; and where we considered the spectrum of possible universes, each with its own set of laws and constants, and the anthropic necessity of finding ourselves in one of the minority of friendly places.

How should we interpret Haldane's 'queerer than we can suppose'? Queerer than can, *in principle*, be supposed? Or just queerer than we can suppose, given the limitation of our brains' evolutionary apprenticeship in Middle World? Could we, by training and practice, emancipate ourselves from Middle

World, tear off our black burka, and achieve some sort of intuitive – as well as just mathematical – understanding of the very small, the very large, and the very fast? I genuinely don't know the answer, but I am thrilled to be alive at a time when humanity is pushing against the limits of understanding. Even better, we may eventually discover that there are no limits.

Afterword
by Daniel Dennett

Four books appeared within a few months of each other a decade ago: Sam Harris's *The End of Faith*, my *Breaking the Spell*, Richard Dawkins' *The God Delusion* and Christopher Hitchens' *God is not Great*. Although the authors knew, or knew of, each other, this near-simultaneous outburst was not planned, but we soon joined forces, informally, and somebody – not one of us – dubbed us the Four Horsemen of the New Atheism. Fame – or notoriety, take your pick – followed, and before long we were joined by a distinguished cadre of other authors who had decisive and well-evidenced cases to present about various problems and failures of religion. Many of these have been well received; but *The God Delusion* has outsold them all, probably by an order of magnitude. Whatever twinges of envy that fact obliges me to experience (I'm only human), they are obliterated by my delight in the fact that his book has outsold all the 'flea' books he mentions in his introduction to this new edition by even wider margins. Those frantically scribbled diatribes – none of which, so far as I know, has attracted favourable attention – are a well-deserved measure of the scale of Richard's impact. And while 'sophisticated theologians' and their friends wanted the world to believe that he failed to engage *serious* religion in his critique, those darn fleas tell a different story: he struck a nerve, and he struck it dead centre.

It was once my distinct pleasure to confront one of these critics directly in print. H. Allen Orr, in 'A mission to convert',[156] reviewing *The God Delusion* and other new books on science and religion, called Dawkins an amateur, not professional, atheist, and claimed he had failed to come to grips with 'religious thought' with its 'meticulous reasoning' in any serious way. I cannot resist quoting from the letter I wrote to the editor in response, since it expresses my considered view better than anything I could write anew:

> He notes that the book is 'defiantly middlebrow,' and I wonder just which highbrow thinkers about religion Orr believes Dawkins should have grappled with. I myself have looked over large piles of recent religious thought in the last few years in the course of researching my own book on these topics, and I have found almost all of it to be so dreadful that ignoring it entirely seemed both the most charitable and most constructive policy. (I devote a scant six pages of *Breaking the Spell* to the arguments for and against the existence of God, while Dawkins devotes roughly a hundred, laying out the standard arguments with admirable clarity and fairness, and skewering them efficiently.) There are indeed *recherché* versions of these traditional arguments that perhaps have not yet been exhaustively eviscerated by scholars, but Dawkins ignores them (as do I) and says why: his book is a consciousness-raiser aimed at the general religious public, not an attempt to contribute to the academic micro-discipline of philosophical theology. The arguments Dawkins exposes and rebuts are the arguments that waft from thousands of pulpits every week and reach millions of television viewers every day, and neither the televangelists nor the authors of best-selling spiritual books pay the slightest heed to the subtleties of the theologians either.
>
> Who does Orr favor? Polkinghorne, Peacocke, Plantinga, or some more recondite thinkers? Orr brandishes the names of two philosophers, William James and Ludwig Wittgenstein, and cites C. S. Lewis's *Mere Christianity*, a fairly nauseating example of middle-brow homiletic in roughly the same

league on the undergraduate hit parade as Lee Strobel's *The Case for Christ* (1998) and transparently evasive when it comes to 'meticulous reasoning.' If it were a book in biology – Orr's discipline – I daresay he'd pounce on it like a pit bull, but like many others he adopts a double standard when the topic is religion. As Orr says, both James and Wittgenstein 'struggled with the question of belief,' in their admirable and entirely different ways, but both also steer clear of the issues that Orr chides Dawkins for oversimplifying. I wonder which themes in these fine thinkers Orr would champion in the current discussion, beyond the speculation he cites from James, that 'the visible world is part of a more spiritual universe.' I'd be curious to know what Orr thinks that means. How should it be clarified and investigated, in his opinion, or does he just want to leave it hanging unchallenged?

Orr ends by wondering why Dawkins – no expert on religion – wrote his book, and he might also wonder why I wrote mine. Didn't we have more intellectually satisfying problems to work on, problems better fitting our training, interests and talents? I'll answer for myself, but I think Dawkins would give much the same answer. Yes, of course I'd much rather have been spending my time working on consciousness and the brain, or on the evolution of cooperation, for instance, or free will, but I felt a moral and political obligation to drop everything for a few years and put my shoulder to the wheel doing a dirty job that I thought somebody had to do. I am aching to get back to my favorite topics, but I still have to do a fair amount of follow-up, apparently, since there are plenty of people like Orr who still want to protect religion from the sort of unflinching scrutiny Dawkins and I (and Sam Harris and Lewis

> Wolpert and others) are calling for. Is this opinion of Orr's just force of habit, or going along with tradition, or has he carefully studied the phenomena and seen that we really mustn't rock the boat, for fear of causing calamity? If the latter, he owes the world a careful and vivid argument to that effect, for it would put Dawkins and the rest of us in our proper place as dangerous intellectual vandals. Such a project would not fit his talents or training, but I should think it would be his duty as a concerned scientist.[157]

Orr never replied, and in the intervening years we have seen many other critics of *The God Delusion*, some of whom have probably actually read the book, do a vanishing act when pressed for details in support of their dire verdicts. The one sentence that is often cited as evidence is Dawkins' delicious catalogue of the (retrospectively obvious) disqualifications of Jehovah as an entity to worship:

> The God of the Old Testament is arguably the most unpleasant character in all fiction: jealous and proud of it; a petty, unjust, unforgiving control-freak; a vindictive, bloodthirsty ethnic cleanser; a misogynistic, homophobic, racist, infanticidal, genocidal, filicidal, pestilential, megalomaniacal, sadomasochistic, capriciously malevolent bully.

This is a startling indictment indeed; but aside from deploring it, who has the temerity to go through it, point by point, and *dispute* it? As Dawkins notes, Dan Barker has now published a rollicking case for the prosecution, *God: The Most Unpleasant Character in All Fiction*, citing chapter and verse for each feature. The ball is in the critics' court, but I don't expect to see a return of service.

We horsemen, and others, have often written about the 'I'm an atheist but' crowd, or the 'faitheists' (a good term coyned by

Jerry Coyne), who claim not to find a need in themselves for religious belief but decry the rudeness with which we impose our scepticism on those who still need a fantasy to live by. It never seems to occur to them how patronizing this complaint is: 'I, like you, see through all the smoke and mirrors, but come on, guys, think of the poor dears who can't handle the truth!' We horsemen and horsewomen (many more than four – perhaps we should be rebranded the Dozen Demons of Doubt or something like that) think we are actually showing respect for our fellow adults, readers all, presumably curious and well informed on matters of importance, by directly challenging their professed opinions. Yes, alas, we are not exactly polite, but consider: there is no polite way of asking somebody to consider the possibility that they have wasted their life on an illusion, is there? And sometimes it is important to ask that very question, however awkward it may be. Thank goodness Dawkins has the style to carry off this rude awakening with elegance.

But is he just preaching to the choir, as some have insisted? Has *The God Delusion* actually converted anybody? Yes, indeed. Several of the clergy Linda LaScola interviewed for our study, *Caught in the Pulpit*,[158] revealed that it was their decision to read *The God Delusion* (on the principle of 'know thine enemy') that eventually led to their still secret apostasy. And I have received many emails from people who tell me that reading Dawkins was either the opening alarm bell or the final straw in their lonely journey to atheism. This has been particularly instructive to me, since I went out of my way to write my own book in a more diplomatic tone. I figured that many readers would be strongly tempted to throw it across the room, but I didn't want to give them anything that seemed like a good excuse. It turns out that if you write with the passion and clarity of Dawkins, you can get some of your readers to retrieve the book they have heaved in anger, finish it, and declare themselves persuaded. I wonder how many readers of *The God Delusion* have time bombs ticking away in their minds, accumulating doubt with every passing day.

Is he ‘angry’? Is he ‘shrill’ and ‘arrogant’? Look closely, and you will see that these familiar charges are without foundation. What leads people to level them is the fact that they have been accustomed their entire lives to having their darling dogmas handled with kid gloves, never challenged, always ‘respected’. I put ‘respected’ in scare quotes because – a dirty little secret that I suspect everyone knows – hardly anybody truly respects the bizarre doctrines of any religion but their own. They just feel obliged to say (in public) that they do, a bit of lip service to ecumenicism. Do you really think that the archbishop respects the angel Gabriel who visited Muhammad in the cave, or the Angel Moroni with the golden plates? Or that the imam respects the transubstantiation of the wafer and wine? As one very sophisticated Episcopalian priest once confided to me: ‘When I found out what my Mormon relatives meant by “God” I rather wished that they *didn’t* believe in God!’

Thanks to the new worldwide transparency opened up by electronic media, and especially the internet,[159] we are now all living in glass houses, and all the diplomatic posturing that concealed this mutual disrespect much of the time (except when fighting bloody wars of religion) is beginning to lose its efficacy; so perhaps it is time to retire the faitheists’ demand for lip service altogether and join Richard Dawkins in a candid exploration of the dreams from which the world is finally awakening.

Appendix

A partial list of friendly addresses, for individuals needing support in escaping from religion

I intend to keep an updated version of this list on the website of the Richard Dawkins Foundation for Reason and Science: www.richarddawkins.net. I apologize for limiting the list below largely to the English-speaking world.

USA

American Atheists
PO Box 158, Cranford, NJ 07016
Voicemail: 1 908 276 7300
www.atheists.org

American Humanist Association
1777 T Street NW, Washington DC 20009-7125
Tel: 202 238 9088
Toll-free: 1 800 837 3792
Fax: 202 238 9003
www.americanhumanist.org

Atheist Alliance International
PO Box 26867, Los Angeles, CA 90026
Toll-free: 1 866 HERETIC
Email: info@atheistalliance.org
www.atheistalliance.org

Atheist community of Austin
www.atheist-community.org

The Brights
PO Box 163418, Sacramento, CA 95816
Email: the-brights@the-brights.net
www.the-brights.net

Camp Quest
Camp Quest National Office
PO Box 2552
Columbus, OH 43216
Tel: 614 441 9534
www.campquest.org

Center For Inquiry Transnational
Council for Secular Humanism
Campus Freethought Alliance
Center for Inquiry – On Campus
African Americans for Humanism
PO Box 741, Amherst, NY 14226
Tel: 716 636 4869
Fax: 716 636 1733
Email: info@centerforinquiry.net

Freedom From Religion Foundation
PO Box 750, Madison, WI 53701
Tel: 608 256 5800
Fax: 608 204 0422
www.ffrf.org

Anti-Discrimination Support Network (ADSN)
Freethought Society of Greater Philadelphia
PO Box 242, Pocopson, PA 19366-0242
Tel: 610 793 2737
Email: fsgp@freethought.org
www.ftsociety.org

The Institute for Humanist Studies
1777T Street NW
Washington DC 20009
Tel: 202 238 9088, ext. 116
www.humaniststudies.org

International Humanist and Ethical Union – USA
Appignani Bioethics Center
PO Box 4104, Grand Central Station, New York, NY 10162
Tel: 212 687 3324
Fax: 212 661 4188

Internet Infidels
711 S. Carson St., Suite 4
Carson City, NV 89701
Fax: 877 501 5113
www.infidels.org

Military Association of Atheists and Freethinkers
888 16th St NW Ste 800
Washington DC 20006
Tel: 202 656 MAAF (6223)
www.militaryatheists.org

Project Reason
www.project-reason.org

The Rational Response Squad
www.rationalresponders.com

Secular Coalition for America
1012 14th Street NW, Suite 205
Washington DC 20005-3429
Tel: 202 299 1091
Fax: 202 293 0922
www.secular.org

Secular Student Alliance
PO Box 2371, Columbus, OH 43216
Tel: 1 614 441 9588
Fax: 1 877 842 9474
Email: ssa@secularstudents.org
www.secularstudents.org

The Skeptics Society
PO Box 338, Altadena, CA 91001
Tel: 626 794 3119
Fax: 626 794 1301
Email: skepticssociety@skeptic.com
www.skeptic.com

Society for Humanistic Judaism
28611 W. 12 Mile Rd, Farmington Hills, MI 48334
Tel: 248 478 7610
Fax: 248 478 3159
Email: info@shj.org
www.shj.org

United Coalition of Reason
1300 Pennsylvania Ave NW, Suite 190-822
Washington DC 20004-3002
Tel: 202 744 1553
www.unitedcor.org

Britain
British Humanist Association
39 Moreland Street, London EC1V 8BB
Tel: 020 7324 3060
Fax: 020 7324 3061
Email: info@humanism.org.uk
www.humanism.org.uk

Conway Hall Ethical Society
Conway Hall, Red Lion Square, London WC1R 4RL
Tel: 020 7405 1818
www.conwayhall.org.uk/ethical-society

GALHA LGBT Humanists
Email: chair@galha.org
www.galha.org
Also known as LGBT Humanists UK

International Humanist and Ethical Union – UK (IHEU)
39 Moreland Street, London EC1V 8BB
Tel: 020 7490 8468
www.iheu.org/

National Secular Society
25 Red Lion Square, London WC1R 4RL
Tel: 020 7404 3126
Fax: 0870 762 8971
Email: enquiries@secularism.org.uk
www.secularism.org.uk/

New Humanist
Merchant's House, 5-7 Southwark Street, London SE1 1RQ
Tel: 020 3117 0630
www.newhumanist.org.uk
www.rationalist.org.uk

Scottish Secular Society
Anne Besant Lodge, c/o Glasgow Theosophical Society Building,
17 Queen's Crescent, St Georges Cross, Glasgow G4 9BL
Email: contact@secularsociety.scot

European Mainland

De Vrije Gedachte
PO Box 398, 3500 AJ Utrecht, Netherlands
Tel: 0031 85 401 6874
Email: info@devrijegedachte.nl
www.devrijegedachte.nl

European Humanist Federation
Campus de la Plaine ULB, Accès 2 cp 237
Boulevard de la Plaine, 1050 Brussels, Belgium
Tel: 32 2 627 68 24
Fax: 32 2 627 68 01
Email: admin@humanistfederation.eu
www.humanistfederation.eu

Giordano-Bruno-Stiftung
Haus WEITBLICK
Auf Fasel 16, 55430 Oberwesel
Tel: 49 06744 7105020
Fax: 49 06744 7105021
Email: info@giordano-bruno-stiftung.de
www.giordano-bruno-stiftung.de

International League of Non-Religious and Atheists
IBKA-Geschäftsstelle
c/o Rainer Ponitka
Tilsiter Str.3
51491 Overath
Tel: 49 02206 8673261
Email: info@ibka.org

Canada
Humanist Canada
45 O'Connor St, Suite 1150, Ottawa, ON, Canada K1P 1A4
Freephone 1 877 486 2671
Tel: 613 739 9569
www.humanistcanada.ca

Movement Laïque Québécois
Casier postal 32132, Succursale Saint-André,
Montréal, (Québec), Canada H2L 4Y5
Tel: 1 514 985 5840
Email: info@mlq.qc.ca

Australia
Australian Skeptics Inc.
PO Box 20, Beecroft, NSW 2119
Tel: 02 8094 1894
04 3271 3195
Fax: 02 8088 4735
Email: nsw@skeptics.com.au
www.skeptics.com.au

Council of Australian Humanist Societies
www.humanist.org.au

Rationalist Society of Australia
Email info@rationalist.com.au
www.rationalist.com.au

New Zealand
Humanist Society of New Zealand
PO Box 3372, Wellington
Email: contact@humanist.org.nz
www.humanist.nz/

New Zealand Skeptics
PO Box 30501, Lower Hutt, 5040.
Email: contact@skeptics.nz
http://skeptics.nz

India
Atheist Centre
Dr. G. Vijayam PhD., Executive Director
Atheist Centre, Benz Circle, Vijayawada 520 010,
Andhra Pradesh, India
Tel: 91 866 247 2330
Fax: 91 866 248 4850
Email: atheistcentre@yahoo.com
www.atheistcentre.in

Rationalist International
PO Box 9110, New Delhi 110091
Tel: 91 11 65 699012
Email: info@rationalistinternational.net
www.rationalistinternational.net/

South-East Asia
SEA-Atheists
www.sea-atheists.org

Brazil
Secular Humanist League of Brazil
Av. Dr. Nilo Peçanha, 1221, conjunto 601.
Trés Figueiras. Porto Alegre / RS. 91330-00
Tel: 55 51 3073 4852
Email: contacto@ligahumanista.org.br
www.lihs.org.br

Islamic
Apostates of Islam
Email: aaoislam@gmail.com
www.alliedapostatesofislam.weebly.com

Dr Homa Darabi Foundation
(To promote the rights of women and children under Islam)
PO Box 11049, Truckee, CA 96162, USA
Tel: 530 582 4197
Fax: 530 582 0156
Email: homa@homa.org
www.homa.org/

FaithFreedom.org
www.faithfreedom.org/index.htm [Archived]

Institute for the Secularization of Islamic Society
www.centreforinquiry.net/isis

Books cited or recommended

Adams, D. (2003). *The Salmon of Doubt*. London: Pan.

Alexander, R. D. and Tinkle, D. W., eds (1981). *Natural Selection and Social Behavior*. New York: Chiron Press.

Anon. (1985). *Life – How Did It Get Here? By Evolution or by Creation?* New York: Watchtower Bible and Tract Society.

Ashton, J. F., ed. (1999). *In Six Days: Why 50 Scientists Choose to Believe in Creation*. Sydney: New Holland.

Atkins, P. W. (1992). *Creation Revisited*. Oxford: W. H. Freeman.

Atran, S. (2002). *In Gods We Trust*. Oxford: Oxford University Press.

Attenborough, D. (1960). *Quest in Paradise*. London: Lutterworth.

Aunger, R. (2002). *The Electric Meme: A New Theory of How We Think*. New York: Free Press.

Baggini, J. (2003). *Atheism: A Very Short Introduction*. Oxford: Oxford University Press.

Barber, N. (1988). *Lords of the Golden Horn*. London: Arrow.

Barker, D. (1992). *Losing Faith in Faith*. Madison, WI: Freedom From Religion Foundation.

Barker, D. (2016). *God: The Most Unpleasant Character in All Fiction*. New York: Sterling.

Barker, E. (1984). *The Making of a Moonie: Brainwashing or Choice?* Oxford: Blackwell.

Barrow, J. D. and Tipler, F. J. (1988). *The Anthropic Cosmological Principle*. New York: Oxford University Press.

Baynes, N. H., ed. (1942). *The Speeches of Adolf Hitler*, vol. 1. Oxford: Oxford University Press.

Behe, M. J. (1996). *Darwin's Black Box*. New York: Simon & Schuster.

Beit-Hallahmi, B. and Argyle, M. (1997). *The Psychology of Religious Behaviour, Belief and Experience*. London: Routledge.

Berlinerblau, J. (2005). *The Secular Bible: Why Nonbelievers Must Take Religion Seriously*. Cambridge: Cambridge University Press.

Blackmore, S. (1999). *The Meme Machine*. Oxford: Oxford University Press.

Blaker, K., ed. (2003). *The Fundamentals of Extremism: The Christian Right in America*. Plymouth, MI: New Boston.

Bouquet, A. C. (1956). *Comparative Religion*. Harmondsworth: Penguin.

Boyd, R. and Richerson, P. J. (1985). *Culture and the Evolutionary Process*. Chicago: University of Chicago Press.

Boyer, P. (2001). *Religion Explained*. London: Heinemann.

Brodie, R. (1996). *Virus of the Mind: The New Science of the Meme*. Seattle: Integral Press.

Buckman, R. (2000). *Can We Be Good Without God?* Toronto: Viking.

Bullock, A. (1991). *Hitler and Stalin*. London: HarperCollins.

Bullock, A. (2005). *Hitler: A Study in Tyranny*. London: Penguin.

Buss, D. M., ed. (2005). *The Handbook of Evolutionary Psychology*. Hoboken, NJ: Wiley.

Cairns-Smith, A. G. (1985). *Seven Clues to the Origin of Life*. Cambridge: Cambridge University Press.

Comins, N. F. (1993). *What if the Moon Didn't Exist?* New York: HarperCollins.

Coulter, A. (2006). *Godless: The Church of Liberalism*. New York: Crown Forum.

Coyne, J. (2015). *Faith versus Fact: Why Science and Religion are Incompatible*. New York: Penguin.

Darwin, C. (1859). *On the Origin of Species by Means of Natural Selection*. London: John Murray.

Dawkins, M. Stamp (1980). *Animal Suffering*. London: Chapman & Hall.

Dawkins, R. (1976). *The Selfish Gene*. Oxford: Oxford University Press.

Dawkins, R. (1982). *The Extended Phenotype*. Oxford: W. H. Freeman.

Dawkins, R. (1986). *The Blind Watchmaker*. Harlow: Longman.

Dawkins, R. (1995). *River Out of Eden*. London: Weidenfeld & Nicolson.

Dawkins, R. (1996). *Climbing Mount Improbable*. New York: Norton.

Dawkins, R. (1998). *Unweaving the Rainbow*. London: Penguin.

Dawkins, R. (2003). *A Devil's Chaplain: Selected Essays*. London: Weidenfeld & Nicolson.

Dennett, D. (1995). *Darwin's Dangerous Idea*. New York: Simon & Schuster.

Dennett, D. C. (1987). *The Intentional Stance*. Cambridge, MA: MIT Press.

Dennett, D. C. (2003). *Freedom Evolves*. London: Viking.

Dennett, D. C. (2006). *Breaking the Spell: Religion as a Natural Phenomenon*. London: Viking.

Dennett, D. and LaScola, L. (2015). *Caught in the Pulpit: Leaving Religion Behind*, new edn. Durham, NC: Pitchstone; first publ. 2013.

Deutsch, D. (1997). *The Fabric of Reality*. London: Allen Lane.

Distin, K. (2005). *The Selfish Meme: A Critical Reassessment*. Cambridge: Cambridge University Press.

Dostoevsky, F. (1994). *The Karamazov Brothers*. Oxford: Oxford University Press.

Dunphy, C. (2015). *From Apostle to Apostate: The Story of The Clergy Project*. Durham, NC: Pitchstone.

Ehrman, B. D. (2003a). *Lost Christianities: The Battles for Scripture and the Faiths We Never Knew*. Oxford: Oxford University Press.

Ehrman, B. D. (2003b). *Lost Scriptures: Books that Did Not Make It into the New Testament*. Oxford: Oxford University Press.

Ehrman, B. D. (2006). *Whose Word Is It*? London: Continuum.

Fisher, H. (2004). *Why We Love: The Nature and Chemistry of Romantic Love*. New York: Holt.

Forrest, B. and Gross, P. R. (2004). *Creationism's Trojan Horse: The Wedge of Intelligent Design*. Oxford: Oxford University Press.

Frazer, J. G. (1994). *The Golden Bough*. London: Chancellor Press.

Freeman, C. (2002). *The Closing of the Western Mind*. London: Heinemann.

Galouye, D. F. (1964). *Counterfeit World*. London: Gollancz.

Glover, J. (2001). *Humanity: A Moral History of the Twentieth Century*. Princeton: Yale University Press.

Glover, J. (2006). *Choosing Children.* Oxford: Oxford University Press.

Goodenough, U. (1998). *The Sacred Depths of Nature.* New York: Oxford University Press.

Goodwin, J. (1994). *Price of Honour: Muslim Women Lift the Veil of Silence on the Islamic World.* London: Little, Brown.

Gould, S. J. (1999). *Rocks of Ages: Science and Religion in the Fullness of Life.* New York: Ballantine.

Grafen, A. and Ridley, M., eds (2006). *Richard Dawkins: How a Scientist Changed the Way We Think.* Oxford: Oxford University Press.

Grand, S. (2000). *Creation: Life and How to Make It.* London: Weidenfeld & Nicolson.

Grayling, A. C. (2003). *What Is Good? The Search for the Best Way to Live.* London: Weidenfeld & Nicolson.

Gregory, R. L. (1997). *Eye and Brain.* Princeton: Princeton University Press.

Halbertal, M. and Margalit, A. (1992). *Idolatry.* Cambridge, MA: Harvard University Press.

Harris, S. (2004). *The End of Faith: Religion, Terror and the Future of Reason.* New York: Norton.

Harris, S. (2006). *Letter to a Christian Nation.* New York: Knopf.

Haught, J. A. (1996). *2000 Years of Disbelief: Famous People with the Courage to Doubt.* Buffalo, NY: Prometheus.

Hauser, M. (2006). *Moral Minds: How Nature Designed our Universal Sense of Right and Wrong.* New York: Ecco.

Hawking, S. (1988). *A Brief History of Time.* London: Bantam.

Henderson, B. (2006). *The Gospel of the Flying Spaghetti Monster.* New York: Villard.

Hinde, R. A. (1999). *Why Gods Persist: A Scientific Approach to Religion.* London: Routledge.

Hinde, R. A. (2002). *Why Good Is Good: The Sources of Morality.* London: Routledge.

Hitchens, C. (1995). *The Missionary Position: Mother Teresa in Theory and Practice.* London: Verso.

Hitchens, C. (2005). *Thomas Jefferson: Author of America.* New York: HarperCollins.

Hodges, A. (1983). *Alan Turing: The Enigma.* New York: Simon & Schuster.

Holloway, R. (1999). *Godless Morality: Keeping Religion out of Ethics.* Edinburgh: Canongate.

Holloway, R. (2001). *Doubts and Loves: What is Left of Christianity.* Edinburgh: Canongate.

Humphrey, N. (2002). *The Mind Made Flesh: Frontiers of Psychology and Evolution.* Oxford: Oxford University Press.

Huxley, A. (2003). *The Perennial Philosophy.* New York: Harper.

Huxley, A. (2004). *Point Counter Point.* London: Vintage.

Huxley, T. H. (1871). *Lay Sermons, Addresses and Reviews.* New York: Appleton.

Huxley, T. H. (1931). *Lectures and Essays.* London: Watts.

Jacoby, S. (2004). *Freethinkers: A History of American Secularism.* New York: Holt.

Jammer, M. (2002). *Einstein and Religion.* Princeton: Princeton University Press.

Jaynes, J. (1976). *The Origin of Consciousness in the Breakdown of the Bicameral Mind.* Boston: Houghton Mifflin.

Juergensmeyer, M. (2000). *Terror in the Mind of God: The Global Rise of Religious Violence.* Berkeley: University of California Press.

Kennedy, L. (1999). *All in the Mind: A Farewell to God.* London: Hodder & Stoughton.

Kertzer, D. I. (1998). *The Kidnapping of Edgardo Mortara.* New York: Vintage.

Kilduff, M. and Javers, R. (1978). *The Suicide Cult.* New York: Bantam.

Krauss, L. M. (2012). *A Universe from Nothing.* New York: Free Press.

Kurtz, P., ed. (2003). *Science and Religion: Are They Compatible?* Amherst, NY: Prometheus.

Kurtz, P. (2004). *Affirmations: Joyful and Creative Exuberance.* Amherst, NY: Prometheus.

Kurtz, P. and Madigan, T. J., eds (1994). *Challenges to the Enlightenment: In Defense of Reason and Science.* Amherst, NY: Prometheus.

Lane, B. (1996). *Killer Cults.* London: Headline.

Lane Fox, R. (1992). *The Unauthorized Version*. London: Penguin.

Levitt, N. (1999). *Prometheus Bedeviled*. New Brunswick, NJ: Rutgers University Press.

Loftus, E. and Ketcham, K. (1994). *The Myth of Repressed Memory: False Memories and Allegations of Sexual Abuse*. New York: St Martin's.

McGrath, A. (2004). *Dawkins' God: Genes, Memes and the Meaning of Life*. Oxford: Blackwell.

Mackie, J. L. (1985). *The Miracle of Theism*. Oxford: Clarendon Press.

Medawar, P. B. (1982). *Pluto's Republic*. Oxford: Oxford University Press.

Medawar, P. B. (1984). *The Limits of Science*. Oxford: Oxford University Press.

Medawar, P. B. and Medawar, J. S. (1977). *The Life Science: Current Ideas of Biology*. London: Wildwood House.

Miller, Kenneth (1999). *Finding Darwin's God*. New York: HarperCollins.

Mills, D. (2006). *Atheist Universe: The Thinking Person's Answer to Christian Fundamentalism*. Berkeley: Ulysses Books.

Mitford, N. and Waugh, E. (2001). *The Letters of Nancy Mitford and Evelyn Waugh*. New York: Houghton Mifflin.

Mooney, C. (2005). *The Republican War on Science*. Cambridge, MA: Basic Books.

Norris, P. and Inglehart, R. (2012). *Sacred and Secular: Religion and Politics Worldwide*, 2nd edn. Cambridge: Cambridge University Press.

Pagels, E. and King, K. L. (2007). *Reading Judas*. London: Viking.

Perica, V. (2002). *Balkan Idols: Religion and Nationalism in Yugoslav States*. New York: Oxford University Press.

Phillips, K. (2006). *American Theocracy*. New York: Viking.

Pinker, S. (1997). *How the Mind Works*. London: Allen Lane.

Pinker, S. (2002). *The Blank Slate: The Modern Denial of Human Nature*. London: Allen Lane.

Plimer, I. (1994). *Telling Lies for God: Reason vs Creationism*. Milsons Point, NSW: Random House.

Polkinghorne, J. (1994). *Science and Christian Belief: Theological Reflections of a Bottom-Up Thinker.* London: SPCK.

Rees, M. (1999). *Just Six Numbers.* London: Weidenfeld & Nicolson.

Rees, M. (2001). *Our Cosmic Habitat.* London: Weidenfeld & Nicolson.

Reeves, T. C. (1996). *The Empty Church: The Suicide of Liberal Christianity.* New York: Simon & Schuster.

Richerson, P. J. and Boyd, R. (2005). *Not by Genes Alone: How Culture Transformed Human Evolution.* Chicago: University of Chicago Press.

Ridley, Mark (2000). *Mendel's Demon: Gene Justice and the Complexity of Life.* London: Weidenfeld & Nicolson.

Ridley, Matt (1997). *The Origins of Virtue.* London: Penguin.

Ruse, M. (1982). *Darwinism Defended: A Guide to the Evolution Controversies.* Reading, MA: Addison-Wesley.

Russell, B. (1957). *Why I Am Not a Christian.* London: Routledge.

Russell, B. (1993). *The Quotable Bertrand Russell.* Amherst, NY: Prometheus.

Russell, B. (1997a). *The Collected Papers of Bertrand Russell*, vol. 2: *Last Philosophical Testament, 1943–1968.* London: Routledge.

Russell, B. (1997b). *Collected Papers*, vol. 11, ed. J. C. Slater and P. Köllner. London: Routledge.

Russell, B. (1997c). *Religion and Science.* Oxford: Oxford University Press.

Ruthven, M. (1989). *The Divine Supermarket: Travels in Search of the Soul of America.* London: Chatto & Windus.

Sagan, C. (1995). *Pale Blue Dot.* London: Headline.

Sagan, C. (1996). *The Demon-Haunted World: Science as a Candle in the Dark.* London: Headline.

Scott, E. C. (2004). *Evolution vs. Creationism: An Introduction.* Westport, CT: Greenwood.

Shennan, S. (2002). *Genes, Memes and Human History.* London: Thames & Hudson.

Shermer, M. (1997). *Why People Believe Weird Things: Pseudoscience, Superstition and Other Confusions of Our Time.* New York: W. H. Freeman.

Shermer, M. (1999). *How We Believe: The Search for God in an Age of Science.* New York: W. H. Freeman.
Shermer, M. (2004). *The Science of Good and Evil: Why People Cheat, Gossip, Care, Share, and Follow the Golden Rule.* New York: Holt.
Shermer, M. (2005). *Science Friction: Where the Known Meets the Unknown.* New York: Holt.
Shermer, M. (2006). *The Soul of Science.* Los Angeles: Skeptics Society.
Silver, L. M. (2006). *Challenging Nature: The Clash of Science and Spirituality at the New Frontiers of Life.* New York: HarperCollins.
Singer, P. (1990). *Animal Liberation.* London: Jonathan Cape.
Singer, P. (1994). *Ethics.* Oxford: Oxford University Press.
Smith, K. (1995). *Ken's Guide to the Bible.* New York: Blast Books.
Smolin, L. (1997). *The Life of the Cosmos.* London: Weidenfeld & Nicolson.
Smythies, J. (2006). *Bitter Fruit.* Charleston, SC: Booksurge.
Spong, J. S. (2005). *The Sins of Scripture.* San Francisco: Harper.
Stannard, R. (1993). *Doing Away with God? Creation and the Big Bang.* London: Pickering.
Steer, R. (2003). *Letter to an Influential Atheist.* Carlisle: Authentic Lifestyle Press.
Stenger, V. J. (2003). *Has Science Found God? The Latest Results in the Search for Purpose in the Universe.* New York: Prometheus.
Stenger, V. J. (2007). *God, the Failed Hypothesis: How Science Shows that God Does Not Exist.* New York: Prometheus.
Susskind, L. (2006). *The Cosmic Landscape: String Theory and the Illusion of Intelligent Design.* New York: Little, Brown.
Swinburne, R. (1996). *Is There a God?* Oxford: Oxford University Press.
Swinburne, R. (2004). *The Existence of God.* Oxford: Oxford University Press.
Taverne, R. (2005). *The March of Unreason: Science, Democracy and the New Fundamentalism.* Oxford: Oxford University Press.
Tiger, L. (1979). *Optimism: The Biology of Hope.* New York: Simon & Schuster.

Toland, J. (1991). *Adolf Hitler: The Definitive Biography*. New York: Anchor.

Trivers, R. L. (1985). *Social Evolution*. Menlo Park, CA: Benjamin/Cummings.

Unwin, S. (2003). *The Probability of God: A Simple Calculation that Proves the Ultimate Truth*. New York: Crown Forum.

Vermes, G. (2000). *The Changing Faces of Jesus*. London: Allen Lane.

Ward, K. (1996). *God, Chance and Necessity*. Oxford: Oneworld.

Warraq, I. (1995). *Why I Am Not a Muslim*. New York: Prometheus.

Weinberg, S. (1993). *Dreams of a Final Theory*. London: Vintage.

Wells, G. A. (1986). *Did Jesus Exist?* London: Pemberton.

Wheen, F. (2004). *How Mumbo-Jumbo Conquered the World: A Short History of Modern Delusions*. London: Fourth Estate.

Williams, W., ed. (1998). *The Values of Science: Oxford Amnesty Lectures 1997*. Boulder, CO: Westview.

Wilson, A. N. (1993). *Jesus*. London: Flamingo.

Wilson, A. N. (1999). *God's Funeral*. London: John Murray.

Wilson, D. S. (2002). *Darwin's Cathedral: Evolution, Religion and the Nature of Society*. Chicago: University of Chicago Press.

Wilson, E. O. (1984). *Biophilia*. Cambridge, MA: Harvard University Press.

Winston, R. (2005). *The Story of God*. London: Transworld/BBC.

Wolpert, L. (1992). *The Unnatural Nature of Science*. London: Faber & Faber.

Wolpert, L. (2006). *Six Impossible Things Before Breakfast: The Evolutionary Origins of Belief*. London: Faber & Faber.

Young, M. and Edis, T., eds (2006). *Why Intelligent Design Fails: A Scientific Critique of the New Creationism*. New Brunswick: Rutgers University Press.

Notes

Preface

1 Wendy Kaminer, 'The last taboo: why America needs atheism', *New Republic*, 14 Oct. 1996; http://www.positiveatheism.org/writ/kaminer.htm.
2 Dr Zoë Hawkins, Dr Beata Adams and Dr Paul St John Smith, personal communication.

Chapter 1: A deeply religious non-believer

Deserved respect

3 The television documentary of which the interview was a part was accompanied by a book (Winston 2005).
4 Dennett (2006).

Undeserved respect

5 The full speech is transcribed in Adams (2003) as 'Is there an artificial God?'
6 Perica (2002). See also http://www.historycooperative.org/journals/ahr/108.5/br_151.html.
7 'Dolly and the cloth heads', in Dawkins (2003).
8 http://www.nbcnews.com/id/11188277/ns/politics/t/court-upholds-church-use-hallucinogenic-tea/#VurOFhirJjQ
9 R. Dawkins, 'The irrationality of faith', *New Statesman* (London), 31 March 1989.
10 *Columbus Dispatch*, 19 Aug. 2005.
11 *Los Angeles Times*, 10 April 2006.
12 http://www.thegatewaypundit.com/2006/02/islamic-society-of-denmark-used-fake-cartoons-to-create-story/
13 http://news.bbc.co.uk/2/hi/south_asia/4686536.stm; http://www.neandernews.com/?cat=6.
14 *Independent*, 5 Feb. 2006.
15 Andrew Mueller, 'An argument with Sir Iqbal', *Independent on Sunday*, 2 April 2006, Sunday Review section, 12–16.

Chapter 2: The God Hypothesis

16 Mitford and Waugh (2001).

Polytheism

17 http://www.newadvent.org/cathen/06608b.htm.

18 http://www.catholicforum.com/forums/showthread.php?24395-IMPORTANT-CHANGE-re-APPARITIONS!

Secularism, the Founding Fathers and the religion of America

19 *Congressional Record*, 16 Sept. 1981.

20 http://www.stephenjaygould.org/ctrl/buckner_tripoli.html.

21 Giles Fraser, 'Resurgent religion has done away with the country vicar', *Guardian*, 13 April 2006.

22 Robert I. Sherman, in *Free Inquiry* 8: 4, Fall 1988, 16.

23 N. Angier, 'Confessions of a lonely atheist', *New York Times Magazine*, 14 Jan. 2001: http://www.nytimes.com/2001/01/14/magazine/the-bush-years-confessions-of-a-lonely-atheist.html?pagewanted=all

24 http://www.ftsociety.org/menu/anti-discrimination-support-network/

25 An especially bizarre case of a man (Larry Hooper) being murdered simply because he was an atheist is recounted in the newsletter of the Freethought Society of Greater Philadelphia for March/April 2006. Go to http://www.ftsociety.org/category/newsletter/

26 http://www.hinduonnet.com/thehindu/mag/2001/11/18/stories/2001111800070400.htm.

The poverty of agnosticism

27 Quentin de la Bédoyère, *Catholic Herald*, 3 Feb. 2006.

28 Carl Sagan, 'The burden of skepticism', *Skeptical Inquirer* 12, Fall 1987.

29 I discussed this case in Dawkins (1998).

30 T. H. Huxley, 'Agnosticism' (1889), repr. in Huxley (1931). The complete text of 'Agnosticism' is also available at http://www.infidels.org/library/historical/thomas_huxley/huxley_wace/part_02.html.

31 Russell, 'Is there a God?' (1952), repr. in Russell (1997b).

32 Andrew Mueller, 'An argument with Sir Iqbal', *Independent on Sunday*, 2 April 2006, Sunday Review section, 12–16.

33 *New York Times*, 29 Aug. 2005. See also Henderson (2006).

34 Henderson (2006).

35 http://uncyclopedia.wikia.com/wiki/Reformed_church_of_Alfredo see also: http://uncyclopedia.wikia.com/wiki/Pastafarian_Schisms

The Great Prayer Experiment

36 H. Benson et al., 'Study of the therapeutic effects of intercessory prayer (STEP) in cardiac bypass patients', *American Heart Journal* 151: 4, 2006, 934–42.

37 Richard Swinburne, in *Science and Theology News*, 7 April 2006, http://users.ox.ac.uk/~orie0087/framesetpdfs.shtml and link to article is http://users.ox.ac.uk/~one0087/pdf-files/Responses%20to%20Controversies/Response%20to%20a%20Statistical%20Study.pdf

38 *New York Times*, 11 April 2006.

The Neville Chamberlain school of evolutionists

39 In court cases, and books such as Ruse (1982). His article in *Playboy* appeared in the April 2006 issue.

40 Jerry Coyne's reply to Ruse appeared in the August 2006 issue of *Playboy*.

41 Madeleine Bunting, *Guardian*, 27 March 2006.

42 Dan Dennett's reply appeared in the *Guardian*, 4 April 2006.

43 http://scienceblogs.com/pharyngula/2006/03/27/the-dawkinsdennett-boogeyman/; http://scienceblogs.com/pharyngula/2006/02/26/our-double-standard/; http://scienceblogs.com/pharyngula/2006/02/23/the-rusedennett-feud/

Little green men

44 Dennett (1995).

Chapter 3: Arguments for God's existence

The ontological argument and other a priori *arguments*

45 http://www.iep.utm.edu/o/ont-arg.htm. William Grey: 'Gasking's proof', Analysis, Vol 60, No 4 (2000), pp. 368–70.

The argument from personal 'experience'

46 The whole subject of illusions is discussed by Richard Gregory in a series of books including Gregory (1997).
47 My own attempt at spelling out the explanation is on pp. 268–9 of Dawkins (1998).
48 http://www.sofc.org/Spirituality/s-of-fatima.htm.

The argument from scripture

49 Tom Flynn, 'Matthew vs. Luke', *Free Inquiry* 25: 1, 2004, 34–45; Robert Gillooly, 'Shedding light on the light of the world', *Free Inquiry* 25: 1, 2004, 27–30.
50 Erhman (2006). See also Ehrman (2003a, b).

The argument from admired religious scientists

51 Beit-Hallahmi and Argyle (1997).
52 E. J. Larson and L. Witham, 'Leading scientists still reject God', *Nature* 394, 1998, 313.
53 http://www.leaderu.com/ftissues/ft9610/reeves.html gives a particularly interesting analysis of historical trends in American religious opinion by Thomas C. Reeves, Professor of History at the University of Wisconsin, based on Reeves (1996).
54 http://www.answersingenesis.org/docs/3506.asp.
55 R. Elisabeth Cornwell and Michael Stirrat, manuscript in preparation, 2006.
56 P. Bell, 'Would you believe it?', *Mensa Magazine*, Feb. 2002, 12–13.

Chapter 4: Why there almost certainly is no God

The Ultimate Boeing 747

57 An exhaustive review of the provenance, usages and quotations of this analogy is given, from a creationist point of view,

by Gert Korthof, at http://wasdarwinwrong.com/kortho46a.htm

Natural selection as a consciousness-raiser

58 Adams (2002), p. 99. My 'Lament for Douglas', written the day after his death, is reprinted as the Epilogue to *The Salmon of Doubt*, and also in *A Devil's Chaplain*, which also has my eulogy at his memorial meeting in the Church of St Martin-in-the-Fields.

59 Interview in *Der Spiegel*, 26 Dec. 2005.

60 Susskind (2006: 17).

The worship of gaps

61 Behe (1996).

62 http://www.millerandlevine.com/km/evol/design2/article.html.

63 This account of the Dover trial, including the quotations, is from A. Bottaro, M. A. Inlay and N. J. Matzke, 'Immunology in the spotlight at the Dover "Intelligent Design" trial', *Nature Immunology* 7, 2006, 433–5.

64 J. Coyne, 'God in the details: the biochemical challenge to evolution', *Nature* 383, 1996, 227–8. The article by Coyne and me, 'One side can be wrong', was published in the *Guardian*, 1 Sept. 2005: http://www.theguardian.com/science/2005/sep/01/schools.research
The quotation from the 'eloquent blogger' is at http://www.religionisbullshit.net/blog/2005_09_01_archive.php.

65 Dawkins (1995).

The anthropic principle: planetary version

66 Carter admitted later that a better name for the overall principle would be 'cognizability principle' rather than the already entrenched term 'anthropic principle': B. Carter, 'The anthropic principle and its implications for biological evolution', *Philosophical Transactions of the Royal Society of London A*, 310, 1983, 347–63. For a book-length discussion of the anthropic principle, see Barrow and Tipler (1988).

67 Comins (1993).

68 I spelled this argument out more fully in *The Blind Watchmaker* (Dawkins 1986).

The anthropic principle: cosmological version

69 Murray Gell-Mann, quoted by John Brockman on the 'Edge' website, http://www.edge.org/3rd_culture/bios/smolin.html.

70 Ward (1996: 99); Polkinghorne (1994: 55).

An interlude at Cambridge

71 J. Horgan, 'The Templeton Foundation: a skeptic's take', *Chronicle of Higher Education*, 7 April 2006. See also http://www.edge.org/3rd_culture/horgan06/horgan06_index.html.

72 P. B. Medawar, review of *The Phenomenon of Man*, repr. in Medawar (1982: 242).

73 Dennett (1995: 155).

Chapter 5: The roots of religion

The Darwinian imperative

74 Quoted in Dawkins (1982: 30).

75 K. Sterelny, 'The perverse primate', in Grafen and Ridley (2006: 213–23).

Group selection

76 N. A. Chagnon, 'Terminological kinship, genealogical relatedness and village fissioning among the Yanomamö Indians', in Alexander and Tinkle (1981: ch. 28).

77 C. Darwin, *The Descent of Man* (New York: Appleton, 1871), vol. 1, 156.

Religion as a by-product of something else

78 Quoted in Blaker (2003: 7).

Psychologically primed for religion

79 See e.g. Buss (2005).

80 Deborah Keleman, 'Are children "intuitive theists"?', *Psychological Science* 15: 5, 2004, 295–301.
81 Dennett (1987).
82 *Guardian*, 31 Jan. 2006.
83 Smythies (2006).
84 http://jmm.aaa.net.au/articles/14223.htm.

Chapter 6: The roots of morality: why are we good?

85 The movie itself, which is very good, can be obtained at http://www.thegodmovie.com/index.php.

A case study in the roots of morality

86 M. Hauser and P. Singer, 'Morality without religion', *Free Inquiry* 26: 1, 2006, 18–19.

If there is no God, why be good?

87 Dostoevsky (1994: bk 2, ch. 6, p. 87).
88 Hinde (2002). See also Singer (1994), Grayling (2003), Glover (2006).

Chapter 7: The 'Good' Book and the changing moral *Zeitgeist*

89 Lane Fox (1992); Berlinerblau (2005).
90 Holloway (1999, 2005). Richard Holloway's 'recovering Christian' line is in a book review in the *Guardian*, 15 Feb. 2003: http://books.guardian.co.uk/reviews/scienceandnature/0,6121,894941,00.html. The Scottish journalist Muriel Gray wrote a beautiful account of my Edinburgh dialogue with Bishop Holloway in the (Glasgow) *Herald*: http://www.sundayherald.com/44517. [Link no longer active, 2016]

The Old Testament

91 For a frightening collection of sermons by American clergymen, blaming hurricane Katrina on human 'sin', see http://universist.org/neworleans.htm.
92 Pat Robertson, reported by the BBC at http://news.bbc.co.uk/2/hi/americas/4427144.stm.

Is the New Testament any better?

93 R. Dawkins, 'Atheists for Jesus', *Free Inquiry* 25: 1, 2005, 9–10.

94 Julia Sweeney is also right on target when she briefly mentions Buddhism. Just as Christianity is sometimes thought to be a nicer, gentler religion than Islam, Buddhism is often cracked up to be the nicest of all. But the doctrine of demotion on the reincarnation ladder because of sins in a past life is pretty unpleasant. Julia Sweeney: 'I went to Thailand and happened to visit a woman who was taking care of a terribly deformed boy. I said to his caretaker, "It's so good of you to be taking care of this poor boy." She said, "Don't say 'poor boy,' he must have done something terrible in a past life to be born this way." '

95 For a thoughtful analysis of techniques used by cults, see Barker (1984). More journalistic accounts of modern cults are given by Lane (1996) and Kilduff and Javers (1978).

96 Paul Vallely and Andrew Buncombe, 'History of Christianity: Gospel according to Judas', *Independent*, 7 April 2006.

97 Vermes (2000).

Love thy neighbour

98 Hartung's paper was originally published in *Skeptic* 3: 4, 1995, but is now most readily available at http://strugglesforexistence.com/?p=article_p&id=13.

99 Smith (1995).

100 *Guardian*, 12 March 2002: http://books.guardian.co.uk/departments/politicsphilosophyandsociety/story/0,,664342,00.html.

101 N. D. Glenn, 'Interreligious marriage in the United States: patterns and recent trends', *Journal of Marriage and the Family* 44: 3, 1982, 555–66.

The moral Zeitgeist

102 http://www.ebonmusings.org/atheism/new10c.html?

103 Huxley (1871).

104 http://www.gutenberg.org/files/2656/2656-h/2656-h.htm?

What about Hitler and Stalin? Weren't they atheists?

105 Bullock (1991).

106 Bullock (2005).

107 http://ffrf.org/legacy/fttoday/1997/march97/holocaust.html. This article by Richard E. Smith, originally published in *Freethought Today*, March 1997, has a large number of relevant quotations from Hitler and other Nazis, giving their sources. Unless otherwise stated, my quotations are from Smith's article.

108 http://homepages.paradise.net.nz/mischedj/ca_hitler.html.

109 Bullock (2005: 96).

110 Adolf Hitler, speech of 12 April 1922. In Baynes (1942: 19–20).

111 Bullock (2005: 43).

112 This quotation, and the following one, are from Anne Nicol Gaylor's article on Hitler's religion, http://ffrf.org/legacy/fttoday/back/hitler.html.

113 http://www.contra-mundum.org/schirrmacher/NS_Religion.pdf.

Chapter 8: What's wrong with religion? Why be so hostile?

Fundamentalism and the subversion of science

114 From 'What is true?', ch. 1.2 of Dawkins (2003).

115 Both my quotations from Wise come from his contribution to the 1999 book *In Six Days*, an anthology of essays by young-Earth creationists (Ashton 1999).

The dark side of absolutism

116 Warraq (1995: 175).

117 John William Gott's imprisonment for calling Jesus a clown is mentioned in *The Indypedia*, published by the *Independent*, 29 April 2006. The attempted prosecution of the BBC for blasphemy is in BBC news, 10 Jan. 2005: http://news.bbc.co.uk/1/hi/entertainment/tv_and_radio/4161109.stm.

118 http://adultthought.ucsd.edu/Culture_War/The_American_Taliban.html.

Faith and homosexuality

119 Hodges (1983).

120 This and the remaining quotations in this section are from the American Taliban site already listed: http://adultthought.ucsd.edu/Culture_War/The_American_ Taliban.html.

121 http://adultthought.ucsd.edu/Culture_War/The_American_Taliban.html.

122 From Pastor Phelps's Westboro Baptist Church official website, godhatesfags.com: http://www.godhatesfags.com/fliers/jan2006/20060131_coretta-scott-king-funeral.pdf.

Faith and the sanctity of human life

123 See Mooney (2005). Also Silver (2006), which arrived when this book was in final proof, too late to be discussed as fully as I would have liked.

124 For an interesting analysis of what makes Texas different in this respect, see http://www.pbs.org/wgbh/pages/frontline/shows/execution/readings/texas.html.

125 http://en.wikipedia.org/wiki/Karla_Faye_Tucker.

126 These Randall Terry quotes are from the same American Taliban site as before: http://adultthought.ucsd.edu/Culture_War/The_American_Taliban.html.

127 Reported on Fox news: http://www.foxnews.com/story/0,2933,96286,00.html.

128 M. Stamp Dawkins (1980).

The Great Beethoven Fallacy

129 Medawar and Medawar (1977).

130 http://www.warroom.com/ethical.htm. [link no longer active, 2016]

How 'moderation' in faith fosters fanaticism

131 Johann Hari's article, originally published in the *Independent*,

15 July 2005, can be found at http://independent.co.uk/2/voices/commentators/johann-hari/johann-hari-the-best-way-to-undermine-the-jihadists-is-to-trigger-a-rebellion-of-muslim-women-5346567.html

132 Village Voice, 18 May 2004: http://villagevoice.com/news/the-jesus-landing-pad-6429003

133 Harris (2004: 29).

134 Nasra Hassan, 'An arsenal of believers', *New Yorker*, 19 Nov. 2001. See also http://www.bintjbeil.com/articles/en/011119_hassan.html.

Chapter 9: Childhood, abuse and the escape from religion

Physical and mental abuse

135 Reported by BBC news: http://news.bbc.co.uk/1/hi/wales/901723.stm.

136 Loftus and Ketcham (1994).

137 See John Waters in the *Irish Times:* http://www.irishtimes.com/news/government-response-to-laffoy-letter-1.374483

138 Associated Press, 10 June 2005: http://www.rickross.com/reference/clergy/clergy426.html [link no longer active, 2016].

139 http://www.av1611.org/hell.html.

In defence of children

140 N. Humphrey, 'What shall we tell the children?', in Williams (1998); repr. in Humphrey (2002).

141 http://www.law.umkc.edu/faculty/projects/ftrials/conlaw/yoder.html.

An educational scandal

142 *Guardian*, 15 Jan. 2005: http://www.guardian.co.uk/weekend/story/0,,1389500,00.html.

143 *Times Educational Supplement*, 15 July 2005.

144 http://www.telegraph.co.uk/comment/personal-view/3574344/Young-Earth-Creationists-teach-bad-science-and-worse-religion.html

145 *Guardian*, 15 Jan. 2005: http://www.guardian.co.uk/weekend/story/0,,1389500,00.html.

146 The text of our letter, drafted by the Bishop of Oxford, was as follows:

> Dear Prime Minister,
>
> We write as a group of scientists and Bishops to express our concern about the teaching of science in the Emmanuel City Technology College in Gateshead. Evolution is a scientific theory of great explanatory power, able to account for a wide range of phenomena in a number of disciplines. It can be refined, confirmed and even radically altered by attention to evidence. It is not, as spokesmen for the college maintain, a 'faith position' in the same category as the biblical account of creation which has a different function and purpose.
>
> The issue goes wider than what is currently being taught in one college. There is a growing anxiety about what will be taught and how it will be taught in the new generation of proposed faith schools. We believe that the curricula in such schools, as well as that of Emmanuel City Technical College, need to be strictly monitored in order that the respective disciplines of science and religious studies are properly respected.
>
> Yours sincerely

147 *British Humanist Association News*, March–April 2006.

148 *Observer*, 22 July 2004: http://observer.guardian.co.uk/magazine/story/0,11913,1258506,00.html.

Consciousness-raising again

149 The Oxford Dictionary takes 'gay' back to American prison slang in 1935. In 1955 Peter Wildeblood, in his famous book *Against the Law*, found it necessary to define 'gay' as 'an American euphemism for homosexual'.

150 http://www.independent.co.uk/news/uk/children-from-all-

faiths-tell-the-christmas-story-pupils-from-many-different-religious-backgrounds-1563455.html [link no longer active. 2016]

Religious education as a part of literary culture

151 Shaheen has written three books, anthologizing biblical references in the comedies, tragedies and histories separately. The summary count of 1,300 is mentioned in http://shakespeareoxfordfellowship.org/biblical-references/

152 http://www.bibleliteracy.org/Secure/Documents/BibleLiteracyReport2005.pdf.

Chapter 10: A much needed gap?

Consolation

153 From memory, I attribute this argument to the Oxford philosopher Derek Parfitt. I have not researched its origins thoroughly because I am using it only as a passing example of philosophical consolation.

154 Reported by BBC News: http://news.bbc.co.uk/1/hi/special_report/1999/06/99/cardinal_hume_funeral/376263.stm.

The mother of all burkas

155 Wolpert (1992).

Afterword by Daniel Dennett

156 *New York Review of Books*, 11 Jan. 2007.

157 *New York Review of Books*, 1 March 2007.

158 For details see bibliography.

159 Daniel Dennett and Deb Roy, 'Our transparent future: no secret is safe in the digital age. The implications for our institutions are downright Darwinian', *Scientific American* 312: 3, March 2015, pp. 64–9.

Index

The Greatest Show on Earth

The Evidence for Evolution

Richard Dawkins

'This is a magnificent book of wonderstanding: Richard Dawkins combines an artist's wonder at the virtuosity of nature with a scientist's understanding of how it comes to be.'
MATT RIDLEY

'A voice of reason in irrational times, Richard Dawkins is both theorist and explainer of one of the greatest discoveries of the human mind.'
THE TIMES

Charles Darwin, whose 1859 masterpiece *On the Origin of Species* shook society to its core, would surely have raised an incredulous eyebrow at the controversy over evolution still raging 150 years later.

The Greatest Show on Earth is a stunning counter-attack on creationists, followers of 'Intelligent Design' and all those who still question evolution as scientific fact. In this brilliant *tour de force* Richard Dawkins pulls together the incontrovertible evidence that underpins it: from living examples of natural selection to clues in the fossil record; from plate tectonics to molecular genetics.

The Greatest Show on Earth comes at a critical time as systematic opposition to the fact of evolution flourishes as never before in many schools worldwide. Dawkins wields a devastating argument against this ignorance while sharing with us his palpable love of science and the natural world. Written with elegance, wit and passion, it is hard-hitting, absorbing and totally convincing.

The Magic of Reality
How We Know What's Really True

Richard Dawkins
Illustrated by Dave McKean

'The force and fluency of a classic . . . a luminous, authoritative prose that transcends age differences.'
THE TIMES

'The clearest and most beautifully written introduction to science I've ever read.'
PHILIP PULLMAN

Magic takes many forms. The ancient Egyptians explained the night by suggesting that the goddess Nut swallowed the sun. The Vikings believed a rainbow was the gods' bridge to earth. The Japanese used to explain earthquakes by conjuring a gigantic catfish that carried the world on its back – and earthquakes occurred each time it flipped its tail. These are magical, extraordinary tales. But there is another kind of magic, and it lies in the exhilaration of discovering the real answers to these questions. It is the magic of reality – science.

Packed with inspiring explanations of space, time and evolution, and with clever thought experiments, *The Magic of Reality* explains a stunningly wide range of natural phenomena. What is stuff made of? How old is the universe? What causes tsunamis? Who was the first man, or woman? This is a page-turning, inspirational detective story that not only mines all the sciences for its clues but primes the reader to think like a scientist too.

Richard Dawkins elucidates the wonders of the natural world to all ages with his inimitable clarity and exuberance in a text that will enlighten and inform for generations to come.

An Appetite for Wonder

The Making of a Scientist

Richard Dawkins

'Richard Dawkins's memoirs are, like their author, honest, perceptive, sometimes ingenuous, always rational and deeply humane.'

MATT RIDLEY

What were the influences that shaped Richard Dawkins's life and who inspired him to become the pioneering scientist and public thinker now famous (and infamous to some) around the world? This is his personal account of a childhood in colonial Africa, life at an English boarding school and his studies at the University of Oxford's dynamic Zoology Department, which sparked his radical new vision of Darwinism, *The Selfish Gene.* Through Dawkins's honest self-reflection, touching reminiscences and witty anecdotes, we are finally able to understand what shaped the man who, more than anyone else in his generation, explained our own origins.

Brief Candle in the Dark

More Reflections on a Life in Science

Richard Dawkins

'One of our greatest living writers . . . the opportunity to eavesdrop on the workings of an extraordinary mind.'

STEVEN PINKER

In *An Appetite for Wonder* Richard Dawkins brought us his engaging memoir of his first thirty-five years. Here he continues his story, painting a vivid picture, coloured with wit, anecdote and digression, of the twenty-five postgraduate years he spent teaching at Oxford, and recalls with characteristic wry humour the idiosyncrasies of an establishment steeped in tradition and ritual. He invites us to share the life of a travelling scientist, from fieldwork on the Panama Canal to conferences in the company of some of the most prominent – and most eccentric – of the world's scientific luminaries. Most important of all, for the first time he reviews with fresh and stimulating insights the evolving narrative of his ideas about science over the course of his highly distinguished career as thinker, teacher and writer.